NONCOMMUTATIVE RATIONAL SERIES WITH APPLICATIONS

The algebraic theory of automata was created by Schützenberger and Chomsky over 50 years ago and there has since been a great deal of development. Classical work on the theory of noncommutative power series has been augmented more recently to areas such as representation theory, combinatorial mathematics and theoretical computer science.

This book presents to an audience of graduate students and researchers a modern account of the subject and its applications. The algebraic approach allows the theory to be developed in a general form of wide applicability. For example, number-theoretic results can now be more fully explored, in addition to applications in automata theory, codes and noncommutative algebra. Much material, for example, Schützenberger's theorem on polynomially bounded rational series, and results on semisimple algebras, appear here for the first time in book form.

In sum, this is an excellent resource and reference for all those working in algebra, theoretical computer science and their areas of overlap.

JEAN BERSTEL is Emeritus Professor of Computer Science in the Gaspard-Monge Institute at the Université Paris-Est.

CHRISTOPHE REUTENAUER is Professor of Mathematics in the Combinatorics and Mathematical Computer Science Laboratory (LaCIM) at the University of Québec, Montréal.

ENCYCLOPEDIA OF MATHEMATICS AND ITS APPLICATIONS

All the titles listed below can be obtained from good booksellers or from Cambridge University Press. For a complete series listing visit

http://www.cambridge.org/uk/series/sSeries.asp?code=EOM

80 O. Stormark *Lie's Structural Approach to PDE Systems*
81 C. F. Dunkl and Y. Xu *Orthogonal Polynomials of Several Variables*
82 J. P. Mayberry *The Foundations of Mathematics in the Theory of Sets*
83 C. Foias *et al. Navier–Stokes Equations and Turbulence*
84 B. Polster and G. F. Steinke *Geometries on Surfaces*
85 R. B. Paris and D. Kaminski *Asymptotics and Mellin–Barnes Integrals*
86 R. J. McEliece *The Theory of Information and Coding, 2nd edn*
87 B. A. Magurn *An Algebraic Introduction to K-Theory*
88 T. Mora *Solving Polynomial Equation Systems I*
89 K. Bichteler *Stochastic Integration with Jumps*
90 M. Lothaire *Algebraic Combinatorics on Words*
91 A. A. Ivanov and S. V. Shpectorov *Geometry of Sporadic Groups II*
92 P. McMullen and E. Schulte *Abstract Regular Polytopes*
93 G. Gierz *et al. Continuous Lattices and Domains*
94 S. R. Finch *Mathematical Constants*
95 Y. Jabri *The Mountain Pass Theorem*
96 G. Gasper and M. Rahman *Basic Hypergeometric Series, 2nd edn*
97 M. C. Pedicchio and W. Tholen (eds.) *Categorical Foundations*
98 M. E. H. Ismail *Classical and Quantum Orthogonal Polynomials in One Variable*
99 T. Mora *Solving Polynomial Equation Systems II*
100 E. Olivieri and M. Eulália Vares *Large Deviations and Metastability*
101 A. Kushner, V. Lychagin and V. Rubtsov *Contact Geometry and Nonlinear Differential Equations*
102 L. W. Beineke and R. J. Wilson (eds.) with P. J. Cameron *Topics in Algebraic Graph Theory*
103 O. J. Staffans *Well-Posed Linear Systems*
104 J. M. Lewis, S. Lakshmivarahan and S. K. Dhall *Dynamic Data Assimilation*
105 M. Lothaire *Applied Combinatorics on Words*
106 A. Markoe *Analytic Tomography*
107 P. A. Martin *Multiple Scattering*
108 R. A. Brualdi *Combinatorial Matrix Classes*
109 J. M. Borwein and J. D. Vanderwerff *Convex Functions*
110 M.-J. Lai and L. L. Schumaker *Spline Functions on Triangulations*
111 R. T. Curtis *Symmetric Generation of Groups*
112 H. Salzmann *et al. The Classical Fields*
113 S. Peszat and J. Zabczyk *Stochastic Partial Differential Equations with Lévy Noise*
114 J. Beck *Combinatorial Games*
115 L. Barreira and Y. Pesin *Nonuniform Hyperbolicity*
116 D. Z. Arov and H. Dym *J-Contractive Matrix Valued Functions and Related Topics*
117 R. Glowinski, J.-L. Lions and J. He *Exact and Approximate Controllability for Distributed Parameter Systems*
118 A. A. Borovkov and K. A. Borovkov *Asymptotic Analysis of Random Walks*
119 M. Deza and M. Dutour Sikirić *Geometry of Chemical Graphs*
120 T. Nishiura *Absolute Measurable Spaces*
121 M. Prest *Purity, Spectra and Localisation*
122 S. Khrushchev *Orthogonal Polynomials and Continued Fractions*
123 H. Nagamochi and T. Ibaraki *Algorithmic Aspects of Graph Connectivity*
124 F. W. King *Hilbert Transforms I*
125 F. W. King *Hilbert Transforms II*
126 O. Calin and D.-C. Chang *Sub-Riemannian Geometry*
127 M. Grabisch *et al. Aggregation Functions*
128 L. W. Beineke and R. J. Wilson (eds.) with J. L. Gross and T. W. Tucker *Topics in Topological Graph Theory*
129 J. Berstel, D. Perrin and C. Reutenauer *Codes and Automata*
130 T. G. Faticoni *Modules over Endomorphism Rings*
131 H. Morimoto *Stochastic Control and Mathematical Modeling*
132 G. Schmidt *Relational Mathematics*
133 P. Kornerup and D. W. Matula *Finite Precision Number Systems and Arithmetic*
134 Y. Crama and P. L. Hammer (eds.) *Boolean Functions*
135 V. Berthé and M. Rigo (eds.) *Combinatorics, Automata and Number Theory*
136 A. Kristály, V. D. Rădulescu and C. Varga *Variational Principles in Mathematical Physics, Geometry, and Economics*

ENCYCLOPEDIA OF MATHEMATICS AND ITS APPLICATIONS

Noncommutative Rational Series with Applications

JEAN BERSTEL
Université Paris-Est

CHRISTOPHE REUTENAUER
University of Québec, Montréal

CAMBRIDGE
UNIVERSITY PRESS

CAMBRIDGE
UNIVERSITY PRESS

University Printing House, Cambridge CB2 8BS, United Kingdom

One Liberty Plaza, 20th Floor, New York, NY 10006, USA

477 Williamstown Road, Port Melbourne, VIC 3207, Australia

314-321, 3rd Floor, Plot 3, Splendor Forum, Jasola District Centre, New Delhi - 110025, India

103 Penang Road, #05-06/07, Visioncrest Commercial, Singapore 238467

Cambridge University Press is part of the University of Cambridge.

It furthers the University's mission by disseminating knowledge in the pursuit of education, learning and research at the highest international levels of excellence.

www.cambridge.org
Information on this title: www.cambridge.org/9780521190220

First published 2011

A catalogue record for this publication is available from the British Library

Library of Congress Cataloging in Publication data
Berstel, Jean, 1941–
Noncommutative rational series with applications / Jean Berstel, Christophe Reutenauer.
p. cm. – (Encyclopedia of mathematics and its applications ; 137)
Includes bibliographical references and index.
ISBN 978-0-521-19022-0 (hardback)
1. Machine theory. 2. Noncommutative algebras. I. Reutenauer, Christophe.
II. Title. III. Series.
QA267.B47 2010
511.3′5 – dc22 2010030398

ISBN 978-0-521-19022-0 Hardback

Pour Anne et Anissa

Contents

Part III Applications

Preface

Formal power series have long been used in all branches of mathematics. They are invaluable in algebra, analysis, combinatorics and in theoretical computer science.

Historically, the work of M.-P. Schützenberger in the algebraic theory of finite automata and the corresponding languages has led him to introduce noncommutative formal power series. This appears in particular in his work with Chomsky on formal grammars. This last point of view is at the origin of this book.

The first part of the book, composed of Chapters 1–4, is especially devoted to this aspect: Formal power series may be viewed as formal languages with coefficients, and finite automata (and more generally weighted automata) may be considered as linear representations of the free monoid. In this sense, via formal power series, algebraic theory of automata becomes a part of representation theory.

The first two chapters contain general results and discuss in particular the equality between rational and recognizable series (Theorem of Kleene–Schützenberger) and the construction of the minimal linear representation. The exposition illustrates the synthesis of linear algebra and syntactic methods inherited from automata theory.

The next two chapters are concerned with the comparison of some typical properties of rational (regular) languages, when they are transposed to rational series. First, Chapter 3 describes the relationship with the family of regular languages studied in theoretical computer science. Next, the chapter contains iteration properties for rational series, also known as pumping lemmas, which are much more involved than those for regular languages. Chapter 4 discusses rational expressions. It contains two main results: the so-called "triviality" of rational identities over a commutative ring and the characterization of the star height of a rational series and its two consequences: the star height is unbounded and the star height over an algebraically closed field is decidable. The same problem for rational languages is known to be extremely difficult.

The second part of the book, composed of Chapters 5–8, is devoted to arithmetic properties of rational series.

Chapter 5 is concerned with automatic sequences and algebraic series. Two main results are the characterization of algebraic series over a finite field by Christol, Kamae, Mendès France and Rauzy, and Fürstenberg's theorem on the diagonal of a rational function.

Chapter 6 gives the proof of a theorem of Pólya characterizing rational series whose set of coefficients have only finitely many distinct prime divisors, and an elementary proof of a theorem of Skolem, Mahler and Lech about vanishing terms in a rational series.

Chapter 7 studies rational series over a principal ring and Fatou extensions. It contains a section on polynomial identities and rationality criteria, and a section on Fatou rings.

Chapter 8 is concerned with rational series with nonnegative coefficients. It contains a simplified proof of Soittola's theorem (Theorem 8.3.1) which is one of the most striking results on these rational series in one variable. Also the presentation of the star height (Theorem 8.4.1) is new.

The third part, composed of the remaining four chapters, is concerned with applications and with the study of important subfamilies of rational series.

Chapter 9 contains some results appearing for the first time in book form. The first section is on the Burnside problem of matrix semigroups. Section 9.2 contains a main result: Schützenberger's theorem on polynomially bounded rational series, one of the most difficult results in the area. The chapter also contains Simon's result on the Burnside problem for the matrix semigroups over the tropical semiring and limitedness of languages.

The next two chapters are devoted to the study of polynomials in noncommutative variables, and to their application to coding theory. Because of noncommutativity, the structure of polynomials is much more complex than in the case of commutativity, and the results are rather delicate to prove. We present here basic properties concerning factorizations. The main purpose of Chapter 10 is to prepare the ground for Chapter 11. The latter contains the generalization of a result of M.-P. Schützenberger concerning the factorization of a polynomial associated with a finite code.

Chapter 12 is a step towards representation theory. It gives results on semisimple syntactic algebras. Main results are the semisimplicity of the syntactic algebra of bifix codes and its converse. The syntactic algebra of a cyclic language is semisimple and its zeta function is rational. The chapter also contains a long appendix on the Rees–Suschkewitsch theorem which describes the structure of the minimal idea of a finite monoid. We included a self-contained exposition for the ease of the reader.

More than 170 exercises are provided, and also short bibliographical notes are given at the end of the chapters.

This book is issued from a previous book of the authors, entitled *Rational Series and their Languages*. The present text is an entirely rewritten and enriched version of this book. An important part of the material presented here appears for the first time in book form.

The text served for advanced courses held several times by the authors, at the University Pierre et Marie Curie, Paris and at the University of Saarbrücken. Parts of the book were also taught at several different levels at other places, such as the University of Québec at Montréal and the University of Marne-la-Vallée.

Many thanks to Sylvain Lavallée, Aaron Lauve, Martin Dagenais, Pierre Bouchard, Franco Saliola, Dominique Perrin and to Christian Mathissen for their help.

Note to the reader

Following usual notation, items such as sections, theorems, corollaries, etc. are numbered within a chapter. When cross-referenced the chapter number is omitted if the item is within the current chapter. Thus "Theorem 1.1" means the first theorem in the first section of the current chapter, and "Theorem 2.1.1" refers to the equivalent theorem in Chapter 2. Exercises are numbered accordingly and the section number should help the reader to find the section relevant to that exercise.

Part I

Rational series

1

Rational series

This chapter contains the definitions of the basic concepts, namely rational and recognizable series in several noncommuting variables.

We start with the definition of a semiring, followed by the notation for the usual objects in free monoids and formal series. The topology on formal series is briefly introduced.

Section 1.4 contains the definition of rational series, together with some elementary properties and the fact that certain morphisms preserve the rationality of series.

Recognizable series are introduced in Section 1.5. An algebraic characterization is given. We also prove (Theorem 5.1) that the Hadamard product preserves recognizability.

Weighted automata are presented in Section 1.6.

The fundamental theorem of Schützenberger (equivalence between rational and recognizable series, Theorem 7.1) is the concern of the last section. This theorem is the starting point for the developments given in the subsequent chapters.

1.1 Semirings

Recall that a *semigroup* is a set equipped with an associative binary operation, and a *monoid* is a semigroup having a neutral element for its law.

A *semiring* is, roughly speaking, a ring without subtraction. More precisely, it is a set K equipped with two operations $+$ and $\cdot$ (sum and product) such that the following properties hold:

(i) $(K, +)$ is a commutative monoid with neutral element denoted by 0.
(ii) $(K, \cdot)$ is a monoid with neutral element denoted by 1.
(iii) The product is distributive with respect to the sum.
(iv) For all a in K, $0a = a0 = 0$.

The last property is not a consequence of the others, as is the case for rings.

A semiring is *commutative* if its product is commutative. A *subsemiring* of K is a subset of K containing 0 and 1, which is stable for the operations of K.

A *semiring morphism* is a function

$$f : K \to K'$$

of a semiring K into a semiring K' that maps the 0 and 1 of K onto the corresponding elements of K' and that respects sum and product.

Let us give some examples of semirings. Among them are, of course, rings and fields. In this text, a *field is always commutative*. Next, the set $\mathbb{N}$ of natural numbers, the sets $\mathbb{Q}_+$ of nonnegative rational numbers and $\mathbb{R}_+$ of nonnegative real numbers are semirings. The *Boolean semiring* $\mathbb{B} = \{0, 1\}$ is completely described by the relation $1 + 1 = 1$ (see Exercise 1.1). If M is a monoid, the set of its subsets is naturally equipped with the structure of a semiring: the sum of two subsets X and Y of M is simply $X \cup Y$ and their product is

$$\{xy \mid x \in X, y \in Y\}.$$

Let K be a semiring and let P, Q be two finite sets. We denote by $K^{P\times Q}$ the set of $P \times Q$-matrices with coefficients in K. The sum of such matrices is defined in the usual way, and if R is a third finite set, a product

$$K^{P\times Q} \times K^{Q\times R} \to K^{P\times R}$$

is defined in the usual manner. In particular, $K^{Q\times Q}$ thus becomes a semiring. If $P = \{1, \ldots, m\}$ and $Q = \{1, \ldots, n\}$, we will write $K^{m\times n}$ for $K^{P\times Q}$; moreover, $K^{1\times 1}$ will be identified with K.

1.2 Formal series

Let A be a finite, nonempty set called *alphabet*. The *free monoid* A^* generated by A is the set of finite sequences

$$a_1 \cdots a_n$$

of elements of A, including the empty sequence denoted by 1. This set is a monoid, the product being the concatenation defined by

$$(a_1 \cdots a_n) \cdot (b_1 \cdots b_p) = a_1 \cdots a_n b_1 \cdots b_p$$

and with neutral element 1. An element of the alphabet is called a *letter*, an element of A^* is a *word*, and 1 is the *empty word*. The *length* of a word

$$w = a_1 \cdots a_n$$

is n; it is denoted by $|w|$. The length $|w|_a$ relative to a letter a is defined to be the number of occurrences of the letter a in w. We denote by A^+ the subsemigroup $A^* \setminus 1$. A *language* is a subset of A^*.

A *formal series* (or formal power series) S is a function

$$A^* \to K.$$

The image by S of a word w is denoted by (S, w) and is called the *coefficient* of w in S. The *support* of S is the language

$$\text{supp}(S) = \{w \in A^* \mid (S, w) \neq 0\}.$$

The set of formal series on A with coefficients in K is denoted by $K\langle\langle A\rangle\rangle$. A semiring structure is defined on $K\langle\langle A\rangle\rangle$ as follows. If S and T are two formal series, their *sum* is given by

$$(S+T, w) = (S, w) + (T, w),$$

and their *product* by

$$(ST, w) = \sum_{xy=w} (S, x)(T, y).$$

Observe that this sum is finite.

Furthermore, two external operations of K on $K\langle\langle A\rangle\rangle$, one acting on the left, the other on the right, are defined, for $k \in K$, by

$$(kS, w) = k(S, w), \quad (Sk, w) = (S, w)k.$$

There is a natural injection of the free monoid into $K\langle\langle A\rangle\rangle$ as a multiplicative submonoid; the image of a word w is still denoted by w. Thus the neutral element of $K\langle\langle A\rangle\rangle$ for the product is the empty word 1. Similarly, there is an injection of K into $K\langle\langle A\rangle\rangle$ as a subsemiring: to each $k \in K$ is associated $k \cdot 1 = 1 \cdot k$, simply denoted by k. Thus we identify A^* and K with their images in $K\langle\langle A\rangle\rangle$.

The *constant term* of a series S is the coefficient of the empty word, that is $(S, 1)$. Note that the mapping $S \mapsto (S, 1)$ from $K\langle\langle A\rangle\rangle$ onto K is a morphism of semirings, see Exercise 2.1.

A *polynomial* is a formal series with finite support. The set of polynomials is denoted by $K\langle A\rangle$. It is a subsemiring of $K\langle\langle A\rangle\rangle$. The *degree* of a nonnull polynomial is the maximal length of the words in its support, and it is $-\infty$ if the polynomial is null.

When $A = \{a\}$ has just one element, one gets the usual sets of formal power series $K\langle\langle a\rangle\rangle = K[[a]]$ and of polynomials $K\langle a\rangle = K[a]$.

For the rest of this chapter, we fix an alphabet A.

1.3 Topology

We have seen that $K\langle\langle A\rangle\rangle$ is the set of functions $A^* \to K$. In other words,

$$K\langle\langle A\rangle\rangle = K^{A^*}.$$

Thus, if K is equipped with the *discrete topology*, the set $K\langle\langle A\rangle\rangle$ can be equipped with the product topology.

This topology can be defined by an *ultrametric distance*. Indeed, let

$$\omega : K\langle\langle A\rangle\rangle \times K\langle\langle A\rangle\rangle \to \mathbb{N} \cup \infty$$

be the function defined by

$$\omega(S, T) = \inf\{n \in \mathbb{N} \mid \exists w \in A^*, |w| = n \text{ and } (S, w) \neq (T, w)\}.$$

For any fixed real number σ with $0 < \sigma < 1$, the function

$$d : K\langle\langle A\rangle\rangle \times K\langle\langle A\rangle\rangle \to \mathbb{R}$$
$$d(S,T) = \sigma^{\omega(S,T)}$$

is an ultrametric distance, that is d is a distance which satisfies the enforced triangular inequality

$$d(S,T) \leq \max(d(S,U), d(U,T)).$$

The function d defines the topology given above (Exercise 3.1). Furthermore, $K\langle\langle A\rangle\rangle$ is *complete* for this topology, and it is a *topological semiring* (that is sum and product are continuous functions).

Let $(S_i)_{i\in I}$ be a family of series. It is called *summable* if there exists a formal series S such that for all $\varepsilon > 0$, there exists a finite subset I' of I such that every finite subset J of I containing I' satisfies the inequality

$$d\Big(\sum_{j\in J} S_j, S\Big) \leq \varepsilon.$$

The series S is then called the *sum* of the family (S_i) and it is unique.

A family $(S_i)_{i\in I}$ is called *locally finite* if for every word w there exists only a finite number of indices $i \in I$ such that $(S_i, w) \neq 0$. It is easily seen that every locally finite family is summable. The sum of such a family can also be defined simply for $w \in A^*$ by

$$(S, w) = \sum_{i\in I} (S_i, w),$$

observing that the support of this sum is finite because the family (S_i) is locally finite (all terms but a finite number in this sum are 0). However, it is not true that a summable family is always locally finite (see Exercise 3.2), but we shall need mainly the second concept.

Let S be a formal series. Then the family of series $((S,w)w)_{w\in A^*}$ clearly is locally finite, since each of these series has a support formed of at most one single word, and these supports are pairwise disjoint. Thus the family is summable, and its sum is S. This justifies the usual notation

$$S = \sum_{w\in A^*} (S, w)w.$$

It follows in particular that $K\langle A\rangle$ is *dense* in $K\langle\langle A\rangle\rangle$, which thus is the completion of $K\langle A\rangle$ for the distance d.

1.4 Rational series

A formal series $S \in K\langle\langle A\rangle\rangle$ is *proper* if its constant term vanishes. In this case, the family $(S^n)_{n\geq 0}$ is locally finite. Indeed, for any word w, the condition $n > |w|$ implies $(S^n, w) = 0$. Thus the family is summable. The sum of this family is denoted by S^*:

$$S^* = \sum_{n \geq 0} S^n ,$$

and is called the *star* of S. Similarly, S^+ denotes the series

$$S^+ = \sum_{n \geq 1} S^n .$$

The fact that $K\langle\langle A\rangle\rangle$ is a topological semiring and the usual properties of summable families imply that

$$S^* = 1 + S^+ \quad \text{and} \quad S^+ = SS^* = S^*S .$$

From these identities, it follows that if K is a ring, then S^* is the inverse of $1 - S$ since $S^*(1 - S) = S^* - S^*S = S^* - S^+ = 1$. This also implies the following classical result: a series is invertible if and only if its constant term is invertible in K (still assuming K to be a ring); see Exercise 4.5.

Let us return to the general case of a semiring.

Lemma 4.1 *Let T and U be formal series, with T proper. Then the unique solution S of the equation $S = U + TS$ ($S = U + ST$) is the series $S = T^*U$ (the series $S = UT^*$, respectively).*

Proof. One has $T^* = 1 + TT^*$, whence $T^*U = U + TT^*U$. Conversely, since T is proper

$$\lim_n T^n = 0 \quad \text{and} \quad \lim_n \sum_{0 \leq i \leq n} T^i = T^* .$$

From $S = U + TS$, it follows that

$$S = U + T(U + TS) = U + TU + T^2S$$

and inductively

$$S = (1 + T + \cdots + T^n)U + T^{n+1}S .$$

Thus, going to the limit, and using the fact that $K\langle\langle A\rangle\rangle$ is a topological semiring, one gets $S = T^*U$. □

Definition The *rational operations* in $K\langle\langle A\rangle\rangle$ are the sum, the product, the two external products of K on $K\langle\langle A\rangle\rangle$ and the star operation. A subset of $K\langle\langle A\rangle\rangle$ is *rationally closed* if it is closed for the rational operations. The smallest subset containing a subset E of $K\langle\langle A\rangle\rangle$ and which is rationally closed is called the *rational closure* of E.

Definition A formal series is *rational* if it is in the rational closure of $K\langle A\rangle$.

When we want to emphasize the underlying semiring, we say that the series is rational *over K* or is *K-rational.*

Observe that if K is a ring, then the rational closure of $K\langle A\rangle$ is the smallest subring of $K\langle\langle A\rangle\rangle$ containing $K\langle A\rangle$ and closed under inversion. In other words, regarding rational closure, the star operation and inversion play equivalent roles.

The *star height* of a rational series $S \in K\langle\langle A\rangle\rangle$ is defined as follows. We define a sequence

$$R_0 \subset R_1 \subset \cdots \subset R_n \subset \cdots$$

of sets of series, such that the union of the R_n is the set of all rational series. The set R_0 is the set of polynomials, and for $S, T \in R_i$, both $S + T$ and ST are in R_i; if $S \in R_i$ is proper, then $S^* \in R_{i+1}$. The star height of a series S is the least integer n with $S \in R_n$.

Definition If L is a language, its *characteristic series* is the formal series

$$\underline{L} = \sum_{w \in L} w\,.$$

In other words, $(\underline{L}, w) = 1$ for $w \in L$, and $(\underline{L}, w) = 0$ if $w \notin L$.

Example 4.1 The series $\underline{A}$ is proper and

$$\underline{A}^* = \sum_{n \geq 0} \underline{A}^n\,.$$

Since $\underline{A}^n$ is the sum of all words of length n, it follows that

$$\underline{A}^* = \sum_{w \in A^*} w$$

is the characteristic series of A^*. Therefore this series is rational. Consider now a letter a. The series $\underline{A}^* a \underline{A}^*$, as a product of $\underline{A}^*$, a and $\underline{A}^*$, is also rational. By the definition of product,

$$(\underline{A}^* a \underline{A}^*, w) = \sum_{xyz = w} (\underline{A}^*, x)(a, y)(\underline{A}^*, z)\,.$$

Since $(a, y) = 0$ unless $y = a$ (and then $(a, y) = 1$), and since $(\underline{A}^*, x) = (\underline{A}^*, z) = 1$, one has $(\underline{A}^* a \underline{A}^*, w) = \sum_{xaz=w} 1$, which is the number of factorizations $w = xaz$, that is the number $|w|_a$ of occurrences of the letter a in w. Thus

$$\underline{A}^* a \underline{A}^* = \sum_{w} |w|_a w$$

is a rational series.

Let B be an alphabet, and let ρ be a function

$$\rho : A \to K\langle\langle B\rangle\rangle\,.$$

Then ρ extends to a morphism of monoids

$$\rho : A^* \to K\langle\langle B\rangle\rangle\,.$$

If K is commutative, then ρ can be extended in a unique manner into a morphism of semirings

$$\rho : K\langle A\rangle \to K\langle\langle B\rangle\rangle$$

with $\rho|_K = \text{id}$. Indeed, it suffices, for any polynomial $P = \sum_{w\in A^*}(P, w)w \in K\langle A\rangle$, to set

$$\rho(P) = \sum_{w\in A^*}(P, w)\rho(w)$$

which is a finite sum since P is a polynomial. Then ρ is K-linear. Moreover, in view of the commutativity of K

$$\begin{aligned}\rho(P)\rho(Q) &= \sum_{x\in A^*}(P, x)\rho(x)\sum_{y\in A^*}(Q, y)\rho(y)\\ &= \sum_{x,y\in A^*}(P, x)\rho(x)(Q, y)\rho(y) = \sum_{x,y\in A^*}(P, x)(Q, y)\rho(x)\rho(y)\\ &= \sum_{x,y\in A^*}(P, x)(Q, y)\rho(xy)\\ &= \rho\Big(\sum_{x,y\in A^*}(P, x)(Q, y)xy\Big) = \rho(PQ)\,.\end{aligned}$$

Assume now that for each letter $a \in A$, the series $\rho(a)$ is proper. Then $\rho : K\langle A\rangle \to K\langle\langle B\rangle\rangle$ is uniformly continuous. Indeed, let P and Q be two polynomials with

$$\omega(P, Q) = n\,.$$

Then, for any word x in B^* of length $< n$,

$$(\rho(P), x) = \sum_{w\in A^*}(P, w)(\rho(w), x) = \sum_{|w|<n}(P, w)(\rho(w), x)$$

since $(\rho(w), x) = 0$ whenever $|w| \geq n$ by the hypothesis on ρ. Thus

$$(\rho(P), x) = \sum_{|w|<n}(Q, w)(\rho(w), x) = (\rho(Q), x)$$

showing that

$$\omega(\rho(P), \rho(Q)) \geq n\,.$$

Since $K\langle\langle A\rangle\rangle$ is the completion of $K\langle A\rangle$ (see Section 1.3), the function ρ extends uniquely to a morphism of semirings

$$K\langle\langle A\rangle\rangle \to K\langle\langle B\rangle\rangle$$

which induces the identity mapping on K and which is continuous.

Proposition 4.2 *Suppose K is commutative. Let $\rho : A \to K\langle\langle B\rangle\rangle$ be a function such that, for all $a \in A$, the series $\rho(a)$ is a proper rational series. Then ρ extends uniquely to a morphism of semirings $K\langle\langle A\rangle\rangle \to K\langle\langle B\rangle\rangle$ which induces the identity on K and which is continuous. Moreover, the image of any rational series is again rational.*

Proof. It remains to show the last claim. First, if P is a polynomial, then $\rho(P) = \sum (P, w)\rho(w)$ is a rational series since $\rho(a)$ is a rational series for each letter a in A and since ρ is multiplicative. Next, if $\rho(S)$ and $\rho(T)$ are rational series, then so are $\rho(S+T)$ and $\rho(ST)$. Finally, if S is a proper series and $\rho(S)$ is rational, then $\rho(S)$ is proper and

$$\rho(S^*) = \rho\Big(\sum_{n\geq 0} S^n\Big) = \sum_{n\geq 0} \rho(S^n) = \rho(S)^*$$

by the continuity of ρ, showing that $\rho(S^*)$ is rational. This proves that ρ preserves rationality. □

1.5 Recognizable series

Definition A formal series $S \in K\langle\langle A\rangle\rangle$ is called *recognizable* if there exists an integer $n \geq 1$ and a morphism of monoids

$$\mu : A^* \to K^{n\times n}$$

($K^{n\times n}$ with its multiplicative structure) and two matrices $\lambda \in K^{1\times n}$ and $\gamma \in K^{n\times 1}$ such that, for all words w,

$$(S, w) = \lambda \mu w \gamma \, .$$

In this case, the triple (λ, μ, γ) is called a *linear representation* of S, and n is its *dimension*. For sake of coherence, we admit the representation of dimension 0, which corresponds to the null series $S = 0$.

As before, when we want to emphasize the underlying semiring, we say that the series is recognizable *over* K or is K*-recognizable*.

We also use the word *representation* or *linear representation* for a morphism of a monoid into a multiplicative monoid of square matrices. If μ is a representation, we say that a series S is *recognized* by μ if S admits a linear representation of the form (λ, μ, γ).

We shall need the notion of module over a semiring. A *left* K*-module* is a commutative monoid M with law denoted by $+$ and neutral element 0, equipped with an external law $K \times M \to M$ denoted by $(k, x) \mapsto kx$ such that, for all k, ℓ in K and x, y in M the following relations hold:

$$k(x + y) = kx + ky \, ,$$

$$(k + \ell)x = kx + \ell x \, ,$$

$$(k\ell)x = k(\ell x) \, ,$$

$$1x = x \, ,$$

$$0x = 0 \, ,$$

$$k0 = 0 \, .$$

A *submodule* of M is a subset of M containing 0 and closed for the operations of M.

A left K-module is *finitely generated* if there exists finitely many elements $x_1, \ldots, x_n \in M$ such that any element in M can be written as a linear combination

$$k_1 x_1 + \cdots + k_n x_n \quad (k_i \in K).$$

The semiring $K\langle\langle A\rangle\rangle$ of formal power series is a left K-module, where the external law $K \times K\langle\langle A\rangle\rangle \to K\langle\langle A\rangle\rangle$ is the law considered in Section 1.2:

$$(k, S) \mapsto kS.$$

We now define an operation of A^* on $K\langle\langle A\rangle\rangle$. To each word x, and to each formal series S, we associate the series denoted by $x^{-1}S$ and defined by

$$x^{-1}S = \sum_{w \in A^*} (S, xw)w.$$

In other terms, for all words x and w, the coefficient of w in the series $x^{-1}S$ is (S, xw), that is,

$$(x^{-1}S, w) = (S, xw).$$

In particular, $(x^{-1}S, 1) = (S, x)$. A combinatorial view of this operation is given in the case where $S = y$ is a single word. Then $x^{-1}y$ vanishes, unless y has x as a *prefix*, that is $y = xy'$. In this case, $x^{-1}y = y'$.

Observe that this defines completely the operation

$$S \to x^{-1}S$$

since the operation is additive, that is

$$x^{-1}(S + T) = x^{-1}S + x^{-1}T,$$

since it commutes with the external operation of K on $K\langle\langle A\rangle\rangle$, that is

$$x^{-1}(kS) = k(x^{-1}S), \; x^{-1}(Sk) = (x^{-1}S)k$$

for all k in K, and since, finally, this operation is continuous.

Example 5.1

$$(ab)^{-1}(a^2 + aba^2 + abab + ab^2 + b) = a^2 + ab + b.$$

Observe that if P is a polynomial, then $x^{-1}P$ is still a polynomial, with degree less than or equal to the degree of P.

Furthermore, this operation of A^* on $K\langle\langle A\rangle\rangle$ is associative in the following sense:

$$(xy)^{-1}S = y^{-1}(x^{-1}S)$$

as is easily verified. Another property is the following formula which holds for any series S:

$$S = (S, 1) + \sum_{a \in A} a(a^{-1}S). \tag{5.1}$$

This formula is indeed easily proved when S is a word, and then extended by linearity and continuity.

A subset M of $K\langle\langle A\rangle\rangle$ is called *stable* if, for all S in M and x in A^*, the series $x^{-1}S$ is in M.

Proposition 5.1 *A formal series $S \in K\langle\langle A\rangle\rangle$ is recognizable if and only if there exists a stable finitely generated left K-submodule of $K\langle\langle A\rangle\rangle$ which contains S.*

Proof. Assume that S is recognizable, and let (λ, μ, γ) be a linear representation of S of dimension n. Consider the formal series $S_1, \ldots, S_n$ defined by

$$(S_i, w) = (\mu w \gamma)_i$$

for all words w. Let M be the left K-module generated by the series S_i. Thus M is finitely generated. It contains S, since

$$(S, w) = \lambda \mu w \gamma = \sum_i \lambda_i (\mu w \gamma)_i = \sum_i \lambda_i (S_i, w),$$

showing that $S = \sum_i \lambda_i S_i$. Next, M is stable. Indeed, let x be a word. Then

$$\begin{aligned}(x^{-1}S_i, w) &= (S_i, xw) = (\mu(xw)\gamma)_i = (\mu x \mu w \gamma)_i \\ &= \sum_j (\mu x)_{i,j} (\mu w \gamma)_j = \sum_j (\mu x)_{i,j} (S_j, w).\end{aligned}$$

Thus $x^{-1}S_i = \sum_j (\mu x)_{i,j} S_j \in M$. Hence M is stable, since the mapping $T \mapsto x^{-1}T$ is K-linear and sends the generators into M.

Conversely, let M be a stable left submodule of $K\langle\langle A\rangle\rangle$ generated by $S_1, \ldots, S_n$ and containing S. Then

$$S = \sum_i \lambda_i S_i$$

for some λ_i in K. Moreover, for any letter a, there exists a matrix $\mu a \in K^{n \times n}$ such that, for all i,

$$a^{-1}S_i = \sum_j (\mu a)_{i,j} S_j.$$

We extend the mapping $\mu : A \to K^{n \times n}$ to a morphism of monoids $A^* \to K^{n \times n}$ still denoted by μ. Then, for any word w,

$$w^{-1}S_i = \sum_j (\mu w)_{i,j} S_j.$$

Indeed, this relation holds for $w = 1$, and if it holds for some word w, then by induction

$$\begin{aligned}(wa)^{-1}S_i &= a^{-1}(w^{-1}S_i) = a^{-1}\Big(\sum_k (\mu w)_{i,k} S_k\Big) \\ &= \sum_k (\mu w)_{i,k}(a^{-1}S_k) = \sum_k (\mu w)_{i,k} \sum_j (\mu a)_{k,j} S_j \\ &= \sum_j \Big(\sum_k (\mu w)_{i,k}(\mu a)_{k,j}\Big) S_j = \sum_j (\mu w a)_{i,j} S_j \,,\end{aligned}$$

and consequently the relation holds for all words.

Set $\gamma_j = (S_j, 1)$ and let $\gamma \in K^{n\times 1}$ be the matrix defined in this way. Then

$$\begin{aligned}(S_i, w) &= (w^{-1}S_i, 1) = \Big(\sum_j (\mu w)_{i,j} S_j, 1\Big) \\ &= \sum_j (\mu w)_{i,j}(S_j, 1) = \sum_j (\mu w)_{i,j}\gamma_j = (\mu w \gamma)_i \,.\end{aligned}$$

Consequently,

$$\lambda \mu w \gamma = \sum_i \lambda_i (\mu w \gamma)_i = \sum_i \lambda_i (S_i, w) = (S, w)\,,$$

showing that S is recognizable. □

Example 5.2 We use Proposition 5.1 to give an example of a recognizable series.

Let $A = \{0, 1\}$ be the alphabet composed of the two "bits" 0 and 1 and let $K = \mathbb{N}$. For each word w over A, let $\nu_2(w)$ be the integer represented by w in base 2. More precisely, if $w = c_{k-1} \cdots c_0$ with $k \geq 0$ and $c_i \in A$, then

$$\nu_2(w) = 2^{k-1}c_{k-1} + \cdots + 2c_1 + c_0 \,.$$

The integer represented by the empty word is 0. We show that the series

$$S = \sum_{w \in A^*} \nu_2(w)\, w$$

is recognizable. Then S starts with

$$\begin{aligned}S &= 1 + 01 + 2\cdot 10 + 3 \cdot 1^2 + 0^2 1 + 2 \cdot 010 + 3 \cdot 01^2 \\ &\quad + 4 \cdot 10^2 + 5 \cdot 101 + 6 \cdot 1^2 0 + 7 \cdot 1^3 + \cdots\end{aligned}$$

Given a word w, one has the relations $(S, 0w) = (S, w)$ and $(S, 1w) = 2^{|w|} + (S, w)$. In other words, $0^{-1}S = S$ and $1^{-1}S = T + S$, where T is the series

$$T = \sum_w 2^{|w|} w \,.$$

Next, $0^{-1}T = 1^{-1}T = 2T$. This shows that the $\mathbb{N}$-submodule M of $\mathbb{N}\langle\langle A \rangle\rangle$ generated by S and T is stable under the operations $U \mapsto a^{-1}U \quad (a \in A)$. Proposition 5.1 shows that S is recognizable.

Corollary 5.2 *Any left or right linear combination of recognizable series is a recognizable series.*

Proof. If M is a stable finitely generated left K-submodule of $K\langle\langle A\rangle\rangle$ containing a series S, then it contains kS for any k in K, hence kS is recognizable. Moreover, the set $Mk = \{Tk \mid T \in M\}$ is a stable finitely generated left K-submodule of $K\langle\langle A\rangle\rangle$ containing Sk; therefore the latter series is recognizable.

Now, let M_1, M_2 be two stable finitely generated left K-submodules of $K\langle\langle A\rangle\rangle$ containing S_1, S_2 respectively. Then the sum of M_1 and M_2, which is $M_1 + M_2 = \{T_1 + T_2 \mid T_i \in M_i\}$, is a stable finitely generated left K-submodule of $K\langle\langle A\rangle\rangle$ containing $S_1 + S_2$; the latter is therefore recognizable.

Thus the corollary follows from Proposition 5.1. □

A direct construction also yields a proof of the corollary. Indeed, if (λ, μ, γ) is a linear representation of S, then kS (resp. Sk) has the linear representation $(k\lambda, \mu, \gamma)$ (resp. $(\lambda, \mu, \gamma k)$). Moreover, if S_i has the linear representation $(\lambda_i, \mu_i, \gamma_i)$ for $i = 1, 2$, then $S_1 + S_2$ has the linear representation (λ, μ, γ) with

$$\lambda = \begin{pmatrix} \lambda_1 & \lambda_2 \end{pmatrix}, \quad \mu = \begin{pmatrix} \mu_1 & 0 \\ 0 & \mu_2 \end{pmatrix}, \quad \gamma = \begin{pmatrix} \gamma_1 \\ \gamma_2 \end{pmatrix}.$$

This is easily verified and left to the reader.

Observe that if M is a stable left K-submodule of $K\langle\langle A\rangle\rangle$ containing a series S, then it contains the series $u^{-1}S$, for $u \in A^*$, and all left K-linear combinations of such series. It follows that the smallest stable left K-submodule containing S is the set of all these linear combinations. Denote it by N.

The left K-submodule N is not always finitely generated, see Exercise 5.5. However, if N is a finitely generated left K-submodule, then it is finitely generated over K by a finite number of series of the form $u^{-1}S$: indeed, N is then generated by finitely many series $S_1, \ldots, S_k$, and each S_i is a linear combination of finitely many series $u_{i,j}^{-1}S$, $j = 1, \ldots, n_i$; thus N is generated by the series $u_{i,j}^{-1}S$, $i = 1, \ldots, k$, $j = 1, \ldots, n_i$.

Corollary 5.4 below describes cases where N is finitely generated.

Recall that a commutative ring K is called *Noetherian* if each submodule of a finitely generated (left or right) K-module is also a finitely generated module. We use the following classical result.

Theorem 5.3 (see (Lang 1984), Cor. IV.2.4 and Prop X.1.4) *Each finitely generated commutative ring is Noetherian.*

Corollary 5.4 *Assume that K is a finite semiring or a commutative ring. Then a series S in $K\langle\langle A\rangle\rangle$ is recognizable if and only if the smallest stable left K-submodule of $K\langle\langle A\rangle\rangle$ containing S is finitely generated. In this case, this submodule is the set of left K-linear combinations of the series $u^{-1}S$, for $u \in A^*$, and in fact of a finite number of them.*

Proof. The "if" part follows directly from Proposition 5.1. Conversely, suppose that S is recognizable. Then, by Proposition 5.1, there is a stable and finitely generated left K-submodule of $K\langle\langle A\rangle\rangle$ containing S. If K is finite, then finitely generated modules and finite modules coincide, hence each submodule of a finitely generated module is finitely generated, and the corollary follows.

Suppose now that K is a commutative ring. Let (λ, μ, γ) be some linear representation of S and let K_1 be the subring generated by the coefficients appearing in the matrices λ, $\mu(a)$ for $a \in A$ and γ. Then K_1 is a finitely generated ring and it is therefore Noetherian by Theorem 5.3, and each submodule of a finitely generated K_1-module is again finitely generated. Since S is recognizable over K_1, it follows from Proposition 5.1 and the fact that K_1 is Noetherian that the K_1-submodule spanned by the series $u^{-1}S$ is finitely generated. Thus, by the remark preceding this corollary, each series $u^{-1}S$ is a K_1-linear combination of finitely many such series. Hence the K-submodule generated by the series $u^{-1}S$ is finitely generated, which proves the corollary. □

Definition The *Hadamard product* of two formal series S and T is the series $S \odot T$ defined by

$$(S \odot T, w) = (S, w)(T, w).$$

Theorem 5.5 (Schützenberger 1962a) *Let K_1 and K_2 be two subsemirings of K such that each element of K_1 commutes with each element of K_2. If S_1 is a K_1-recognizable series and S_2 is a K_2-recognizable series, then $S_1 \odot S_2$ is K-recognizable.*

Proof. We apply Proposition 5.1. Let M_1 (M_2) be a left submodule of $K_1\langle\langle A\rangle\rangle$ (of $K_2\langle\langle A\rangle\rangle$) which contains S_1 (S_2), is stable, and is generated by the series $T_1^1, \ldots, T_1^n \in K_1\langle\langle A\rangle\rangle$ (the series $T_2^1, \ldots, T_2^m \in K_2\langle\langle A\rangle\rangle$ respectively).

Let M be the left K-submodule of $K\langle\langle A\rangle\rangle$ generated by $M_1 \odot M_2 = \{T_1 \odot T_2 \mid T_1 \in M_1, T_2 \in M_2\}$. Clearly, $S_1 \odot S_2$ is in M. Moreover, M is finitely generated. Indeed, if $T_1 = \sum_{1\le i\le n} k_i T_1^i \in M_1$ with $k_i \in K_1$ and $T_2 = \sum_{1\le j\le m} \ell_j T_2^j \in M_2$ with $\ell_j \in K_2$, then for any word w,

$$\begin{aligned}(T_1 \odot T_2, w) = (T_1, w)(T_2, w) &= \sum_{i,j} k_i (T_1^i, w)\ell_j (T_2^j, w)\\ &= \sum_{i,j} k_i \ell_j (T_1^i, w)(T_2^j, w)\end{aligned}$$

since (T_1^i, w) and ℓ_j commute. Thus

$$T_1 \odot T_2 = \sum_{i,j} k_i \ell_j T_1^i \odot T_2^j,$$

showing that M is generated, as a left K-module, by the series $T_1^i \odot T_2^j$.

Finally, M is stable, since for any word x, and for series $T_1 \in M_1$, $T_2 \in M_2$,

$$x^{-1}(T_1 \odot T_2) = (x^{-1}T_1) \odot (x^{-1}T_2) \in M . \qquad \square$$

Example 5.3 For $n \in \mathbb{N}$, we denote by n the element $1 + \cdots + 1$ (n times) of K. Let a be a letter. Then the series $\sum_w |w|_a w$ is recognizable (it is also rational, as seen in Example 4.1). Indeed the series admits the linear representation (λ, μ, γ) defined by $\lambda = (1, 0)$, $\mu a = \begin{pmatrix} 1 & 1 \\ 0 & 1 \end{pmatrix}$, $\mu b = \begin{pmatrix} 1 & 0 \\ 0 & 1 \end{pmatrix}$, for $b \in A \setminus a$, and $\gamma = \begin{pmatrix} 0 \\ 1 \end{pmatrix}$. It is indeed easily seen that for any word w,

$$\mu w = \begin{pmatrix} 1 & |w|_a \\ 0 & 1 \end{pmatrix} .$$

As an application, let $P(t_1, \ldots, t_n)$ be a *commutative* polynomial with coefficients in K. Then the formal series (over the alphabet $A = \{a_1, \ldots, a_n\}$)

$$S = \sum_{w \in A^*} P(|w|_{a_1}, \ldots, |w|_{a_n})w$$

is recognizable. This follows from Theorem 5.5, Corollary 5.2 and from the recognizability of $\sum |w|_a w$.

1.6 Weighted automata

We present now the notion of *weighted finite automaton* which is a graphical way to represent a linear representation. Its advantage is that it shows the relation with usual finite automata, and helps in understanding some constructions.

Let K be a semiring, and let A be an alphabet.

Definition A *weighted* (finite) *automaton* $\mathcal{A} = (Q, I, E, T)$ with weights in K, or a K-automaton over A is composed of a (finite) set Q of *states*, and of three mappings

$$I : Q \to K, \quad E : Q \times A \times Q \to K, \quad T : Q \to K .$$

A triple (p, a, q) such that $E(p, a, q) \neq 0$ is an *edge*, p and q are its *start* and *end states*, the letter a is its *label* and $E(p, a, q)$ is its *weight*. A *path* is a sequence

$$c = (q_0, a_1, q_1)(q_1, a_2, q_2) \cdots (q_{n-1}, a_n, q_n)$$

of edges. The *weight* of the path c is the product

$$E(c) = E(q_0, a_1, q_1)E(q_1, a_2, q_2) \cdots E(q_{n-1}, a_n, q_n)$$

of the weights of its edges. Its *label* is the word $w = a_1 a_2 \cdots a_n$. The series S recognized by $\mathcal{A}$ is defined by

$$(S, w) = \sum_{q_0, \ldots, q_n \in Q} I(q_0)E(q_0, a_1, q_1) \cdots E(q_{n-1}, a_n, q_n)T(q_n) . \qquad (6.1)$$

It is useful to call a state q *initial* (*final*) if $I(q) \neq 0$ ($T(q) \neq 0$). The coefficient (S, w) is the sum of the weights of all paths c labeled w from an initial state p to a final state q, each weight being multiplied on the left by the coefficient of the initial state and on the right by the coefficient of the final state.

If $K = \mathbb{B}$, a weighted automaton is just a usual nondeterministic automaton. In this case, I, E and T may be represented by subsets of Q, $Q \times A \times Q$ and Q respectively, which is the usual way of representing an automaton. Note also that the automaton is *deterministic* if for any p in Q and $a \in A$, there is at most one q in Q such that $E(p, a, q) \neq 0$, and if moreover there is exactly one initial state.

A weighted automaton is represented by a graph. Each state is a vertex, and each edge carries an expression ka, where k is its weight and a is its label. Whenever the weight is 1, its value is understood. Each initial (final) state q is distinguished by an incoming (outgoing) edge which carries the weight $I(q)$ ($T(q)$). Again, when the weight is 1, it is omitted.

Example 6.1 Consider the series S over $\mathbb{Z}$ on $A = \{a, b\}$ defined by

$$(S, w) = \begin{cases} 2^n & \text{if } w = a^n, n \geq 1 \\ -3 \cdot 2^n & \text{if } w = a^n b, n \geq 0 \\ 0 & \text{otherwise.} \end{cases}$$

In other words

$$S = \sum_{n \geq 1} 2^n a^n - 3 \sum_{n \geq 0} 2^n a^n b \,.$$

The support of S is the set $a^+ \cup a^* b$. The series is recognized by the following $\mathbb{Z}$-automaton

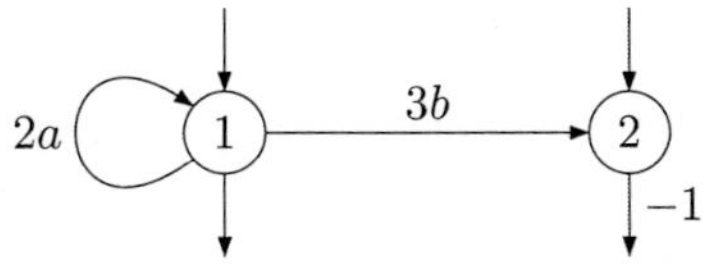

Indeed, for a^n with $n > 0$ there is a unique path with label a^n: it goes from state 1 to state 1 and its weight is 2^n. Similarly, for $a^n b$ with $n \geq 0$ there is a unique path, from 1 to 2 with weight $2^n \cdot 3$, so the coefficient of $a^n b$ in the series recognized by the automaton is $-2^n \cdot 3$. There are two paths labeled with the empty word, the first through state 1, and the second through state 2. The first path contributes 1 to the coefficient of the empty word, and the second path contributes -1, so the coefficient of the empty word in the series recognized by the automaton is 0.

Proposition 6.1 *A series is recognized by a finite weighted automaton if and only if it is recognizable.*

Proof. Assume S is recognized by the automaton $\mathcal{A} = (Q, I, E, T)$. One may suppose $Q = \{1, \ldots, n\}$. Then S is recognized by the linear representation (λ, μ, γ), where $\lambda \in K^{1\times n}$, $\mu : A^* \to K^{n\times n}$, $\gamma \in K^{n\times 1}$ are defined by $\lambda_p = I(p)$, $(\mu a)_{p,q} = E(p, a, q)$, $\gamma_q = T(q)$ for $1 \le p, q \le n$. Indeed, for $w = a_1 \cdots a_m$,

$$(\mu(w))_{p,q} = \sum_{p_1,\ldots,p_{m-1}} E(p, a_1, p_1)E(p_1, a_2, p_2) \cdots E(p_{m-1}, a_m, q)$$

is the sum of the weights of the paths from p to q labeled w. Therefore (S, w), which is given by Equation (6.1), is equal to $\lambda\mu w\gamma$.

Conversely, let (λ, μ, γ) be a linear representation recognizing S, and define a weighted automaton $\mathcal{A} = (Q, I, E, T)$ by setting $I(p) = \lambda_p$, $E(p, a, q) = (\mu(a))_{p,q}$, $T(q) = \gamma_q$. Then $\mathcal{A}$ recognizes S. □

The proof shows that there is a complete equivalence between the notion of a weighted automaton and that of a linear representation: they are called *associated* to each other.

Example 6.2 The automaton of the previous example corresponds to the linear representation

$$\lambda = (1 \quad 1) \quad \mu(a) = \begin{pmatrix} 2 & 0 \\ 0 & 0 \end{pmatrix} \quad \mu(b) = \begin{pmatrix} 0 & 3 \\ 0 & 0 \end{pmatrix} \quad \gamma = \begin{pmatrix} 1 \\ -1 \end{pmatrix}.$$

Observe that in particular

$$\mu(a^n) = \begin{pmatrix} 2^n & 0 \\ 0 & 0 \end{pmatrix}, \quad \mu(a^n b) = \begin{pmatrix} 0 & 3\cdot 2^n \\ 0 & 0 \end{pmatrix}.$$

1.7 The fundamental theorem

Theorem 7.1 (Schützenberger 1961a) *A formal series is recognizable if and only if it is rational.*

We start with several lemmas which will be needed for the proof.

Lemma 7.2 *Let S and T be formal series, and let a be a letter. Then*

$$a^{-1}(ST) = (a^{-1}S)T + (S, 1)(a^{-1}T).$$

If S is proper, then

$$a^{-1}(S^*) = (a^{-1}S)S^*.$$

Proof. For any word w,

$$(a^{-1}(ST), w) = (ST, aw) = \sum_{uv=aw} (S, u)(T, v)$$

$$= (S, 1)(T, aw) + \sum_{uv=w} (S, au)(T, v)$$

$$= (S, 1)(T, aw) + \sum_{uv=w} (a^{-1}S, u)(T, v)$$

$$= (S, 1)(a^{-1}T, w) + ((a^{-1}S)T, w)\,.$$

This proves the first relation.

For the second relation, observe that $S^* = 1 + SS^*$, whence, using the first relation, $a^{-1}(S^*) = (a^{-1}S)S^*$, since $(S, 1) = 0$. □

Let m be an $n \times n$-matrix with coefficients in $K\langle\langle A\rangle\rangle$:

$$m \in K\langle\langle A\rangle\rangle^{n\times n}\,.$$

The matrix is *proper* if, for all indices i and j, the series $m_{i,j}$ is proper. In this case, the *star* of m can be defined as

$$m^* = \sum_{k\geq 0} m^k\,.$$

The existence of m^* can be verified by considering the product topology induced by $K\langle\langle A\rangle\rangle$ on $K\langle\langle A\rangle\rangle^{n\times n}$ (the details are left to the reader). It is easily seen that

$$m^* = 1 + mm^*\,, \tag{7.1}$$

where 1 is the identity matrix.

Lemma 7.3 *If m is a proper matrix with elements in $K\langle\langle A\rangle\rangle$, then all coefficients of m^* are in the rational closure of the coefficients of m.*

Proof. Let m be an $n \times n$-matrix. If $n = 1$, the result is clear. Arguing by induction on n, assume $n > 1$ and consider a decomposition into blocks

$$m = \begin{pmatrix} a & b \\ c & d \end{pmatrix}$$

where a and d are square matrices, and set

$$m^* = \begin{pmatrix} \alpha & \beta \\ \gamma & \delta \end{pmatrix}$$

where the blocks have the same dimensions as the corresponding blocks in m.

By Eq. (7.1), we get

$$\alpha = 1 + a\alpha + b\gamma \qquad \beta = a\beta + b\delta$$

$$\gamma = c\alpha + d\gamma \qquad \delta = 1 + c\beta + d\delta\,.$$

Observe that Lemma 4.1 extends to matrix equations; thus we have

$$\beta = a^* b\delta, \quad \gamma = d^* c\alpha\,,$$

whence

$$\alpha = 1 + a\alpha + bd^* c\alpha = 1 + (a + bd^* c)\alpha$$

$$\delta = 1 + ca^* b\delta + d\delta = 1 + (ca^* b + d)\delta\,.$$

Again, Lemma 4.1 gives

$$\alpha = (a + bd^* c)^*$$

$$\delta = (ca^* b + d)^*\,.$$

Finally

$$\beta = a^* b(ca^* b + d)^*$$

$$\gamma = d^* c(a + bd^* c)^*\,.$$

By the induction hypothesis, all coefficients of a^*, d^* are in the rational closure of the coefficients of m. The same holds for the coefficients of $a + bd^* c$ and $ca^* b + d$, and using again the induction hypothesis, the coefficients of α, δ, and also those of β and γ, are in this rational closure. □

Proof of Theorem 7.1. In order to show that any rational series is recognizable, we use Proposition 5.1. If P is a polynomial, then $w^{-1}P = 0$ for any word w of length greater than $\deg(P)$. Consequently, the set $\{w^{-1}P \mid w \in A^*\}$ is finite. Since it is stable, it generates a stable submodule which, moreover, is finitely generated and also contains P (because $1^{-1}P = P$). Thus P is recognizable.

If S and T are recognizable, then there exist stable finitely generated submodules M and N of $K\langle\langle A\rangle\rangle$ with $S \in M$ and $T \in N$. Then $M + N$ contains $S + T$, is finitely generated and is stable, showing that $S + T$ is recognizable.

Next, let P be the submodule $P = MT + N$. Clearly, P contains ST, and according to Lemma 7.2, P is stable. It is finitely generated because M and N are finitely generated. Hence ST is recognizable.

Assume now that S is proper. Let Q be the submodule $Q = K + MS^*$. Then Q contains $S^* = 1 + SS^*$, and Q is stable since, by Lemma 7.2,

$$a^{-1}(S'S^*) = (a^{-1}S')S^* + (S', 1)(a^{-1}S)S^*$$

is in Q for all S' in M. Finally, Q is finitely generated. Hence S^* is recognizable.

Conversely, let S be a recognizable series and let (λ, μ, γ) be a linear representation of S of dimension n. Consider the proper matrix

$$m = \sum_{a \in A} \mu a a \in K^{n \times n}\langle\langle A\rangle\rangle\,.$$

We use below the natural isomorphism between $K^{n\times n}\langle\langle A\rangle\rangle$ and $K\langle\langle A\rangle\rangle^{n\times n}$. Then

$$m^* = \sum_{k\geq 0} m^k = \sum_{k\geq 0}\Big(\sum_{a\in A}\mu a a\Big)^k = \sum_{k\geq 0}\sum_{w\in A^k}\mu w w = \sum_{w\in A^*}\mu w w\,.$$

Thus

$$m^*_{i,j} = \sum_w (\mu w)_{i,j} w$$

is rational in view of Lemma 7.3. Since

$$S = \sum_{i,j}\lambda_i m^*_{i,j}\gamma_j\,,$$

the series S is rational. □

Exercises for Chapter 1

1.1 Let $K = \{0, 1\}$ be a semiring composed of two elements. Show that, according to the value of $1 + 1$, K is either the field with two elements or the Boolean semiring.

1.2 Let K be a semiring. A *congruence* in K is an equivalence relation $\equiv$ which is compatible with the laws of K, that is for all $a, b, c, d \in K$,

$$a \equiv b, c \equiv d \implies a + c \equiv b + d,\ ac \equiv bd\,.$$

(a) Show that $K/\!\equiv$ has a natural structure of a semiring. Such a semiring is called a *quotient* of K.

(b) Show that if K is a ring then there is a bijection between congruences and two-sided ideals in K.

(c) Show that any quotient semiring of $\mathbb{N}$ which is not isomorphic to $\mathbb{N}$ is finite.

1.3 The *prime* subsemiring of a semiring K is the semiring L generated by 1. Show that every element in L commutes with every element in K and that L either is isomorphic to $\mathbb{N}$ or is finite.

1.4 Let K be a commutative semiring.

(a) Define two operations on $K \times K$ by

$$(a, b) + (a', b') = (a + a', b + b')\,,$$
$$(a, b)(a', b') = (aa' + bb', ab' + ba')\,.$$

Show that these operations make $K \times K$ a commutative semiring with zero $(0, 0)$ and unity $(1, 0)$. Show that

$$i : a \mapsto (a, 0)$$

is an injection of K into $K \times K$. Show that the relation $\equiv$ defined by

$$(a, b) \equiv (a', b') \iff \exists c : a + b' + c = a' + b + c$$

is a congruence on $K \times K$. Show that $L = K \times K/\equiv$ is a ring.

(b) Denote by p the canonical surjection

$$p : K \times K \to L .$$

Show that $p \circ i : K \to L$ is injective if and only if for all $a, b, c \in K$

$$a + b = a + c \implies b = c .$$

A semiring having this property is called *regular*. Show that K can be embedded into a ring if and only if it is regular.

(c) Assume that K is regular. Show that the ring L is without zero divisors if and only if for all $a, b, c, d \in K$, the following condition holds:

$$ac + bd = ad + bc \implies a = b \text{ or } c = d .$$

Show that K can be embedded into a field if and only if K is regular and this condition is satisfied.

(d) K is *simplifiable* if for all $a, b, c \in K$

$$ab = ac \implies b = c \text{ or } a = 0 .$$

Show that if K can be embedded into a field, then it is regular and simplifiable.

(e) Let a, b, c, d be commutative indeterminates and let I be the ideal of $\mathbb{Z}[a, b, c, d]$ generated by $(a - b)(c - d)$. Show that the image K of $\mathbb{N}[a, b, c, d]$ in the quotient $\mathbb{Z}[a, b, c, d]/I$ is a regular and simplifiable semiring, but that K cannot be embedded into any field.

2.1 Verify that the mapping $S \mapsto (S, 1)$ is a semiring morphism $K\langle\langle A\rangle\rangle \to K$.

3.1 Give complete proofs for the claims in Section 1.3.

3.2 Let $\mathbb{B}$ be the Boolean semiring and for all $n \in \mathbb{N}$, let $S_n = 1$. Show that the family $(S_n)_{n\in\mathbb{N}}$ is summable, but not locally finite.

3.3 Let K, L be two semirings, and let A, B be two alphabets. A function

$$f : K\langle\langle A\rangle\rangle \to L\langle\langle B\rangle\rangle$$

is a *morphism of formal series* if f is a morphism of semirings and moreover is uniformly continuous.

(a) Show that the mapping

$$L\langle\langle B\rangle\rangle \to L$$
$$S \mapsto (S, 1)$$

is a continuous morphism of semirings. Show that if

$$f : K\langle\langle A\rangle\rangle \to L\langle\langle B\rangle\rangle$$

is a morphism of semirings which is continuous at 0, then
(i) for all $k \in K$ and $a \in A$, the elements $f(k)$ and $f(a)$ commute,
(ii) the multiplicative subsemigroup of L generated by

$$\{(f(a), 1) \mid a \in A\}$$

is nilpotent, that is, for some n, each product of n elements of this semigroup is null.
(b) Let $f : A \cup K \to L\langle\langle B\rangle\rangle$ be a function satisfying conditions (i) and (ii) of (a). Show that f extends in a unique manner to a morphism of formal series

$$K\langle\langle A\rangle\rangle \to L\langle\langle B\rangle\rangle .$$

3.4 Let M be a commutative monoid, with law denoted additively, having an ultrametric distance d which is *subinvariant* with respect to translation (that is such that $d(a + c, b + c) \leq d(a, b)$ for $a, b, c \in M$). Show that every series that converges in M converges commutatively.

3.5 Assume that K is a field. Recall that for any K-vector space E, for any subspace F and any vector v in $E \setminus F$, there exists a linear function h on E such that $h(E) = 0$ and $h(v) \neq 0$. We use here the identification of $K\langle\langle A\rangle\rangle$ and of the dual of $K\langle A\rangle$ (see beginning of Chapter 2).
(a) For each subspace V of $K\langle A\rangle$ (subspace W of $K\langle\langle A\rangle\rangle$), define its *orthogonal* in $K\langle\langle A\rangle\rangle$ (in $K\langle A\rangle$) to be given by

$$V^{\perp} = \{S \in K\langle\langle A\rangle\rangle \mid \forall P \in V, (S, P) = 0\}$$

$$(W^{\perp} = \{P \in K\langle A\rangle \mid \forall S \in W, (S, P) = 0\}, \text{respectively.})$$

Show that if V is a subspace of $K\langle A\rangle$, then $V^{\perp\perp} = V$.
(b) Show that if a linear function h on $K\langle\langle A\rangle\rangle$ is continuous (for the discrete topology on K and the product topology on K^{A^*}) then Ker(h) contains all but a finite number of elements of A^*. Show that the topological dual space of $K\langle\langle A\rangle\rangle$ can be identified with $K\langle A\rangle$.
(c) Show that for any *closed* subspace W of $K\langle\langle A\rangle\rangle$, and for any formal series S not in W, there exists a *continuous* linear function h on $K\langle\langle A\rangle\rangle$ such that $h(S) \neq 0$ and $h(W) = 0$. Show from this that for any subspace W of $K\langle\langle A\rangle\rangle$, $W^{\perp\perp}$ is the adherence of W.

4.1 Let $S \in K\langle\langle A\rangle\rangle$, let c be its constant term and let T be a proper series with $S = c + T$.
(a) Show that if $\sum S^n$ converges in $K\langle\langle A\rangle\rangle$, then $\sum c^n$ also converges in K for the discrete topology.

(b) Show that if $\sum c^n$ converges in K, then $\sum S^n$ converges in $K\langle\langle A\rangle\rangle$, and then

$$\sum_{n\geq 0} S^n = \Big(\big(\sum_{n\geq 0} c^n\big)T\Big)^*\big(\sum_{n\geq 0} c^n\big).$$

(c) Show that if S is rational and if $\sum S^n$ converges, then $\sum S^n$ is rational.

(d) Show that if $f : K\langle\langle A\rangle\rangle \to L\langle\langle B\rangle\rangle$ is a morphism of formal series (see Exercise 3.3) such that $f(S)$ is rational for all $S \in K \cup A$, then f preserves rationality.

4.2 Let (S_n) be a sequence of proper series. Show that if $\lim S_n = S$, then S is proper and $\lim S_n^* = S^*$.

4.3 Recall that an element a of a ring K is called *quasi-regular* (in the sense of Jacobson) if there exists some $b \in K$ such that $a + b + ab = 0$. Recall also that the radical R of K is the greatest two-sided ideal of K having only quasi-regular elements (it exists by (Herstein 1968) Th. 1.2.3).

(a) Show that $S \in K\langle\langle A\rangle\rangle$ is quasi-regular in $K\langle\langle A\rangle\rangle$ if and only if its constant term is quasi-regular in K.

(b) Show that the radical of $K\langle\langle A\rangle\rangle$ is

$$\{S \in K\langle\langle A\rangle\rangle \mid (S, 1) \in R\}.$$

4.4 Let $k \geq 2$ be an integer and let $A = \{0, \ldots, k-1\}$. For any word w over A, we denote by $\nu_k(w)$ the integer represented by w in base k. For example $\nu_k(0111) = k^2 + k + 1$. We write $\underline{c}$ for c when we need to distinguish the symbol $\underline{c}$ from the number c. Let S and T be the series defined by

$$S = \sum_w \nu_k(w)\, w,\ T = \sum_w k^{|w|} w.$$

Show that $T = 1 + k\underline{A}T$ and that $S = PT + \underline{A}S$. Deduce that

$$S = \underline{A}^* P(k\underline{A})^*,$$

where $P = 1 + 2\underline{2} + \cdots (k-1)\underline{k-1}$.

4.5 Assume that K is a ring. Show that a series is invertible in $K\langle\langle A\rangle\rangle$ if and only if its constant term is invertible in K. Show that if K is a field, the set of proper series is the unique maximal ideal of $K\langle\langle A\rangle\rangle$.

5.1 (a) Suppose that K is a field with absolute value $|\,|$. Show that if $S \in K\langle\langle A\rangle\rangle$ is recognizable, then there is a constant $C \in \mathbb{R}$ such that for all $w \in A^*$

$$|(S, w)| \leq C^{1+|w|}.$$

(b) Suppose that K is a commutative integral domain with quotient field F. Show that if $S \in F\langle\langle A\rangle\rangle$ is recognizable and has a linear representation (λ, μ, γ), then for some $C \in K \setminus 0$ the series $\sum_w C^{2+|w|}(S, w)w$ is in $K\langle\langle A\rangle\rangle$, is K-recognizable and has the linear representation $(C\lambda, C\mu, C\gamma)$ over K ("Eisenstein's criterion").

5.2 Verify that a series in $K\langle\langle A\rangle\rangle$ is Hadamard-invertible if and only if no coefficient in this series is 0 (we assume that K is a field).

Show that the inverse of a rational series is in general not rational, by considering the series $\sum_{n\geq 0} 1/(n+1)a^n$ in $\mathbb{Q}\langle\langle a\rangle\rangle$ (use Eisenstein's criterion).

5.3 Let $w = a_1 \cdots a_n$ be a word ($a_i \in A$). For any subset $I = \{i_1 < \cdots < i_k\}$ of $\{1, \ldots, n\}$, define $w|I$ to be the word $a_{i_1} \cdots a_{i_k}$. Given two words x and y of length n and p respectively, define their *shuffle* product $x \sqcup\!\sqcup\, y$ to be the polynomial

$$x \sqcup\!\sqcup\, y = \sum w(I, J),$$

where the sum is over all couples (I, J) with $\{1, 2, \ldots, n+p\} = I \cup J$, $|I| = n$, $|J| = p$, and where $w(I, J)$ is defined by $w(I, J)|I = x$, $w(I, J)|J = y$. Moreover, $1 \sqcup\!\sqcup\, y = y \sqcup\!\sqcup\, 1 = y$. For example,

$$ab \sqcup\!\sqcup\, ac = abac + 2a^2bc + 2a^2cb + acab\,.$$

Let K be a commutative semiring. Extend the shuffle product to $K\langle\langle A\rangle\rangle$ by linearity and continuity, that is

$$S \sqcup\!\sqcup\, T = \sum_{x,y\in A^*} (S, x)(T, y)x \sqcup\!\sqcup\, y\,.$$

Show that the shuffle product is commutative and associative. Show that the operator

$$S \mapsto a^{-1}S \quad (a \in A)$$

is a derivation for the shuffle, that is

$$a^{-1}(S \sqcup\!\sqcup\, T) = (a^{-1}S) \sqcup\!\sqcup\, T + S \sqcup\!\sqcup\, (a^{-1}T)\,. \tag{*}$$

Show that the shuffle product of two recognizable series is still recognizable. (*Hint*: Proceed as in the proof of Theorem 5.5 and use Eq. (*).)

5.4 To show that for each $k \geq 2$, the series $\sum n^{k-1}a^n$ over one letter a is recognizable without using the Hadamard product, consider the matrix representation of order k defined by

$$\mu(a)_{i,j} = \binom{k-i}{k-j}.$$

For instance, for $k = 4$, one gets

$$\mu(a) = \begin{pmatrix} 1 & 3 & 3 & 1 \\ 0 & 1 & 2 & 1 \\ 0 & 0 & 1 & 1 \\ 0 & 0 & 0 & 1 \end{pmatrix}.$$

Show that $\mu(a^n)_{1,k} = n^{k-1}$. Compare the dimension k of this representation to the dimension of the $k-1$-th Hadamard power of the series $\sum na^n$.

5.5 Show that, although the series $S = \sum_{n\geq 0} na^n$ is recognizable over the semiring $\mathbb{N}$, the smallest stable $\mathbb{N}$-submodule of $\mathbb{N}\langle\langle a\rangle\rangle$ containing S is not finitely generated over $\mathbb{N}$. (*Hint*: Otherwise, for some $n_1, \ldots, n_k$ in $\mathbb{N}$, each series $(a^\ell)^{-1}S$ is a $\mathbb{N}$-linear combination of the series $(a^{n_1})^{-1}S, \ldots, (a^{n_k})^{-1}S$.)

5.6 Denote by $\bar{S}$ the smallest stable left K-submodule of $K\langle\langle A\rangle\rangle$ containing S. Show that if K is a ring and S is invertible, then $\overline{S^{-1}} + K = S^{-1}(\bar{S} + K)$.

7.1 Let S have the representation (λ, μ, γ) of dimension n over K. Let S_i have the representations (e_i, μ, γ), where e_i is the i-th canonical vector. Show that $S = \sum \lambda_i S_i$. Show that $S_1, \ldots, S_n$ satisfy

$$a^{-1}S_i = \sum_j (\mu a)_{i,j} S_j$$

for any letter a. Show that they satisfy the system of linear equations

$$S_i = (S_i, 1) + \sum_{j=1}^{n} \Big(\sum_{a \in A} (\mu a)_{i,j} a\Big) S_j \,.$$

7.2 Let $P_{i,j}$, Q_j be series, with each $P_{i,j}$ proper. Use iteratively Lemma 4.1 to show how to solve the system of linear equations

$$S_i = Q_i + \sum_{j=1}^{n} P_{i,j} S_j \,, \qquad i = 1, \ldots, n \,,$$

where the S_i are unknown series. Deduce from this and from Exercise 7.1 another proof of the fact that a recognizable series is rational.

7.3 Show how to construct algorithmically a rational expression representing a recognizable series given by some representation. (*Hint*: use Lemma 7.3 or Exercises 7.1 and 7.2.)

Notes to Chapter 1

Theorem 7.1 showing the equivalence between rationality and recognizability was first proved by Kleene (1956) for languages (which may be seen as series with coefficients in the Boolean semiring) and later extended by Schützenberger (1961a, 1962a,b) to arbitrary semirings.

Here we have derived Kleene's theorem from Schützenberger's (see Chapter 3). The condition "recognizable" $\Longrightarrow$ "rational," which is essentially Lemma 7.3, is proved here by using an argument of Conway (1971). Other proofs are also given in Eilenberg (1974) and Salomaa and Soittola (1978). The characterization of recognizable series (Proposition 5.1) is taken from Jacob (1975) who extends to semirings a Hankel-like property given by Fliess (1974a) for fields. A generalization of Schützenberger's theorem 7.1 to free partially commutative monoids has been given by Droste and Gastin (1999), see also Berstel and Reutenauer (2008b); this generalization has a long history, starting with the Boolean case, see Droste and Gastin (1999) for details.

It is well-known that if K is a field, then $K\langle\langle A\rangle\rangle$ can be embedded in the field $K((\Gamma))$ of Malcev–Neumann series over the free group Γ generated by A, see for instance Cohn (1985, Cor. 8.7.6) or Sakarovitch (2009a, Th. IV.4.7). The subfield of $K((\Gamma))$ generated by $K\langle A\rangle$, that is the subfield of rational elements of $K((\Gamma))$, is isomorphic to the free (skew) field of Cohn, according to a theorem of Lewin (1974), see also Reutenauer (1999). Fliess (1970) has shown that the intersection of the free field and of $K\langle\langle A\rangle\rangle$ is the set of rational series in $K\langle\langle A\rangle\rangle$, see also Duchamp and Reutenauer (1997, Cor. 13).

There exists a detailed study of semirings according to their behavior with respect to the star operation. A semiring K which is equipped with an additional unary operation denoted by * is called a *starsemiring*. A *Conway semiring* is a starsemiring K in which the equations $(a+b)^* = (a^*b)^*a^*$ and $(ab)^* = 1 + a(ba)^*b$ hold for all a, b in K. The main property of interest of Conway semirings is that Lemma 7.3 holds in these semirings. For a recent exposition, see Droste and Kuich (2009).

Formal power series, viewed as functions from A^* into K, can also be generalized to structures other than free monoids. One may consider functions from a – not necessarily free – monoid M into K. The product ST of two such functions S, T should satisfy the usual identity

$$(ST, z) = \sum_{xy=z} (S, x)(T, y)\,. \tag{7.2}$$

This is well defined provided the sum (7.2) exists in the semiring K. This holds in particular when the number of terms in the sum is finite. One way to ensure this is to require that the monoid M is *graded*, that is M is equipped with a length function $|\cdot| : M \to \mathbb{N}$ such that $|mm'| = |m| + |m'|$ and $|m| = 0$ if and only if m is the neutral element in M. It is easily seen that, in a finitely generated graded monoid, all sets $\{(x, y) | xy = z\}$ are finite. For an exposition, see Sakarovitch (2009a,b). Note that a graded monoid M is free if and only if Levi's lemma holds for M, that is if $xy = zt$ implies that $x = ze$, $ey = t$ or $xe = z$, $y = et$ for some $e \in M$.

Another extension is formal power series on trees. These have been considered by several authors. A comprehensive survey paper is Ésik and Kuich (2003).

As for formal languages, there are close connections between rational formal power series and logic theories, see Droste and Gastin (2007), Droste *et al.* (2008).

There is also an extension of the characterization of aperiodic languages to partially commutative formal power series, by Droste and Gastin (2008).

Closure under shuffle product (Exercise 5.3) is due to Fliess (1974b) and has many applications in Control Theory, see Fliess (1981). Exercise 5.6 is from Bacher (2008, Prop. 5.5). We do not consider algebraic formal series in this book; the reader may consult Salomaa and Soittola (1978) or Kuich and Salomaa (1986). Bacher (2009) gives a closure property of rational series over a finite field.

Applications to a large variety of domains, especially in computer science, are reported in the recent handbook of weighted automata, Droste *et al.* (2009).

2

Minimization

This chapter gives a presentation of results concerning the minimization of linear representations of recognizable series. A central concept of this study is the notion of syntactic algebra, which is introduced in Section 2.1. Rational series are characterized by the fact that their syntactic algebras are finite dimensional (Theorem 1.2). The syntactic right ideal leads to the notion of rank and of Hankel matrix; the quotient by this ideal is the analogue for series of the minimal automaton for languages.

Section 2.2 is devoted to the detailed study of minimal linear representations. The relations between representations and syntactic algebra are given. Two minimal representations are always similar (Theorem 2.4), and an explicit form of the minimal representation is given (Corollary 2.3).

The minimization algorithm is presented in Section 2.3. We start with a study of prefix sets. The main tool is a description of bases of right ideals of the ring of noncommutative polynomials (Theorem 3.2).

Several important consequences are given. Among them are Cohn's result on the freeness of right ideals, the Schreier formula for right ideals and linear recurrence relations for the coefficients of a rational series. A detailed description of the minimization algorithm completes the chapter.

2.1 Syntactic ideals

We start by assuming that K is a commutative ring. The algebra of polynomials $K\langle A\rangle$ is a free K-module having as a basis the free monoid A^*. Consequently, the set $K\langle\langle A\rangle\rangle$ of formal series can be identified with the dual of $K\langle A\rangle$. Each formal series S defines a linear function

$$\begin{aligned} K\langle A\rangle &\to K, \\ P &\mapsto (S, P) = \sum_{w\in A^*}(S, w)(P, w), \end{aligned}$$

the sum having a finite support because P is a polynomial. Thus, one may consider the kernel of S, denoted by $\mathrm{Ker}(S)$:

$$\mathrm{Ker}(S) = \{P \in K\langle A\rangle \mid (S, P) = 0\}.$$

Next, any multiplicative morphism $\mu : A^* \to \mathfrak{M}$, where $\mathfrak{M}$ is a K-algebra, can be extended uniquely to a morphism of algebras

$$K\langle A\rangle \to \mathfrak{M}.$$

This extension will also be denoted by μ. We shall use this convention tacitly in the sequel. Clearly

$$\mu(P) = \sum_{w \in A^*} (P, w)\mu(w).$$

Definition The *syntactic ideal* of a formal series $S \in K\langle\langle A\rangle\rangle$ is the greatest two-sided ideal of $K\langle A\rangle$ contained in the kernel of S. It is denoted by I_S.

This ideal always exists, since it is the sum of all ideals contained in Ker(S),

$$I_S = \sum_{I \subset \mathrm{Ker}(S)} I.$$

Lemma 1.1 *The syntactic ideal of a series S is equal to*

$$\begin{aligned} I_S &= \{Q \in K\langle A\rangle \mid \forall P, R \in K\langle A\rangle, (S, PQR) = 0\} \\ &= \{Q \in K\langle A\rangle \mid \forall x, y \in A^*, (S, xQy) = 0\}. \end{aligned}$$

Proof. Exercise 1.1. □

Definition The *syntactic algebra* of a formal series $S \in K\langle\langle A\rangle\rangle$, denoted by $\mathfrak{M}_S$, is the quotient algebra of $K\langle A\rangle$ by the syntactic ideal of S,

$$\mathfrak{M}_S = K\langle A\rangle / I_S.$$

The canonical morphism $K\langle A\rangle \to \mathfrak{M}_S$ is denoted by μ_S. Since $\mathrm{Ker}(\mu_S) = I_S \subset \mathrm{Ker}(S)$, the series S induces on $\mathfrak{M}_S$ a linear function denoted ϕ_S. Consequently

$$S = \phi_S \circ \mu_S.$$

Theorem 1.2 (Reutenauer 1978, 1980a) *A formal series is rational if and only if its syntactic algebra is a finitely generated module over K.*

Proof. If S is rational, S is recognizable (by Theorem 1.7.1) and has a linear representation (λ, μ, γ), with $\mu : A^* \to K^{n\times n}$ a morphism. Since A is finite, the subring L of K generated by the coefficients of λ, $\mu(a)$, $(a \in A)$ and γ is a finitely generated ring. Thus L is Noetherian and therefore each submodule of a finitely generated L-module is finitely generated by Theorem 1.5.3.

Since $L^{n\times n}$ is a finitely generated module over L, this implies that so is $\mu(L\langle A\rangle)$. In other words, for w in A^* long enough, μw is a L-linear combination of $\mu(v)$ for shorter words v. This implies in turn that $\mu(K\langle A\rangle)$ is a finitely generated K-module.

Now Ker(μ) is an ideal contained in Ker(S). Thus by definition $\mathrm{Ker}(\mu) \subset I_S$, and $\mathfrak{M}_S$ is a quotient of $\mu(K\langle A\rangle)$. Hence it is a finitely generated module over K.

Conversely, suppose that the syntactic algebra of S is a finitely generated module over K. Consider, for each word w in A^*, the K-endomorphism νw of $\mathfrak{M}_S$ defined by

$$m \mapsto \nu w(m) = \mu_S(w)m\,.$$

The function

$$\nu : A^* \to \mathrm{End}(\mathfrak{M}_S)$$

is a monoid morphism, and moreover

$$(S, w) = \phi_S \circ \mu_S(w) = \phi_S(\mu_S(w)) = \phi_S(\nu w(1))\,.$$

In order to conclude, it suffices to apply the following lemma and Theorem 1.7.1. □

Lemma 1.3 (This lemma is true for any semiring K, even noncommutative.) *Let $\mathfrak{M}$ be a finitely generated right K-module, let ϕ be a K-linear function on $\mathfrak{M}$, let m_0 be an element of $\mathfrak{M}$ and let ν be a monoid morphism $A^* \to \mathrm{End}(\mathfrak{M})$. Then the formal series*

$$S = \sum_{w \in A^*} \phi(\nu w(m_0))w$$

is recognizable. Moreover, if $\mathfrak{M}$ has a generating system of n elements, then S admits a linear representation of dimension n.

Proof. Let $m_1, \ldots, m_n$ be generators of $\mathfrak{M}$. Then for each letter $a \in A$, and each j in $\{1, \ldots, n\}$, there exist coefficients $\alpha^a_{i,j}$ in K such that

$$\nu a(m_j) = \sum_i m_i \alpha^a_{i,j}\,.$$

The matrices $(\alpha^a_{i,j})_{i,j} \in K^{n\times n}$ define a function $\mu : A \to K^{n\times n}$, $a \mapsto (\alpha^a_{i,j})_{i,j}$, which extends to a morphism $\mu : A^* \to K^{n\times n}$. A straightforward induction shows that for any word w,

$$\nu w(m_j) = \sum_i m_i(\mu w)_{i,j}\,.$$

Let $\lambda \in K^{1\times n}$ and $\gamma \in K^{n\times 1}$ be given by $\lambda_i = \phi(m_i)$ and $m_0 = \sum_j m_j\gamma_j$. Then

$$\nu w(m_0) = \nu w\Big(\sum_j m_j\gamma_j\Big) = \sum_j \sum_i m_i(\mu w)_{i,j}\gamma_j\,,$$

thus

$$\phi(\nu w(m_0)) = \sum_{i,j} \lambda_i(\mu w)_{i,j}\gamma_j = \lambda\mu w\gamma\,,$$

which completes the proof. □

Definition The *syntactic right ideal* of a formal series $S \in K\langle\langle A\rangle\rangle$ is the greatest right ideal of $K\langle A\rangle$ contained in Ker(S). It is denoted I^r_S.

The existence of I^r_S is shown in the same manner as that of I_S.

We now introduce an operation of $K\langle A\rangle$ on $K\langle\langle A\rangle\rangle$ on the right. Recall that, since $K\langle\langle A\rangle\rangle$ is the dual of $K\langle A\rangle$, each endomorphism f of the K-module $K\langle A\rangle$ defines an endomorphism ${}^t f$ of the K-module $K\langle\langle A\rangle\rangle$, called the *adjoint* morphism, by the relation

$$(S, f(P)) = ({}^t f(S), P)$$

for every series S and polynomial P. The function $f \mapsto {}^t f$ is an antimorphism:

$${}^t(g \circ f) = {}^t f \circ {}^t g\,. \tag{1.1}$$

Given a polynomial P, we consider the endomorphism $Q \mapsto PQ$ of $K\langle A\rangle$ and its adjoint morphism, denoted by $S \mapsto S \circ P$. Thus

$$(S, PQ) = (S \circ P, Q)\,.$$

In particular, for words x, y,

$$(S, xy) = (S \circ x, y)\,. \tag{1.2}$$

Consequently,

$$S \circ x = x^{-1} S$$

with the notation of Section 1.5. Observe that the operation $\circ$ is already defined by Equation (1.2); it suffices to extend it by linearity. In view of Equation (1.1), one obtains

$$(S \circ P) \circ Q = S \circ (PQ)\,. \tag{1.3}$$

Thus $K\langle\langle A\rangle\rangle$ is a right $K\langle A\rangle$-module.

Proposition 1.4 *The syntactic right ideal of a series S is*

$$I_S^r = \{P \in K\langle A\rangle \mid S \circ P = 0\}\,.$$

Proof. Since the operation $\circ$ defines on $K\langle\langle A\rangle\rangle$ a structure of right $K\langle A\rangle$-module, it is clear that the right-hand side of the equation is a right ideal of $K\langle A\rangle$. It is contained in Ker(S) because $S \circ P = 0$ implies $(S, P) = (S \circ P, 1) = 0$. It is the greatest right ideal with that property since, given a polynomial P, the relation $PK\langle A\rangle \subset \mathrm{Ker}(S)$ implies $(S \circ P, Q) = (S, PQ) = 0$ for all polynomials Q, whence $S \circ P = 0$. □

Corollary 1.5 *$K\langle A\rangle / I_S^r$ is isomorphic to $S \circ K\langle A\rangle$ as a right $K\langle A\rangle$-module.* □

This module is the analogue for series of the *minimal automaton* of a formal language.

We suppose from now on that K is a field.

Definition The *rank* of a formal series S is the dimension of the space $S \circ K\langle A\rangle$.

Definition The *Hankel matrix* of a formal series S is the matrix H indexed by $A^* \times A^*$ defined by

$$H(x, y) = (S, xy)$$

for all words x, y.

Theorem 1.6 (Carlyle and Paz 1971, Fliess 1974a) *The rank of a formal series S is equal to the codimension of its syntactic right ideal, and is equal to the rank of its Hankel matrix. The series S is rational if and only if this rank is finite and in this case, its rank is equal to the minimum of the dimensions of the linear representations of S.*

The theorem shows that the rank of a formal series could have been defined by an operation of $K\langle A\rangle$ on $K\langle\langle A\rangle\rangle$ on the left (analogue to $\circ$), or also by means of the syntactic left ideal (whose definition is straightforward). This follows from the left-right symmetry of the Hankel matrix.

Recall that the *rank* of a matrix (even an infinite one) can be defined to be the greatest dimension of a nonvanishing subdeterminant, and that it is equal to the rank of the rows and to the rank of the columns.

Proof. The first equality, namely $\text{rank}(S) = \text{codim}(I_S^r)$ is a direct consequence of Corollary 1.5. Next, the space $S \circ K\langle A\rangle$ has as set of generators $\{S \circ x \mid x \in A^*\}$. Thus $\text{rank}(S)$ is equal to the rank of this set. Since each $S \circ x$ can be identified with the row of index x in the Hankel matrix of S, the rank of S is equal to the rank of this matrix.

If S is rational, it has a linear representation (λ, μ, γ) of dimension n. The right ideal

$$J = \{P \in K\langle A\rangle \mid \lambda\mu(P) = 0\}$$

is contained in $\text{Ker}(S)$, and its codimension is $\leq n$. Consequently, J is contained in I_S^r, showing that $\text{rank}(S) = \text{codim}(I_S^r) \leq \text{codim}(J) \leq n$.

Conversely, let $n = \text{rank}(S) = \dim(S \circ K\langle A\rangle)$. Let ϕ be the linear form

$$S \circ K\langle A\rangle \to K,$$
$$T \mapsto (T, 1).$$

Then for any word w,

$$(S, w) = (S \circ w, 1) = \phi(S \circ w). \tag{1.4}$$

Let μw be the matrix of the endomorphism of $S \circ K\langle A\rangle$ which maps a series T on $T \circ w$, in some basis of $S \circ K\langle A\rangle$. (Each element of $S \circ K\langle A\rangle$ is represented by a vector $K^{1\times n}$, and each endomorphism of $S \circ K\langle A\rangle$ is represented by a matrix in $K^{n\times n}$; then $K^{n\times n}$ acts *on the right* on $K^{1\times n}$.) In view of Eq. (1.3), one has $(\mu x)(\mu y) = \mu(xy)$ for any words x and y. Let λ be the row vector representing S in the chosen basis, and let γ be the column representing ϕ. Then Equation (1.4) can be expressed as

$$(S, w) = \lambda\mu w\gamma$$

showing that S is recognizable, with a linear representation of dimension n. □

The theorem justifies the following definition.

Definition A *minimal linear representation* of a rational series S is a linear representation of S with minimal dimension among all its representations.

Example 1.1 The only series of rank 0 is the null series.

Example 1.2 Let S be a series of rank 1. It admits a representation (λ, μ, γ), with $\mu : K\langle A\rangle \to K$ a morphism of algebras and $\lambda, \gamma \in K$. Set $\alpha_a = \mu(a)$ for each letter a. For $w = a_1 \cdots a_n (a_i \in A)$, this gives

$$\mu(w) = \alpha_{a_1} \cdots \alpha_{a_n} = \prod_{a\in A} \alpha_a^{|w|_a} .$$

Consequently,

$$(S, w) = \lambda\gamma \prod_{a\in A} \alpha_a^{|w|_a} .$$

Such a series is called *geometric*. It follows that

$$S = \lambda\gamma\Big(\sum_{a\in A} \alpha_a a\Big)^* = \lambda\gamma\Big(1 - \sum_{a\in A} \alpha_a a\Big)^{-1} .$$

An example of a geometric series is the characteristic series of A^*:

$$S = \sum_{w\in A^*} w = \Big(\sum_{a\in A} a\Big)^* = \Big(1 - \sum_{a\in A} a\Big)^{-1} .$$

Example 1.3 The series $S = \sum_{w\in A^*} |w|_a w$ has rank 2. Indeed, it has a linear representation of dimension 2 (see Example 1.5.3). Next, the subdeterminant of its Hankel matrix corresponding to the rows and columns 1 and a is

$$\begin{vmatrix} 0 & 1 \\ 1 & 2 \end{vmatrix} = -1 .$$

Thus, S has rank ≥ 2. In view of Theorem 1.6, the rank of S is 2.

2.2 Minimal linear representations

K denotes a field.

Proposition 2.1 *A linear representation (λ, μ, γ) of dimension n of a series S is minimal if and only if, setting $\mathfrak{M} = \mu(K\langle A\rangle)$,*

$$\lambda\mathfrak{M} = K^{1\times n} \quad \text{and} \quad \mathfrak{M}\gamma = K^{n\times 1} .$$

In this case,

$$I_S^r = \{P \mid \lambda\mu P = 0\}.$$

Proof. Suppose that (λ, μ, γ) is minimal, and let $J = \{P \mid \lambda\mu P = 0\}$. Then J is a right ideal of $K\langle A\rangle$ and $\operatorname{codim}(J) = \dim(\lambda\mathfrak{M}) \leq n$. Since $J \subset \operatorname{Ker}(S)$, one has $J \subset I_S^r$ and $\operatorname{codim}(J) \geq \operatorname{codim}(I_S^r) = n$ (Theorem 1.6). Consequently $\operatorname{codim}(J) = n$, $J = I_S^r$ and $\lambda\mathfrak{M} = K^{1\times n}$. The equality $\mathfrak{M}\gamma = K^{n\times 1}$ is derived symmetrically.

Conversely, assume $\lambda\mathfrak{M} = K^{1\times n}$ and $\mathfrak{M}\gamma = K^{n\times 1}$. Then there exist words $x_1, \ldots, x_n$ $(y_1, \ldots, y_n)$ such that $\lambda\mu x_1, \ldots, \lambda\mu x_n$ $(\mu y_1\gamma, \ldots, \mu y_n\gamma)$ is a basis of $K^{1\times n}$ (of $K^{n\times 1}$). Consequently

$$\det(\lambda(\mu x_i y_j)\gamma)_{1\leq i,j\leq n} \neq 0.$$

Since $\lambda(\mu x_i y_j)\gamma = (S, x_i y_j)$, the Hankel matrix of S has rank $\geq n$. In view of Theorem 1.6, the representation (λ, μ, γ) is minimal. □

Corollary 2.2 *If the linear representation (λ, μ, γ) of the formal series S is minimal, then the kernel of μ is exactly the syntactic ideal of S, and consequently $\mu(K\langle A\rangle)$ is isomorphic to the syntactic algebra of S.*

Proof. Since $\operatorname{Ker}(\mu)$ is contained in $\operatorname{Ker}(S)$, it is contained in I_S. Conversely let $P \in I_S$. Then QPR is in I_S for all polynomials Q, R, and consequently $(S, QPR) = 0$. It follows that $\lambda(\mu QPR)\gamma = 0$ and thus $\lambda\mu(K\langle A\rangle)\mu P\mu(K\langle A\rangle)\gamma = 0$. In view of Proposition 2.1, this implies $\mu P = 0$, whence $P \in \operatorname{Ker}(\mu)$. □

Corollary 2.3 (Schützenberger 1961a) *If (λ, μ, γ) is a minimal representation of dimension n of a formal series S, then there exist polynomials $P_1, \ldots, P_n$, $Q_1, \ldots, Q_n$ such that, for every word w,*

$$\mu w = ((S, P_i w Q_j))_{1\leq i,j\leq n}.$$

Proof. In view of Proposition 2.1, there are polynomials $P_1, \ldots, P_n, Q_1, \ldots, Q_n$ such that $(\lambda\mu P_i)_{1\leq i\leq n}$ is the canonical basis of $K^{1\times n}$ and similarly $(\mu Q_j\gamma)_{1\leq j\leq n}$ is that of $K^{n\times 1}$. Thus

$$(\mu w)_{i,j} = \lambda\mu P_i \mu w \mu Q_j \gamma = (S, P_i w Q_j).$$ □

Two linear representations (λ, μ, γ) and $(\lambda', \mu', \gamma')$ are called *similar* if there exists an invertible matrix m such that $\lambda' = \lambda m$, $\mu' w = m^{-1}\mu w m$ (for all words w), $\gamma' = m^{-1}\gamma$. Clearly they recognize the same series.

Theorem 2.4 (Schützenberger 1961a, Fliess 1974a) *Two minimal linear representations are similar.*

Proof. Let (λ, μ, γ) be a minimal linear representation of a series S. Since, by Propositions 1.4 and 2.1,

$$I_S^r = \{P \in K\langle A\rangle \mid \lambda\mu P = 0\} = \{P \in K\langle A\rangle \mid S \circ P = 0\},$$

the two right $K\langle A\rangle$-modules $S \circ K\langle A\rangle$ and $K^{1\times n} = \lambda\mu(K\langle A\rangle)$ (with the action on $K^{1\times n}$ defined by $(v, P) \mapsto v\mu(P)$) are isomorphic. Consequently, there exists a K-isomorphism

$$f : K^{1\times n} \to S \circ K\langle A\rangle$$

such that, for any polynomial P, and any $v \in K^{1\times n}$,

$$f(v\mu P) = f(v) \circ P$$

and, moreover

$$f(\lambda) = S\,.$$

Next, consider the linear function ϕ on $S \circ K\langle A\rangle$ defined by $\phi(T) = (T, 1)$. Then for $v = \lambda\mu P$, one gets $\phi(f(v)) = \phi(f(\lambda\mu P)) = \phi(f(\lambda) \circ P) = \phi(S \circ P) = (S \circ P, 1) = (S, P) = \lambda\mu P\gamma = v\gamma$, which shows that

$$\phi \circ f = \gamma$$

if γ is set to be the linear function $v \to v\gamma$.

If $(\lambda', \mu', \gamma')$ is another minimal linear representation, there exists an analogous isomorphism f'. Thus there exists an isomorphism

$$\psi = f^{-1} \circ f' : K^{1\times n} \to K^{1\times n}$$

such that

$$\psi(v\mu' P) = \psi(v)\mu P, \ \psi(\lambda') = \lambda, \ \gamma' = \gamma \circ \psi\,.$$

It suffices to write ψ in matrix form to obtain the announced result. □

Corollary 2.5 (Schützenberger 1961a) *Let (λ, μ, γ) and $(\lambda', \mu', \gamma')$ be two linear representations of some series S, and assume the second representation is minimal. Then there exists a representation $(\bar\lambda, \bar\mu, \bar\gamma)$ similar to (λ, μ, γ) and having a block decomposition of the form*

$$\bar\lambda = (\times, \lambda', 0), \quad \bar\mu = \begin{pmatrix} \mu_1 & 0 & 0 \\ \times & \mu' & 0 \\ \times & \times & \mu_2 \end{pmatrix}, \quad \bar\gamma = \begin{pmatrix} 0 \\ \gamma' \\ \times \end{pmatrix}.$$

Proof. 1. Assume first that (λ, μ, γ) has the block decomposition

$$\lambda = (\lambda_1, \lambda_2, 0), \quad \mu = \begin{pmatrix} \mu_1 & 0 & 0 \\ \times & \mu_2 & 0 \\ \times & \times & \mu_3 \end{pmatrix}, \quad \gamma = \begin{pmatrix} 0 \\ \gamma_2 \\ \gamma_3 \end{pmatrix}$$

for some morphisms $\mu_i : A^* \to K^{n_i \times n_i}$, with the conditions

(i) $\lambda\mu(K\langle A\rangle) = K^{n_1} \times K^{n_2} \times \{0\}^{n_3}$ (we write here K^r for $K^{r\times 1}$, the set of row vectors), and
(ii) if $v \in K^{n_2}$ and $(0, v, 0)\mu(K\langle A\rangle)\gamma = 0$, then $v = 0$.

By using the block decomposition, we see that $\lambda\mu w\gamma = \lambda_2\mu_2 w\gamma_2$, so that $(\lambda_2, \mu_2, \gamma_2)$ is a representation of S, of dimension n_2. We show that it is minimal, by using Proposition 2.1.

Using again the block decomposition, we obtain $\lambda\mu(P) = (\times, \lambda_2\mu_2(P), 0)$ for P in $K\langle A\rangle$. Thus (i) implies that $\lambda_2\mu_2(K\langle A\rangle) = K^{n_2}$. Now, let $v \in K^{n_2}$ be such that $v\mu_2(K\langle A\rangle)\gamma_2 = 0$. Then, since $(0, v, 0)\mu(P)\gamma = v\mu_2(P)\gamma_2$, we see by (ii) that $v = 0$. This implies that $\mu_2(K\langle A\rangle)\gamma_2 = K^{n_2\times 1}$, and Proposition 2.1 now shows that $(\lambda_2, \mu_2, \gamma_2)$ is minimal. Applying Theorem 2.4, we deduce the corollary in this case.

2. Now consider any representation (λ, μ, γ) of S. Define $V_1 = \lambda\mu(K\langle A\rangle) \cap \{v \mid v\mu(K\langle A\rangle)\gamma = 0\}$. Let V_2 be a subspace of $K^{1\times n}$ such that $V_1 \oplus V_2 = \lambda\mu(K\langle A\rangle)$ and V_3 such that $V_1 \oplus V_2 \oplus V_3 = K^{1\times n}$. The subspaces V_1 and $V_1 \oplus V_2$ are both stable under the right action of the matrices in $\mu(K\langle A\rangle)$. Moreover λ is in $V_1 \oplus V_2$ and $V_1\gamma = 0$. This shows that, by a change of basis (which amounts to similarity), we may assume that (λ, μ, γ) is of the form in 1. We verify that (i) and (ii) hold. Condition (i) is implied by the very definition of V_1 and V_2. For (ii), let $w \in V_2$ be such that $w\mu(K\langle A\rangle)\gamma = 0$; then $w \in V_1$, so that $w = 0$. □

2.3 The minimization algorithm

We now give an effective procedure for computing a minimal linear representation of a recognizable series.

Definition A *prefix set* is a subset C of A^* such that $x, xy \in C$ implies $y = 1$ for all words x and y. It is *right complete* if CA^* meets every right ideal of A^*.

In other words, C is right complete if for every word w in A^*, wA^* meets CA^*. Equivalently, each word w either has a prefix in C, or is a prefix of some word in C.

Definition A subset P of A^* is *prefix-closed* if $xy \in P$ implies $x \in P$ for all words x and y.

In other words, a prefix-closed set contains all the proper prefixes of its elements, while a prefix set contains none of them.

Proposition 3.1 *There exists a natural bijection between prefix sets and prefix-closed sets: to a prefix set C is associated the prefix-closed set $P = A^* \setminus CA^*$, and the inverse bijection is defined by $C = I \setminus IA^+$, with $I = A^* \setminus P$. The prefix set $C = \{1\}$ and the prefix-closed set $P = \emptyset$ correspond each to another. In all other* *cases,* $C = PA \setminus P$. *Furthermore, finite right complete prefix sets correspond to finite prefix-closed sets.*

Proof. The prefix order $u \leq v$ on A^* is defined by the condition that u is a prefix of v. Clearly, a right ideal I of A^* is generated, as a right ideal, by the set of its minimal elements for the prefix order. Evidently, this set is a prefix set. On the other hand, the complement of a right ideal is a prefix-closed set, and conversely. This proves the existence of the bijection.

This shows also that if the prefix-closed set P and the prefix set C correspond to each other under this bijection, then $P = A^* \setminus CA^*$ and $I = A^* \setminus P = CA^*$.

Note that if $P = \emptyset$, then $C = 1$ and conversely. We assume now that $C \neq 1$. Let $w \in C$; then $w \neq 1$ and w is minimal in I, hence $w = ua, a \in A$ and $u \in A^* \setminus I = P$, implying $C \subset PA$. The fact that $P = A^* \setminus CA^*$ implies that P and C are disjoint, hence $C \subset PA \setminus P$. Conversely, if $w \in PA \setminus P$, then $w \in A^* \setminus P \implies w \in CA^*$. Thus $w = xu = pa$, $a \in A$, $x \in C$. Then x cannot be a prefix of p (otherwise I meets P), hence p is a proper prefix of x and this implies $|pa| \leq |x|$, therefore $u = 1$, hence $w \in C$.

If P is finite, then $C = 1$ or $C = PA \setminus P$ is finite. Moreover $A^* = P \cup CA^*$, hence each long enough word is in CA^*, implying that C is right complete. Conversely, suppose that C is right complete and finite. Let n be the length of the longest words in C. Since $CA^* \cap wA^* \neq \emptyset$, any word w of length at least n is in CA^*, hence not in P. Thus P is finite. □

Remark In order to illustrate Proposition 3.1, let us consider the *tree representation* of the free monoid A^*. Let for instance $A = \{a, b\}$. Then A^* is represented by

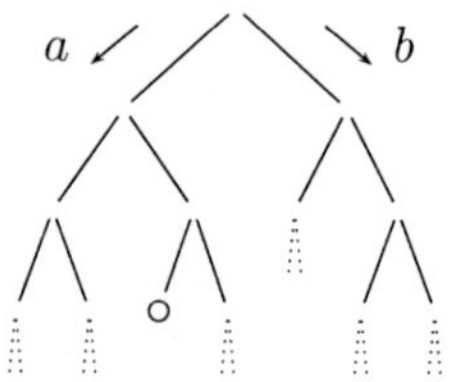

Here, the circled node corresponds to aba. A finite right complete prefix set C then is represented by a finite tree of the shape

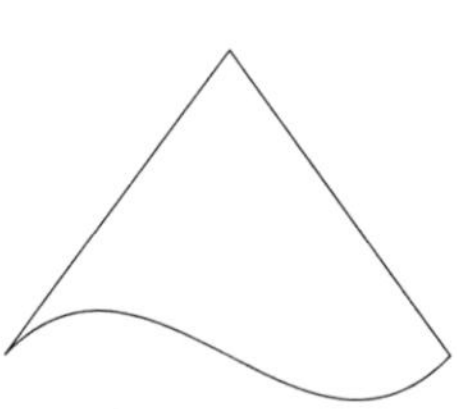

with the elements of the set being the tree's leaves, and the prefix-closed set associated with C being represented by its interior nodes.

Example 3.1 The tree

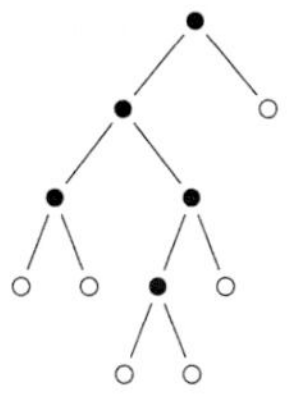

represents the prefix set

$$C = a^3 + a^2b + aba^2 + abab + ab^2 + b\,,$$

with

$$P = 1 + a + a^2 + ab + aba\,.$$

The white circles $\circ$ represent the elements of the set, and the black circles $\bullet$ the elements of P. This representation helps in understanding the proof.

In the following statement, K is assumed to be a field.

Theorem 3.2 *Let I be a right ideal of $K\langle A\rangle$. There exists a prefix set C with associated prefix-closed set P, and coefficients $\alpha_{c,p}(c \in C,\, p \in P)$, such that the polynomials $P_c = c - \sum_{p\in P} \alpha_{c,p}\, p\ (c \in C)$ generate freely I as a right $K\langle A\rangle$-module and such that P defines a K-basis in $K\langle A\rangle/I$.*

Proof. Let

$$\phi : K\langle A\rangle \to M = K\langle A\rangle/I$$

be the canonical morphism. Let P be a prefix-closed subset of A^* such that the elements $\phi(p)$, for $p \in P$, are K-linearly independent in M, and maximal among the subsets of A^* having this property.

Let C be the prefix set corresponding to P by Proposition 3.1. For each $c \in C$, the set $P \cup c$ is prefix-closed: indeed, either $P = \emptyset$ and $c = 1$ or $C = PA \setminus P$ by Proposition 3.1. By the maximality of P, $\phi(c)$ is in the subspace of $\mathfrak{M}$ spanned by $\phi(P)$. Thus there exist coefficients $\alpha_{c,p} \in K$ such that

$$P_c = c - \sum_{p\in P} \alpha_{c,p}\, p \in I\,. \tag{3.1}$$

We now show that any polynomial R can be written as

$$R = \sum_{c\in C} P_c Q_c + \sum_{p\in P} \beta_p p \tag{3.2}$$

for some polynomials Q_c $(c \in C)$ and coefficients β_p $(p \in P)$. It suffices to prove this for the case where $R = w$ is a word, and even in the case where $w \notin P$. But then $w = cx$ $(c \in C)$ since $A^* \setminus P = CA^*$ by Proposition 3.1. We argue by induction on the length of the word x. First, observe that by Equation (3.1),

$$w = P_c x + \sum_{p} \alpha_{c,p}\, px\,.$$

Each of the words px is either in P or of the form $c'x'$; in the latter case, c' cannot be a prefix of p (since $P \cap CA^* = \emptyset$), hence $|p| < |c'|$, whence $|x'| < |x|$. Thus the induction hypothesis completes the proof.

If the polynomial R of Equation (3.2) is in I, then

$$0 = \phi(R) = \sum_p \beta_p \phi(p)\,.$$

Consequently, $\beta_p = 0$ for all p and

$$R = \sum_{c \in C} P_c Q_c\,,$$

which shows that the right ideal I is generated by the P_c.

Let $\sum P_c Q_c = 0$ be a relation of $K\langle A\rangle$-dependency between the P_c, and assume that not all Q_c vanish. Then

$$\sum_c c Q_c = \sum_{c,p} \alpha_{c,p} p Q_c\,. \tag{3.3}$$

Consider a word w for which there is a $c_0 \in C$ with $(Q_{c_0}, w) \neq 0$, and which is a word of maximal length. For this word w, the coefficient of $c_0 w$ on the left-hand side of Equation (3.3) is $(Q_{c_0}, w) \neq 0$ because C is a prefix set. Thus

$$0 \neq (Q_{c_0}, w) = \sum_{c,p} \alpha_{c,p}(pQ_c, c_0 w)\,.$$

However, $px = c_0 w$ implies that p is a proper prefix of c_0, thus $c_0 = py$ for some $y \neq 1$ and $x = yw$. Consequently, the right-hand side of the previous equality is

$$\sum_{y \neq 1, c_0 = py} \alpha_{c,p}(Q_c, yw) = 0$$

in view of the maximality of w, a contradiction. □

Corollary 3.3 (Cohn 1969) *Each right ideal of $K\langle A\rangle$ is a free right $K\langle A\rangle$-module.* □

Corollary 3.4 (Lewin 1969) *Let I be a right ideal of $K\langle A\rangle$ of codimension n and rank d as a right $K\langle A\rangle$-module. Let r be the cardinality of A. Then*

$$d = n(r-1) + 1\,.$$

Proof. Indeed, if P is a nonempty finite prefix-closed set, with associated prefix set C, then by Proposition 3.1, $C = PA \setminus P$. Now, each nonempty word in P is in PA. Therefore we have the equality with disjoint unions: $C \cup P = PA \cup \{1\}$. Observe that this holds also if $P = \emptyset$. Thus in all cases $\mathrm{Card}(C) + \mathrm{Card}(P) = \mathrm{Card}(P)\mathrm{Card}(A) + 1$, implying $d + n = nr + 1$. □

We also obtain *linear recurrence relations* for rational series which generalize those for one-variable series (see Chapter 6).

Corollary 3.5 *For any rational series S of rank n, there exist a prefix-closed set P with n elements, with an associated prefix set C, and coefficients $\alpha_{c,p}$ ($c \in C$, $p \in P$) such that, for all words w and all $c \in C$,*

$$(S, cw) = \sum_{p \in P} \alpha_{c,p}(S, pw)\,. \tag{3.4}$$

Proof. It suffices to apply Theorem 3.2 to the syntactic right ideal of S which has codimension n. □

Corollary 3.6 *Let S be a rational series of rank $\leq n$, such that $(S, w) = 0$ for all words w of length $\leq n - 1$. Then $S = 0$.*

Proof. This is a consequence of Corollary 3.5. Indeed, $|p| \leq n - 1$ and therefore $(S, p) = 0$ for all $p \in P$. Assume $S \neq 0$, and let w be a word with $(S, w) \neq 0$. Then $w = cx$ for some $c \in C$. We choose w in such a way that the corresponding word x has minimal length. By Equation (3.4),

$$(S, cx) = \sum_{p \in P} \alpha_{c,p}(S, px)\,,$$

and by the choice of x, one has $(S, px) = 0$ for all $p \in P$: indeed, either $px \in P$, or $px = c'y$ for some $c' \in C$ and y shorter than x. Thus $(S, cx) = 0$, a contradiction. □

A subset T of A^* is *suffix-closed* if $xy \in T$ implies $y \in T$ for all words x and y.

Corollary 3.7 *Let S be a rational series of rank n. There exists a prefix-closed set P and a suffix-closed set T, both with n elements, such that*

$$\det((S, pt))_{p \in P, t \in T} \neq 0\,.$$

Proof. Let (λ, μ, γ) be a minimal linear representation of S. It has dimension n. In view of Theorem 3.2, applied to the right ideal $\{P \in K\langle A\rangle \mid \lambda\mu P = 0\}$, which is of codimension n by Proposition 2.1, there exists a prefix-closed set P such that $\lambda\mu(P)$ is a basis of $K^{1\times n}$, and symmetrically, there is a suffix-closed set T such that $\mu(T)\gamma$ is a basis of $K^{n\times 1}$. Thus the determinant of the matrix

$$(\lambda\mu p\mu t\gamma)_{p,t}$$

does not vanish. This proves the corollary. □

A careful analysis of the preceding proofs shows how to compute effectively a minimal linear representation of a rational series S given by any of its linear representations.

Indeed, let (λ, μ, γ) be such a representation, of dimension $n \geq 1$. The first step consists in reducing the representation to satisfy $K^{1\times n} = \lambda\mu(K\langle A\rangle)$. To do this, consider a prefix-closed subset P of A^* such that the vectors $\lambda\mu p$, for $p \in P$, are linearly independent, and which is maximal for this property. Then for each c in the prefix set $C = PA \setminus P$, there are coefficients $\alpha_{c,p}$ such that

$$\lambda\mu c = \sum_p \alpha_{c,p}\lambda\mu p .$$

Consider, for each letter a, the matrix $\mu' a \in K^{P\times P}$ defined by

$$(\mu' a)_{p,q} = \begin{cases} 1 & \text{if } pa = q \\ \alpha_{c,q} & \text{if } pa = c \in C \\ 0 & \text{otherwise.} \end{cases}$$

In other words, $\mu' a$ is the matrix, in the basis $\lambda\mu P$ of $\lambda\mu(K\langle A\rangle)$, of the endomorphism $v \mapsto v\mu a$. In this basis the matrix for λ is λ' defined by $\lambda'_1 = 1$, and $\lambda'_p = 0$ for $p \neq 1$; the matrix for γ is γ' defined by $\gamma'_p = \lambda\mu p\gamma = (S, p)$. Then $(\lambda', \mu', \gamma')$ is a linear representation of S, since for any word w, one has $\lambda\mu w \in \lambda\mu(K\langle A\rangle)$, whence $\lambda\mu w\gamma = \lambda'\mu' w\gamma'$. Moreover, the representation $(\lambda', \mu', \gamma')$ satisfies $K^{1\times P} = \lambda'\mu'(K\langle A\rangle)$. Indeed, since $\lambda'\mu' p$ represents the vector $\lambda\mu p$ in the basis $\lambda\mu(P)$, one has $\lambda'\mu' p = (\delta_{p,q})_{q\in P}$, which shows that $\lambda'\mu'(K\langle A\rangle)$ contains the canonical basis of $K^{1\times P}$.

If in the preceding construction, we assume moreover that $\mu(K\langle A\rangle)\gamma = K^{n\times 1}$, then also $\mu'(K\langle A\rangle)\gamma' = K^{P\times 1}$. Indeed, the first equality implies that every linear function on the space $\lambda'\mu'(K\langle A\rangle)$ is represented by a matrix of the form $\mu(R)\gamma$ for some $R \in K\langle A\rangle$. In the new basis $\lambda'\mu'(P)$ of $\lambda'\mu'(K\langle A\rangle)$, this matrix becomes $\mu'(R)\gamma'$. Thus any linear function on $K^{1\times P} = \lambda'\mu'(K\langle A\rangle)$ is represented as some $\mu'(R)\gamma'$, which proves the claim.

Now the work is almost done. In a first step, one reduces the representation to satisfy the condition $\mu(K\langle A\rangle)\gamma = K^{n\times 1}$, using a construction which is symmetric to the preceding one, based on suffix sets and suffix-closed sets. In a second step, the representation is transformed to satisfy in addition $\lambda\mu(K\langle A\rangle) = K^{1\times n}$, and (λ, μ, γ) is minimal by Proposition 2.1.

Exercises for Chapter 2

1.1 Prove Lemma 1.1. (*Hint*: The second set is an ideal and it contains each ideal which is contained in Ker$S$.)

1.2 Show that $I_S^r = \{P \in K\langle A\rangle \mid \forall Q \in K\langle A\rangle, (S, PQ) = 0\} = \{P \in K\langle A\rangle \mid \forall x \in A^*, (S, Px) = 0\}$.

1.3 The *reversal* of a word w, denoted by $\tilde{w}$, is defined as follows. If $w = 1$, then $\tilde{w} = 1$; if $w = a_1 \cdots a_n$ ($a_i \in A$), then $\tilde{w} = a_n \cdots a_1$. A word w is a *palindrome* if it is equal to its reversal. Let L be the set of palindromes.

(a) Assume Card$(A) \geq 2$. Show that if $x, x_1, \ldots, x_n$ are words with $|x| \leq |x_1|, \ldots, |x_n|$, and $x \neq x_1, \ldots, x_n$, then there exists y such that $xy \in L$,

$x_1 y, \dots, x_n y \notin L$. (*Hint*: Take $y = a^p bba^p \tilde{x}$, where a and b are distinct letters and $p = \sup\{|x_i| - |x|\}$.)

(b) Let $S \in K\langle\langle A \rangle\rangle$ be such that $(S, w) = 1$ if $w \in L$ and $(S, w) = 0$ for $w \notin L$. Show that all syntactic ideals of S are null (see Reutenauer (1980a)).

(c) (K is a commutative semiring.) Let $S \in K\langle\langle A \rangle\rangle$ be a recognizable series. Show that $S' = \sum_w (S, \tilde{w})w$ is recognizable.

1.4 (K is a commutative ring.) Let S be a formal series, let $\mathfrak{A}$ be an algebra, let $\mu : K\langle A \rangle \to \mathfrak{A}$ be an algebra morphism, and let φ be a linear mapping $\mathfrak{A} \to K$ such that $(S, w) = \varphi(\mu w)$ for any word w. Show that the syntactic algebra of S is a quotient of the algebra $\mu(\mathfrak{A})$.

1.5 (K is a field.) A finitely generated K-algebra $\mathfrak{M}$ is *syntactic* if there exists a formal series S whose syntactic algebra is isomorphic to $\mathfrak{M}$.

(a) Show that $\mathfrak{M}$ is syntactic if and only if it contains a hyperplane which contains no nonnull two-sided ideal.

(b) Let $\mathfrak{M} = K \cdot 1 \oplus K \cdot \alpha \oplus K \cdot \beta$, with multiplication defined by

$$\alpha^2 = \alpha\beta = \beta\alpha = \beta^2 = 0\,.$$

Show that $\mathfrak{M}$ is not syntactic.

(c) Show that $K\langle A \rangle$ is syntactic (use Exercise 1.3).

1.6 Show that the converse of Lemma 1.3 holds, and that $\mathfrak{M}$ may be chosen to be a free right K-module (K is any semiring).

1.7 (K is a field.) For any rational series S, define $N(S) = \dim(K + S \circ K\langle A \rangle) - 1$. Show that if S, T have constant term 1, then $N(ST) \le N(S) + N(T)$ with equality if S and T are polynomials; show that $N(S^{-1}) = N(S)$ and that $N(S) = 0$ if and only if $S = 1$. Show that if S is a series in one variable, written as a quotient P/Q of two relatively prime polynomials, then $N(S) = \max(\deg P, \deg Q)$. (*Hint*: Use Exercise 1.5.6 and Section 6.1.)

1.8 Show that Theorem 1.2 is no longer true for semirings. (*Hint*: Use the example of Exercise 1.5.5.)

1.9 Let K be a field. Show that the mapping $S \circ K\langle A \rangle \times K\langle A \rangle \circ S \to K$ given by $(S \circ P, Q \circ S) \mapsto (S, PQ)$ is well-defined and defines a nondegenerate duality between the spaces $S \circ K\langle A \rangle$ and $K\langle A \rangle \circ S$ (For $Q \in K\langle A \rangle$, the series $Q \circ S$ is defined by $(Q \circ S, w) = (S, wQ)$ for any $w \in A^*$).

2.1 Let K be a field and let Γ be the free group generated by A. It is well-known that the elements of Γ are uniquely represented by reduced words on the alphabet $A \cup A^{-1}$ (such a word has by definition no factor aa^{-1} or $a^{-1}a$ with $a \in A$). Let E denote the set of edges of the Cayley graph of Γ. By definition, E is the set of $\{\gamma, \gamma x\}$

with $\gamma \in \Gamma$, $x \in A \cup A^{-1}$, and no simplification occurs in the product γx. Define a mapping $F : \Gamma \to E \cup K$ by $F(1) = 0$ and $F(\gamma_1) = \{\gamma, \gamma x\}$ if $\gamma_1 = \gamma x$ and $\gamma, \gamma x$ are as above.

(a) Show that Γ acts on the left on E, that is $\gamma_1\{\gamma, \gamma x\} = \{\gamma_1\gamma, \gamma_1\gamma x\}$ is in E. For a set V, denote by KV (resp. $\overline{KV}$) the set of (resp. of infinite) K-linear combinations of elements of V; F extends naturally to linear mappings $K\Gamma \to KE$ and $\overline{K\Gamma} \to \overline{KE}$, still denoted F.

(b) Let $S \in \overline{K\Gamma}$. Show that S defines by left multiplication linear mappings $K\Gamma \to \overline{K\Gamma}$ and $KE \to \overline{KE}$. We denote them by S.

(c) Let $S \in \overline{K\Gamma}$. Define the linear mapping $D = FS - SF : K\Gamma \to \overline{KE}$. Show that if the image of D is finite dimensional, then the series $\text{red}(S) \in K\langle\langle A \cup A^{-1}\rangle\rangle$ is recognizable, where $\text{red}(S)$ is obtained from S by replacing each $\gamma \in \Gamma$ by its reduced word.

(d) Conversely, show that if $S \in \overline{K\Gamma}$ and $\text{red}(S)$ is recognizable, then $\text{Im}(D)$ has finite dimension.

2.2 Let K be a commutative semiring. Denote by $K\langle\langle A\rangle\rangle \bar{\otimes} K\langle\langle A\rangle\rangle$ the *complete tensor product*, which is the set of infinite linear combinations over K of the elements $u \otimes v$ with $u, v \in A^*$. If $S, T \in K\langle\langle A\rangle\rangle$, then $S \otimes T$ denotes the element

$$S \otimes T = \sum_{u,v \in A^*} (S, u)(T, v) u \otimes v\,.$$

Define a mapping $\Delta : K\langle\langle A\rangle\rangle \to K\langle\langle A\rangle\rangle \bar{\otimes} K\langle\langle A\rangle\rangle$ by

$$\Delta(S) = \sum_{u,v \in A^*} (S, uv) u \otimes v\,.$$

(a) Show that the series S is recognizable if and only if $\Delta(S)$ is a finite sum $\sum_{1 \le i \le r} S_i \otimes T_i$, with $S_i, T_i \in K\langle\langle A\rangle\rangle$. Show that the smallest possible r in such a sum is the smallest number of generators of all stable submodules of $K\langle\langle A\rangle\rangle$ containing S, and also the smallest dimension of a representation of S.

(b) Determine the series where $r = 1$. A series is *group-like* if $\Delta(S) = S \otimes S$. Determine these series.

2.3 Let K be a field and let (λ, μ, γ) be a minimal linear representation of a series S. Show that S is a polynomial if and only if $\mu w = 0$ for each word of length n, where n is the rank of S. (*Hint*: Show that if S is a polynomial of degree d, then the polynomials $u^{-1}S$ are linearly independent, for suitable words u of length $0, \ldots, d$; deduce that $n \ge d + 1$ by using Theorem 1.6 and Corollary 1.5. From Corollary 2.3, deduce that $\mu w = 0$ for each word of length n.)

2.4 A (right) *serial module* is a triple (ℓ, M, c) where M is a right $K\langle A\rangle$-module, ℓ is an element of M and $c : M \to K$ is a K-linear mapping. Its *dimension* is $\dim_K(M)$. It *recognizes* the series $S = \sum_{w \in A^*} c(\ell w) w$. A *morphism* $\sigma : (\ell, M, c) \to (\ell', M', c')$ between two serial modules is a right $K\langle A\rangle$-linear morphism $\sigma : M \to M'$ such that $\sigma\ell = \ell'$ and $c'\sigma = c$.

The *canonical serial module* of S is (ℓ_S, M_S, c_S), where $M_S = S \circ K\langle A\rangle$, $\ell_S = S$ and $c_S(T) = (T, 1)$.

(a) Associate to each serial module (ℓ, M, c) with an n element basis and recognizing S, a linear representation of dimension n of S. Show that this defines a bijective correspondence.

(b) Show that any serial right module recognizing the rational series S which is of minimal dimension is isomorphic to the canonical serial module of S.

(c) Deduce from a) and b) another proof of Theorem 2.4.

(d) Give a formulation of Proposition 2.1 in terms of serial modules.

(e) Do the same for Corollary 2.5.

3.1 Show that if P and C correspond each to another under the bijection of Proposition 3.1, and $C \neq 1$, then each word has a unique factorization $x_1 \cdots x_n p$ with $n \geq 0$, $x_i \in C$, $p \in P$. Show that one has the following equalities of formal series $\underline{A}^* = \underline{C}^* \underline{P}$ and $\underline{C} - 1 = \underline{P}(\underline{A} - 1)$.

3.2 Show that it is decidable whether two rational series are equal. (*Hint*: Use Corollary 3.6.)

3.3 Show that the recurrence relations of Corollary 3.5, together with the *initial values* (S, p), for $p \in P$, allow each coefficient of S to be computed explicitly.

3.4 Let C, P and P_c be as in Theorem 3.2. Show that the right ideal I generated by the polynomials P_c is freely generated by them. Show that P mod I is a K-basis in $K\langle A\rangle/I$. (*Hint*: Use the ideas of the proof of Theorem 3.2, in particular prove Equation 3.2 and its uniqueness.)

Notes to Chapter 2

The notions of syntactic ideal and algebra are introduced in Reutenauer (1978, 1980a), which also contains Theorem 1.2.

The notions of Hankel matrix and rank of a formal series, which are classical in the case of one variable, were introduced by Carlyle and Paz (1971) and Fliess (1974a).

Bacher (2008) computes the polynomials in q that count the number of rational series of fixed rank and on a fixed alphabet, when the field of scalars has q elements.

The minimal linear representations of a rational series are studied in Inagaki *et al.* (1972), Turakainen (1972) and Fliess (1974a). They were however first considered by Schützenberger (1961a,b), mainly in connection with the linear recurrence relations (Corollary 3.4). His methods are used here to prove Theorem 3.2 and the minimization algorithm. Observe that this construction is similar to Schreier's construction of a basis of a subgroup of a free group (see Lyndon and Schupp (1977), Proposition I.3.7).

The introduction of serial modules allows Fliess (1974a) to give the good minimization theory of the linear representations of a given series. The results are essentially contained, without the terminology, in Theorem 2.4 and Corollary 2.5 and their proofs. Serial modules are the analogues for series of minimal automata for automata, see Exercise 2.4.

Cobham (1978) shows that a rational series S of rank n may be expressed as a sum of two series, each of rank less than n, if and only if the right $K\langle A\rangle$-module $S \circ K\langle A\rangle$ (or equivalently $K\langle A\rangle / I_S^r$, or $K^{1\times n}$ with right action of $K\langle A\rangle$ via μ, for some minimal linear representation (λ, μ, γ) of S) contains two submodules, neither of which contains the other.

Fliess (1974a) shows that a rational series S of rank n is a sum $S = S_1 + \cdots + S_k$ of rational series of rank $n_1, \ldots, n_k$ with $n_1 + \cdots + n_k = n$ if and only if the right $K\langle A\rangle$-module $S \circ K\langle A\rangle$ is a direct sum of k submodules of K-dimension $n_1, \ldots, n_k$ (see also the corollary in Cobham (1978)). He shows that such a maximal decomposition is unique and corresponds to a maximal decomposition of $S \circ K\langle A\rangle$ as a sum of indecomposable submodules (Krull–Schmidt theorem); at most one of the series S_i above is a polynomial (see also Cohn and Reutenauer (1999)).

The operators F and D defined in Exercise 2.1 are due to Connes (1994). The exercise is from Duchamp and Reutenauer (1997). Exercise 1.7 is from Bacher (2008).

3

Series and languages

This chapter describes the relations between rational series and languages.

We start with Kleene's theorem, presented as a consequence of Schützenberger's theorem. Then we describe the cases where the support of a rational series is a rational language. The most important result states that if a series has finite image, then its support is a rational language (Theorem 2.10).

The family of languages which are supports of rational series have closure properties given in Section 3.4. The iteration theorem for rational series is proved in Section 3.5. The last section is concerned with an extremal property of supports which forces their rationality; to prove it, we use a remarkable characterization of rational languages due to Ehrenfeucht, Parikh and Rozenberg.

3.1 Kleene's theorem

Definition A *congruence* in a monoid is an equivalence relation which is compatible with the operation in the monoid. A language L is *recognizable* if there exists a congruence with finite index in A^* that *saturates* L (that is L is a union of equivalence classes).

It is equivalent to say that L is recognizable if there exists a finite monoid M, a morphism of monoids $\phi : A^* \to M$ and a subset P of M such that $L = \phi^{-1}(P)$.

The *product* of two languages L_1 and L_2 is the language $L_1L_2 = \{xy \mid x \in L_1, y \in L_2\}$. If L is a language, the submonoid generated by L is $\cup_{n\geq 0}L^n$.

Definition The set of *rational languages* over A is the smallest set of subsets of A^* containing the finite subsets and closed under union, product, and submonoid generation.

Rational languages are also often called *regular* languages.

Theorem 1.1 (Kleene 1956) *A language is rational if and only if it is recognizable.*

We will obtain this theorem as a consequence of Schützenberger's Theorem 1.7.1.

Lemma 1.2 *Let K, L be two semirings, and let $\phi : K \to L$ be a morphism of semirings. If $S \in K\langle\langle A\rangle\rangle$ is recognizable, then $\phi(S) = \sum \phi((S, w))w \in L\langle\langle A\rangle\rangle$ is recognizable.*

Proof. If indeed S has a linear representation (λ, μ, γ), then $\phi(S)$ admits the linear representation $(\phi(\lambda), \phi \circ \mu, \phi(\gamma))$, where we still denote ϕ the extension of ϕ to matrices. □

Lemma 1.3 *A language L is recognizable if and only if it is the support of some recognizable series $S \in \mathbb{N}\langle\langle A\rangle\rangle$.*

Proof. If L is recognizable, there exists a finite monoid M, a morphism of monoids $\phi : A^* \to M$ and a subset P of M such that $L = \phi^{-1}(P)$. Consider the *right regular representation* of M

$$\psi : M \to \mathbb{N}^{M \times M}$$

defined by

$$\psi(m)_{m_1,m_2} = \begin{cases} 1 & \text{if } m_1 m = m_2, \\ 0 & \text{otherwise.} \end{cases}$$

Then ψ is a morphism of monoids. Define $\lambda \in \mathbb{N}^{1 \times M}$ and $\gamma \in \mathbb{N}^{M \times 1}$ by

$$\lambda_m = \delta_{m,1},$$

$$\gamma_m = \begin{cases} 1 & \text{if } m \in P, \\ 0 & \text{otherwise.} \end{cases}$$

Then $\psi(m)_{1,m'} = 1$ if and only if $m = m'$, and consequently $\lambda\psi(m)\gamma = 1$ if $m \in P$, and $= 0$ otherwise. Now let

$$\mu = \psi \circ \phi : A^* \to \mathbb{N}^{M \times M}$$

and let S be the recognizable series with representation (λ, μ, γ). Then $S = \sum_{w \in L} w$, whence $L = \operatorname{supp}(S)$.

Conversely, assume that $S \in \mathbb{N}\langle\langle A\rangle\rangle$ is recognizable and let $L = \operatorname{supp}(S)$. Consider the Boolean semiring $\mathbb{B} = \{0, 1\}$ with $1 + 1 = 1$. Then the function

$$\phi : \mathbb{N} \to \mathbb{B}$$

defined by $\phi(0) = 0$ and $\phi(r) = 1$ for $r \geq 1$ is a morphism of semirings. By Lemma 1.2, the series $\phi(S) = \sum \phi((S, w))w \in \mathbb{B}\langle\langle A\rangle\rangle$ is $\mathbb{B}$-recognizable.

Thus there exists a linear representation (λ, μ, γ) of $\phi(S)$ with

$$\mu : A^* \to \mathbb{B}^{n \times n}.$$

Let $M = \mathbb{B}^{n \times n}$, and $P = \{m \in M \mid \lambda m \gamma = 1\}$. Since M is finite, the language

$$\{w \mid \mu(w) \in P\}$$

is recognizable, but this language is exactly $\operatorname{supp}(\phi(S)) = \operatorname{supp}(S) = L$. □

Lemma 1.4 *A language L over A is rational if and only if it is the support of some rational series $S \in \mathbb{N}\langle\langle A \rangle\rangle$.*

Proof. The following relations hold for series S and T in $\mathbb{N}\langle\langle A \rangle\rangle$:

$$\operatorname{supp}(S+T) = \operatorname{supp}(S) \cup \operatorname{supp}(T),$$

$$\operatorname{supp}(ST) = \operatorname{supp}(S)\operatorname{supp}(T),$$

$$\operatorname{supp}(S^*) = (\operatorname{supp}(S))^* \quad \text{if } S \text{ is proper.}$$

It follows easily that the support of a rational series in $\mathbb{N}\langle\langle A \rangle\rangle$ is a rational language.

For the converse, one can use the same relations, provided one has proved that any rational language can be obtained from finite sets by union, product and submonoid generation restricted to *proper* languages (that is languages not containing the empty word). We shall prove a stronger result, namely that for any rational language L, the language $L \setminus 1$ can be obtained from the finite subsets of $A^+ = A^* \setminus 1$ by union, product and generation of subsemigroup (that is $A \mapsto A^+ = \bigcup_{n \geq 1} A^n = AA^*$).

Indeed, if L_1 and L_2 have this property, then clearly so does $L_1 \cup L_2$ also, since $(L_1 \cup L_2) \setminus 1 = L_1 \setminus 1 \cup L_2 \setminus 1$; moreover $L_1 L_2$ has the property, since $L_1 L_2 \setminus 1 = (L_1 \setminus 1)(L_2 \setminus 1) \cup K$, where $K = \emptyset$, $= L_1 \setminus 1$, $= L_2 \setminus 1$, $= (L_1 \setminus 1) \cup (L_2 \setminus 1)$ according to the four cases: $1 \notin L_1 \cup L_2$, $1 \in L_2 \setminus L_1$, $1 \in L_1 \setminus L_2$, $1 \in L_1 \cap L_2$. Finally, if L has the announced property, then so does L^*, since $L^* \setminus 1 = (L \setminus 1)^* \setminus 1 = (L \setminus 1)^+$. □

Kleene's Theorem 1.1 is now an immediate consequence of Lemmas 1.3, 1.4 and of Theorem 1.7.1.

Corollary 1.5 *The family of rational languages is closed under Boolean operations.*

Proof. If L and L' are saturated by a congruence with finite index, then $L \cup L'$ and $L \cap L'$ are saturated by the congruence whose classes are intersections of classes of the congruences. This congruence has finite index. If L is saturated by a congruence with finite index, then $A^* \setminus L$ is saturated by the same congruence. □

3.2 Series and rational languages

Proposition 2.1 *Over any semiring, the characteristic series of a rational language is a rational series.*

Proof. This follows from the first part of the proof of Lemma 1.3, with "recognizable" replaced by "rational", which can be done in view of Theorem 1.1 and Theorem 1.7.1. Indeed, the right regular representation may be defined over any semiring. □

Given a language $L \subset A^*$, we call *generating function* of L the series $\sum_{n \geq 0} \alpha_n x^n$, where $\alpha_n = \operatorname{Card}(L \cap A^n)$.

Corollary 2.2 *A series $\sum_{n\geq 0} \alpha_n x^n$ in $\mathbb{Z}[[x]]$ is the generating function of some rational language if and only if it is rational over the semiring $\mathbb{N}$ and has constant term* 0 *or* 1.

In particular, the α_n satisfy a linear recurrence relation, see Chapter 6.

Proof. Suppose that $\sum \alpha_n x^n$ is the generating function of the rational language L. By Proposition 2.1, the characteristic series $\underline{L}$ of L is rational over $\mathbb{N}$. By sending each letter a of A onto x, we obtain a morphism $K\langle\langle A\rangle\rangle \to K[[x]]$ which sends $\underline{L}$ onto an $\mathbb{N}$-rational series in $\mathbb{N}[[x]]$ by Proposition 1.4.2. Clearly, this series is the generating series of L, which therefore is $\mathbb{N}$-rational.

Conversely, let S be an $\mathbb{N}$-rational series in $\mathbb{N}[[x]]$. It is obtained from elements in $\mathbb{N}[x]$ by the rational operations. It has therefore a rational expression involving these operations. We may assume that the only scalar in the expression is 1 (by replacing n by $1 + 1 \cdots + 1$). We now replace in the expression each monomial x^d by $a_1 a_2 \cdots a_d$, where a_i are distinct letters, distinct also from the letters for each monomial. An inductive argument then shows that this rational expression defines an $\mathbb{N}$-rational series T with coefficients 0 and 1. Hence T is the characteristic series of some rational language, whose generating series is S. □

Example 2.1 Let $S = (x + x^2)^* = \sum_{n\geq 0} F_n x^n$, where the F_n are the *Fibonacci numbers* ($F_0 = F_1 = 1$, $F_{n+2} = F_{n+1} + F_n$ for $n \geq 0$). Then S is the generating function of the rational language $(a \cup bc)^*$.

Similarly, $(x + 2x^2)^*(1 + 2x) + x$ is the generating function of the rational language $(a \cup bc \cup de)^*(1 \cup f \cup g) \cup h$ over the alphabet $\{a, b, c, d, e, f, g, h\}$.

Corollary 2.3 *If S is a rational series over the semiring K and L is a rational language, then $S \odot \underline{L} = \sum_{w\in L}(S, w)w$ is a rational series.*

Proof. Let K_1 be the prime semiring of K, that is the subsemiring generated by 1. Then by Proposition 2.1, the series $\underline{L}$ is K_1-rational. Since the elements of K_1 and K commute, it suffices to apply Theorem 1.5.5. □

Let S be a formal series, and let V be a subset of K. We denote by $S^{-1}(V)$ the language $S^{-1}(V) = \{w \in A^* \mid (S, w) \in V\}$.

Proposition 2.4 *If K is finite and if $S \in K\langle\langle A\rangle\rangle$ is rational, then $S^{-1}(V)$ is a rational language for any subset V of K. In particular,* supp(S) *is rational.*

Proof. Since S is recognizable, it admits a linear representation (λ, μ, γ). Since K is finite, $K^{n\times n}$ is finite, and $S^{-1}(V)$ is saturated by a congruence with finite index. Thus $S^{-1}(V)$ is recognizable, hence rational. □

Corollary 2.5 *A language is rational (or recognizable) if and only if its characteristic series over the Boolean semiring is so.*

Proof. Similar to that of Lemma 1.3. □

Corollary 2.6 *If $S \in \mathbb{Z}\langle\langle A \rangle\rangle$ is a rational series and $a, b \in \mathbb{Z}$, $b \neq 0$, then $S^{-1}(a + b\mathbb{Z})$ is a rational language.*

Proof. Let $\phi : \mathbb{Z} \to \mathbb{Z}/b\mathbb{Z}$ be the canonical morphism. Then $\phi(S)$ is rational by Lemma 1.2. Since $S^{-1}(a + b\mathbb{Z}) = \phi(S)^{-1}(\phi(a))$, the result follows from Proposition 2.4. □

Corollary 2.7 *If $S \in \mathbb{N}\langle\langle A \rangle\rangle$ is rational and if $a \in \mathbb{N}$, then the languages $S^{-1}(a)$, $S^{-1}(\{n \mid n > a\})$, $S^{-1}(\{n \mid n \leq a\})$ are rational.*

Proof. Let $\sim$ be the congruence of the semiring $\mathbb{N}$ generated by the relation $a + 1 \sim a + 2$; in this congruence, all integers $n \geq a + 1$ are in a single class, and each $n \leq a$ is alone in its class. Let K be the quotient semiring and let $\phi : \mathbb{N} \to K$ be the canonical morphism. Then $\phi(S)$ is rational by Lemma 1.2, and it suffices to apply Proposition 2.4, K being finite. □

Corollary 2.8 *A language L over A is rational if and only if the set of languages $\{w^{-1}L \mid w \in A^*\}$ is finite, with $w^{-1}L = \{x \in A^* \mid wx \in L\}$.*

Proof. By Corollary 2.5, this is a consequence of Proposition 1.5.1. □

Corollary 2.9 *Let $S \in \mathbb{Z}\langle\langle A \rangle\rangle$ be a rational series. If there is a nonzero integer $d \in \mathbb{N}$ which divides none of the nonzero coefficients of S, then the support of S is a rational language.*

Proof. If this is true, then $\operatorname{supp}(S) = A^* \setminus S^{-1}(d\mathbb{Z})$ and it suffices to apply Corollaries 2.6 and 1.5. □

We denote by $\operatorname{Im}(S)$ the set of coefficients of S. It is called the *image* of S.

Theorem 2.10 (Schützenberger 1961a, Sontag 1975) *Assume that K is a commutative ring. If $S \in K\langle\langle A \rangle\rangle$ is a rational series with finite image, then $S^{-1}(V)$ is rational for any $V \subset K$. Thus in particular the support of S is rational.*

Proof. (i) Arguing as in the proof of Theorem 2.1.2, we may assume that K is a Noetherian ring. Then, using Corollary 1.5.4 and the remarks before it, we see that there is some integer N such that for each word w, the series $w^{-1}S$ is a K-linear combination of the series $u^{-1}S$ with $|u| \leq N - 1$. Let $C = A^N$ and $P = 1 \cup A \cup \cdots \cup A^{N-1}$. We deduce that, for some coefficients $\alpha_{c,p}$ in K, $c \in C$, $p \in P$, one has, for any word w,

$$(S, cw) = \sum_{p \in P} \alpha_{c,p}(S, pw)\,. \tag{2.1}$$

(ii) We now consider the set E of sequences of words of the form $(pw)_{p \in P}$. For each word x, define a function f_x from E into E by

$$f_x((pw)_p) = (pxw)_p\,.$$

Then $f_y \circ f_x = f_{yx}$ since indeed $f_y \circ f_x((pw)_p) = f_y((pxw)_p) = (pyxw)_p = f_{yx}((pw)_p)$.

Consider the image of E by S, that is the set F of sequences $((S, pw))_{p \in P}$. The functions f_x induce functions on F (still denoted f_x); indeed if $((S, pw))_{p \in P} = ((S, pw'))_{p \in P}$ (which means that $(S, pw) = (S, pw')$ for all $p \in P$), then one has also $((S, pxw))_{p \in P} = ((S, pxw'))_{p \in P}$. It suffices to prove this claim for $x = a \in A$. In this case, either $pa \in P$ and then $(S, paw) = (S, paw')$, or $pa = c \in C$, and $(S, paw) = (S, paw')$ by Equation (2.1).

(iii) We have defined a morphism of monoids of A^* into the monoid M of functions from F into F by

$$x \mapsto f_x\,.$$

We now apply the hypothesis. Since Im(S) is finite, the set F is finite, and consequently M is finite. Let Q be the subset of M composed of those functions that map the sequence $((S, p))_{p \in P}$ onto an element F of the form $(\beta_p)_p$ with $\beta_1 \in V$. Since $f_x(((S, p))_{p \in P}) = ((S, px))_{p \in P}$, we have

$$f_x \in Q \iff (S, x) \in V \iff x \in S^{-1}(V)\,.$$

This shows that $S^{-1}(V)$ is recognizable, whence rational. □

3.3 Syntactic algebras and syntactic monoids

Let L be a language. The *syntactic congruence* of L, denoted by $\sim_L$, is the congruence on A^* defined by

$$u \sim_L v \text{ if and only if } \forall x, y \in A^*,\ xuy \in L \iff xvy \in L\,.$$

It is easily verified that this is indeed a congruence on A^*. Moreover, the syntactic congruence saturates L. In other words, if $u \sim_L v$, then $u \in L$ if and only if $v \in L$.

If $\sim$ is another congruence that saturates L, then $u \sim v$ implies $xuy \sim xvy$ (since $\sim$ is a congruence), therefore $xuy \in L$ if and only if $xvy \in L$. This shows that $u \sim v$ implies $u \sim_L v$. Thus the syntactic congruence of L is the coarsest congruence of A^* which saturates L. The monoid $M_L = A^*/\sim_L$ is called the *syntactic monoid* of L. In view of the definition of recognizable languages and of Theorem 1.1, we have the following result.

Proposition 3.1 *A language is rational if and only if its syntactic monoid is finite.* □

Given a language L, we call *syntactic algebra* of L the syntactic algebra of its characteristic series $\underline{L}$ (and we do similarly for other objects associated to the series). Here we take for K a commutative ring.

Proposition 3.2 *Let L be a language and let $\mathfrak{A}$ be its syntactic algebra, with the natural algebra homomorphism $\mu : K\langle A\rangle \to \mathfrak{A}$. Then $u \sim_L v$ if and only if $\mu(u) = \mu(v)$, and $\mu(A^*)$ is the syntactic monoid of L.*

Proof. Let $S = \underline{L}$. By definition of the syntactic algebra and Lemma 2.1.2, we have

$$\begin{aligned}\mu(u) = \mu(v) &\iff u - v \in I_S \\ &\iff (S, x(u - v)y) = 0 \text{ for all } x, y \in A^* .\end{aligned}$$

This latter condition is equivalent to $(S, xuv) = (S, xvy)$ for all $x, y \in A^*$. This is seen to be equivalent to $u \sim_L v$.

This proves the first statement, and the second follows. □

Recall that the *monoid algebra* KM of a monoid M is the K-module of formal K-linear combinations of elements of M, with K-bilinear product extending that of M. In particular, $K\langle A\rangle$ is the monoid algebra of the monoid A^*.

Proposition 3.3 *Let L be a language, let M be its syntactic monoid and $\mathfrak{A}$ its syntactic algebra. There are natural surjective algebra morphisms such that the following diagram is commutative.*

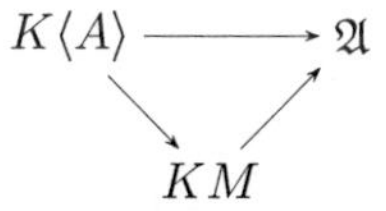

In particular, $\mathfrak{A}$ is a quotient of KM.

In general, $\mathfrak{A}$ is not isomorphic with KM, see Exercise 3.1.

Proof. We have an algebra morphism $\bar{\rho} : K\langle A\rangle \to KM$ which extends the syntactic monoid morphism $\rho : A^* \to M$. There is a subset P of M such that $L = \rho^{-1}(P)$. Define the linear mapping $\varphi : KM \to K$ by $\varphi(m) = 1$ if $m \in P$, and $\varphi(m) = 0$ otherwise. Then $(\underline{L}, w) = \varphi \circ \bar{\rho}(w)$ for any word w. Hence the ideal $\mathrm{Ker}(\bar{\rho})$ is contained in $\mathrm{Ker}(\underline{L})$ and therefore $\mathrm{Ker}(\bar{\rho})$ is contained in the syntactic ideal $I_{\underline{L}}$ of $\underline{L}$. From this, we deduce the algebra morphism $KM \to \mathfrak{A}$ which makes the diagram commutative. □

3.4 Support

In this and the next two sections, we study properties of languages which are supports of rational series. These languages strongly depend on the underlying semiring. Thus we have seen in Sections 3.1 and 3.2 that the rational languages are exactly the supports of rational series when the semiring is $\mathbb{N}$ or is finite. This is not generally true.

Example 4.1 Let $K = \mathbb{Z}$, $A = \{a, b\}$, and let S be the series

$$S = \sum_w (|w|_a - |w|_b)w .$$

This series is rational (Example 1.5.3). Its support is the language

$$\text{supp}(S) = \{w \in A^* \mid |w|_a \neq |w|_b\}$$

and its complement is

$$L = \{w \in A^* \mid |w|_a = |w|_b\} .$$

We shall prove that L is not a support of a rational series over $\mathbb{Z}$. This shows that L is not a rational language, by Proposition 2.1, and shows also that supp(S) is not rational, by Corollary 1.5.

Arguing by contradiction, we assume that $L = \text{supp}(T)$ for some rational series T having a linear representation (λ, μ, γ) of dimension n. Then the matrix μa^n is a linear combination of the matrices $\mu 1, \mu a, \ldots, \mu a^{n-1}$, and

$$\mu a^n = \alpha_1 \mu 1 + \cdots + \alpha_n \mu a^{n-1} .$$

Multiplying on the left by λ and on the right by $\mu b^n \gamma$, one gets

$$(T, a^n b^n) = \alpha_1 (T, b^n) + \cdots + \alpha_n (T, a^{n-1} b^n) .$$

Since $a^i b^n \notin L$ for $i \neq n$, the right-hand side of this equation vanishes, and the left-hand side is not zero, a contradiction.

Example 4.2 Recall that a *palindrome* w is a word which is equal to its reversal, that is $w = \tilde{w}$ (see Exercise 2.1.3). We show that the language $L = \{w \in A^* \mid w \neq \tilde{w}\}$ of words which are not palindromes is the support of a rational series over $\mathbb{Z}$.

Assume for simplicity that $A = \{a_0, a_1\}$, and consider the series

$$\sum_w \langle w \rangle w ,$$

where $\langle w \rangle$ is the integer represented by w in base 2. This series is rational (see Example 1.5.2). Consequently the series

$$\sum_w \langle \tilde{w} \rangle w$$

also is rational (see Exercise 2.1.3). Thus the series

$$\sum_w (\langle w \rangle - \langle \tilde{w} \rangle) w$$

is rational, and its support is L. Note that, by a technique analogous to that of Example 4.1, one can show that the set of palindromes is not a support of a rational series.

For the rest of this section, we fix a subsemiring K of the field $\mathbb{R}$ of real numbers. We denote by $\mathfrak{K}$ the family of languages which are supports of rational series, that is $L \subset A^*$ is in $\mathfrak{K}$ if and only if $L = \mathrm{supp}(S)$ for some rational series $S \in K\langle\langle A \rangle\rangle$.

We shall see that $\mathfrak{K}$ has all the closure properties usually considered in formal language theory, excepting complementation, as follows from Example 4.1.

The morphisms considered in the next statement are morphisms from one free monoid into another.

Theorem 4.1 (Schützenberger 1961a, Fliess 1971) *The family $\mathfrak{K}$ contains the rational languages. Moreover, $\mathfrak{K}$ is closed under finite union, intersection, product, submonoid generation, direct and inverse morphism.*

Proof. The first claim is a consequence of Proposition 2.1. Consider now a language $L \subset A^*$ in $\mathfrak{K}$, and let $S \in K\langle\langle A \rangle\rangle$ be a rational series with $L = \mathrm{supp}(S)$. If $\phi : B^* \to A^*$ is a morphism, then

$$\phi^{-1}(S) = \sum_{w \in B^*} (S, \phi(w))w$$

is rational. Indeed, if (λ, μ, γ) is a linear representation of S, then clearly $(\lambda, \mu \circ \phi, \gamma)$ is a linear representation of $\phi^{-1}(S)$. Consequently $\phi^{-1}(L) = \mathrm{supp}(\phi^{-1}(S))$ is in $\mathfrak{K}$.

Next, let $L' \subset A^*$ be another language in $\mathfrak{K}$, with $L' = \mathrm{supp}(S')$, and S' rational. Then $L \cap L' = \mathrm{supp}(S \odot S')$ is also in $\mathfrak{K}$, by Theorem 1.5.5.

In order to show that the submonoid L^* generated by L is also in $\mathfrak{K}$, observe first that $L^* = (L \setminus 1)^*$ and that $L \setminus 1 = L \cap A^+$ is in $\mathfrak{K}$. Thus we may assume $1 \notin L$, that is $(S, 1) = 0$. Next, we may suppose that S has only nonnegative coefficients, by considering $S \odot S$ instead of S, which is possible in view of Theorem 1.5.5. Under these conditions,

$$L^* = \mathrm{supp}(S^*)\,,$$

showing that L^* is in $\mathfrak{K}$. It is easily seen that $\mathfrak{K}$ is closed by union and product, using the formulas

$$\mathrm{supp}(S + S') = \mathrm{supp}(S) \cup \mathrm{supp}(S')\,,$$

$$\mathrm{supp}(SS') = \mathrm{supp}(S)\,\mathrm{supp}(S')\,,$$

which hold if S and S' have nonnegative coefficients.

Finally, consider a morphism $\phi : A^* \to B^*$.

(i) First we assume that $\phi(A) \subset B^+$. In this case, the family $\big((S, w)\phi(w)\big)_{w \in A^*}$ of series with each of these series reduced to a monomial, is locally finite, and its sum, the series

$$\phi(S) = \sum_{w \in A^*} (S, w)\phi(w)$$

is rational by Proposition 1.4.2. If moreover S has nonnegative coefficients, then

$$\mathrm{supp}(\phi(S)) = \phi(L)\,,$$

showing that $\phi(L)$ is in $\mathfrak{K}$.

(ii) Next, we assume that $A = B \cup \{a\}$, with $a \notin B$, and that ϕ is the projection $A^* \to B^*$, that is $\phi|_B = \mathrm{id}, \phi(a) = 1$. Let n be the dimension of a linear representation (λ, μ, γ) of S, and set

$$P = A^* \setminus A^* a^n A^* .$$

We claim that

$$\phi(L) = \phi(L \cap P) . \tag{4.1}$$

Let indeed $w \in L$. If $w \notin P$, then $w = xa^n y$ for some words x and y. Using the Cayley–Hamilton theorem for μa, we see that $(S, xa^n y)$ is a linear combination of the $(S, xa^i y)$ with $0 \le i \le n-1$. Consequently, there is such an i with $(S, xa^i y) \neq 0$, whence $xa^i y \in L$. Since $\phi(w) = \phi(xa^i b)$, induction on the length completes the proof.

Let $\psi : B^* \to K\langle A \rangle$ be the morphism of monoids defined by

$$\psi(b) = (1 + \cdots + a^{n-1})b(1 + \cdots + a^{n-1}) .$$

Furthermore, recall that we may assume that S has nonnegative coefficients. Let $T \in K\langle\langle B \rangle\rangle$ be the rational series with the linear representation $(\lambda, \mu \circ \psi, \gamma)$, with μ extended to $K\langle A \rangle$ by linearity.

Let $w = b_1 \cdots b_m \in B^*$. The coefficient of w in T is $\lambda(\mu \circ \psi w)\gamma$. Since ψw is an $\mathbb{N}$-linear combination of words of the form

$$a^{i_0} b_1 a^{i_1} \cdots b_m a^{i_m} \tag{4.2}$$

and since, by definition of ψ, *any* word of the form given by Equation (4.2) with $i_0, \ldots, i_m \in \{0, \ldots, n-1\}$ appears in ψw, it follows that (T, w) is an $\mathbb{N}$-linear combination of coefficients of the form

$$(S, a^{i_0} b_1 a^{i_1} \cdots y_m a^{i_m}) .$$

In view of Equation (4.1), and by the fact that all coefficients are nonnegative, this implies that

$$\phi(\mathrm{supp}(S)) = \mathrm{supp}(T) .$$

(iii) Consider finally an arbitrary morphism $\phi : A^* \to B^*$ and L in $\mathfrak{K}$. We may assume that A and B are disjoint. Then $\phi = \phi_2 \circ \phi_1$, where $\phi_1 : A^* \to (A \cup B)^*$ is defined by $\phi_1(a) = a\phi(a)$ for each letter a, and with $\phi_2 : (A \cup B)^* \to B^*$ defined by $\phi_2(a) = 1$ for $a \in A$, and $\phi_2(b) = b$ for $b \in B$. In view of (i), $\phi_1(L) \in \mathfrak{K}$. Moreover, ϕ_2 can be factorized into a sequence of morphisms of the type considered in (ii). Thus $\phi_2(\phi_1(L)) \in \mathfrak{K}$, and $\phi(L) \in \mathfrak{K}$. □

3.5 Iteration

In this section, we assume that K is a *field*. We prove the following.

Theorem 5.1 (Jacob 1980) *Let L be a language which is the support of a rational series. There exists an integer N such that for any word w in L, and for any factorization $w = xuy$ satisfying $|u| \geq N$, there exists a factorization $u = pvs$ such that the language*

$$L \cap xpv^*sy$$

is infinite.

We need a definition and a lemma.

Definition A *quasi-power of order* 0 is any nonempty word. A *quasi-power of order* $n+1$ is a word of the form xyx, where x is a quasi-power of order n.

Example 5.1 If $x \neq 1$, then $xyxzxyx$ is a quasi-power of order 2.

Lemma 5.2 (Schützenberger 1961b) *Let A be a (finite) alphabet. There exists a sequence of integers (c_n) such that any word on A of length at least c_n has a factor which is a quasi-power of order n.*

Proof. Let $d = \mathrm{Card}(A)$, $c_0 = 1$ and inductively

$$c_{n+1} = c_n(1 + d^{c_n}).$$

Suppose that any word of length c_n contains a factor which is a quasi-power of order n. Let w be a word of length at least $c_{n+1} = c_n(1 + d^{c_n})$. Then w has a factor of the form $x_1x_2 \cdots x_r$, with each x_i of length c_n and $r = 1 + d^{c_n}$. Since there are only d^{c_n} distinct words of length c_n on A, two of the x_i's are identical, and w has a factor xyx with $|x| = c_n$. By the induction hypothesis, $x = zx't$ with x' a quasi-power of order n. Thus w has as a factor $x'tyzx'$ which is a quasi-power of order $n+1$. □

Proof of Theorem 5.1. Let S be a rational series with $L = \mathrm{supp}(S)$, let (λ, μ, γ) be a linear representation of S, of dimension n. Set $N = c_n$ where c_n has the meaning of Lemma 5.2. Consider a word $w = xuy \in L$, with $|u| \geq N$. Then u contains a quasi-power of order n. Thus there exist words $1 \neq x_0, x_1, \ldots, x_n, y_1, \ldots, y_n$ such that x_n is a factor of u and, for each $i = 1, \ldots, n$, $x_i = x_{i-1}y_ix_{i-1}$. Next

$$n \geq \mathrm{rank}(\mu x_{i-1}) \geq \mathrm{rank}(\mu x_{i-1}y_ix_{i-1}) = \mathrm{rank}(\mu x_i).$$

Since $w \in \mathrm{supp}(S)$, we have $\mu(x_n) \neq 0$, thus there is an integer i such that $\mathrm{rank}(\mu x_{i-1}) = \mathrm{rank}(\mu x_{i-1}y_ix_{i-1})$. Set $p = \mu x_{i-1}$ and $q = \mu y_i$. Let these matrices act *on the right* on $K^{1\times n}$. From $\mathrm{rank}(p) = \mathrm{rank}(pqp)$, it follows that

$$\mathrm{Im}(p) \cap \mathrm{Ker}(qp) = 0. \tag{5.1}$$

Moreover,

$$\mathrm{rank}(p) \geq \mathrm{rank}(qp) \geq \mathrm{rank}(pqp) = \mathrm{rank}(p),$$

showing that $\text{rank}(p) = \text{rank}(qp)$, and since $\text{Im}(qp) \subset \text{Im}(p)$, it follows that $\text{Im}(qp) = \text{Im}(p)$. By Equation (5.1), this gives

$$\text{Im}(qp) \cap \text{Ker}(qp) = 0\,.$$

Since $n = \dim \text{Ker}(qp) + \dim \text{Im}(qp)$, the space $K^{1\times n}$ is the direct sum of $\text{Im}(qp)$ and $\text{Ker}(qp)$. In a basis adapted to this direct sum, the matrix qp has the form

$$\begin{pmatrix} m & 0 \\ 0 & 0 \end{pmatrix}$$

where m is an invertible matrix. Consequently the minimal polynomial $P(t)$ of qp is not divisible by t^2. We deduce that u can be factorized into $u = pvs$, with $v \neq 1$, and where the minimal polynomial

$$P(t) = t^r - a_1 t^{r-1} - \cdots - a_{r-1} t - a_r$$

of μv has at least one of the coefficients a_{r-1} or a_r nonnull. Consider the sequence of numbers (b_k) defined by

$$b_k = (S, xpv^k sy) = \lambda \mu(xp)(\mu v)^k \mu(sy)\gamma\,.$$

For all $k \geq 0$, the following relation holds:

$$b_{k+r} = a_1 b_{r+k-1} + \cdots + a_{r-1} b_{k+1} + a_r b_k\,.$$

Since $w \in L$, one has $b_1 = (S, xpvsy) = (S, w) \neq 0$. The condition $a_{r-1} \neq 0$ or $a_r \neq 0$ implies that there exist infinitely many k for which $b_k \neq 0$, whence $xpv^k sy \in L$. □

3.6 Complementation

In this section, K is a *field*. We have seen that the complement of the support of a rational series is not the support of a rational series, in general. However, the following result holds.

Theorem 6.1 (Restivo and Reutenauer 1984) *If the complement of the support of a rational series is also the support of a rational series, then it is a rational language.*

For the proof, we use the following theorem.

Theorem 6.2 (Ehrenfeucht *et al.* 1981) *Let L be a language, and let n be an integer such that for any word w and any factorization $w = ux_1 \cdots x_n v$, there exist i, j with $0 \leq i < j \leq n$ such that*

$$w \in L \iff ux_1 \cdots x_i x_{j+1} \cdots x_n v \in L\,.$$

Then L is a rational language.

The condition means that, given a word w in L (resp. w not in L) with n consecutive factors, one may remove in w some word which is a product of some of these factors, obtaining a word w' in L (resp. w' not in L).

Proof of Theorem 6.1. Let $L = \mathrm{supp}(S)$ and let $L' = A^* \setminus L = \mathrm{supp}(T)$ be two complementary languages which are supports of the rational series S and T respectively. Consider linear representations (λ, μ, γ) and $(\lambda', \mu', \gamma')$ of S and T. Further, let n be an integer greater than the dimension of both representations.

Let $w = ux_1 \cdots x_n v \in A^*$.

(i) Assume that w is in L. Then $0 \neq \lambda\mu(ux_1 \cdots x_n v)\gamma$ and consequently $\lambda\mu u \neq 0$. The $n+1$ vectors

$$\lambda\mu u, \lambda\mu u x_1, \ldots, \lambda\mu u x_1 \cdots x_n$$

belong to a space of dimension at most n. Consequently, there is an integer j with $1 \leq j \leq n$ such that $\lambda\mu u x_1 \cdots x_j$ is a linear combination of the vectors $\lambda\mu u x_1 \cdots x_i$ $(0 \leq i < j)$, say

$$\lambda(\mu u x_1 \cdots x_j) = \sum_{0 \leq i < j} \alpha_i \lambda\mu(ux_1 \cdots x_i)$$

with $\alpha_i \in K$. Multiplying on the right by $\mu(x_{j+1} \cdots x_n v)\gamma$, one gets

$$(S, w) = \sum_{0 \leq i < j} \alpha_i (S, ux_1 \cdots x_i x_{j+1} \cdots x_n v)\,.$$

Since $(S, w) \neq 0$, there exists i with $0 \leq i < j$ such that

$$(S, ux_1 \cdots x_i x_{j+1} \cdots x_n v) \neq 0$$

and hence $ux_1 \cdots x_i x_{j+1} \cdots x_n v \in L$.

(ii) Assume now that $w \notin L$, that is $w \in L'$. A similar proof, this time with $(\lambda', \mu', \gamma')$, shows that there are integers i, j $(0 \leq i < j \leq n)$ such that $(T, ux_1 \cdots x_i x_{j+1} \cdots x_n v) \neq 0$, showing that $ux_1 \cdots x_i x_{j+1} \cdots x_n v \in L'$, whence

$$ux_1 \cdots x_i x_{j+1} \cdots x_n v \notin L\,.$$

Thus we have shown that the language L satisfies the conditions of Theorem 6.2. Consequently, L is rational. □

For the proof of Theorem 6.2, we use without proof the well-known theorem of Ramsey. In order to state it simply, we introduce the following notation: For any set E, we denote by $E(p)$ the set of subsets of p elements of E.

Theorem 6.3 (Ramsey; see e.g. Ryser 1963 or Harrison 1978) *For any integers m, p, r, there exists an integer $N = N(m, p, r)$ such that for any set E of N elements and for any partition $E(p) = X_1 \cup \cdots \cup X_r$, there exists a subset F of E with m elements, such that $F(p)$ is contained in one of the X_i's.*

Proof of Theorem 6.2. Let n be a fixed integer, and let $\mathbf{L}$ be the set of all languages L over A satisfying the conditions of Theorem 6.2 for this n. We prove below that $\mathbf{L}$ is finite. It is not difficult to show that for any $L \in \mathbf{L}$ and any word w, the language

$$w^{-1}L = \{x \in A^* \mid wx \in L\}$$

is still in $\mathbf{L}$. In view of Corollary 2.8, any language in $\mathbf{L}$ is rational.

In order to show that $\mathbf{L}$ is finite, we use Ramsey's theorem for $m = 1 + n$, $p = 2$, $r = 2$. Let $N = N(m, 2, 2)$. Let L and K be two languages in $\mathbf{L}$ such that for all w of length $< N - 1$,

$$w \in L \iff w \in K\,. \tag{6.1}$$

We prove that then $L = K$. This clearly implies that $\mathbf{L}$ is finite. To prove the equality, we argue by induction on the lengths of words in A^*. Let w be a word of length $\geq N - 1$, let

$$w = a_1 a_2 \cdots a_{N-1} s \quad (a_i \in A)$$

and $E = \{0, 1, \ldots, N - 1\}$. Consider the partition

$$E(2) = X \cup Y\,,$$

with

$$X = \{(i, j) \mid 0 \leq i < j \leq N - 1 \text{ and } a_1 \cdots a_i a_{j+1} \cdots a_{N-1} s \in L\}\,,$$
$$Y = E(2) \setminus X\,.$$

Observe that by the induction hypothesis,

$$X = \{(i, j) \mid 0 \leq i < j \leq N - 1 \text{ and } a_1 \cdots a_i a_{j+1} \cdots a_{N-1} s \in K\}\,.$$

By Ramsey's theorem, there exists a subset F of E with $m = n + 1$ elements such that

$$F(2) \subset X \quad \text{or} \quad F(2) \subset Y\,.$$

Let $F = \{f_1 < f_2 < \cdots < f_m\}$ and let $u = a_1 \cdots a_{f_1}$, $x_1 = a_{f_1+1} \cdots a_{f_2}, \ldots, x_n = a_{f_n+1} \cdots a_{f_{n+1}}$ and $v = a_{f_{n+1}+1} \cdots a_{N-1} s$. Then we obtain a factorization

$$w = u x_1 \cdots x_n v$$

such that

(i) either, for all $0 \leq i < j \leq n$, the word $u x_1 \cdots x_i x_{j+1} \cdots x_n v$ is both in L and K;
(ii) or, for all $0 \leq i < j \leq n$, the word $u x_1 \cdots x_i x_{j+1} \cdots x_n v$ is neither in L nor in K.

Since L and K are in $\mathbf{L}$, the first condition implies that $w \in L$ and $w \in K$, and the second condition that $w \notin L$ and $w \notin K$. Thus Equation (6.1) is satisfied and the proof is complete. □

Theorem 6.1 is a special case of the following open problem.

Open problem Let L and K be disjoint languages which are both supports of some rational series. Does there exist two disjoint rational languages L' and K' such that

$$K \subset K', \quad L \subset L'$$

(that is K and L are *rationally separated*)?

Exercises for Chapter 3

1.1 Show that a subset of a^* (where a is a letter) is rational if and only if it is the union of a finite set and of a finite set of arithmetic progressions (we identify $a^* = \{a^n \mid n \in \mathbb{N}\}$ with $\mathbb{N}$).

1.2 For subsets X, Y of A^*, set $X^{-1}Y = \{x^{-1}y \mid x \in X, y \in Y\}$. Show that whatever is X, if Y is a rational language, then $X^{-1}Y$ is a rational language. (*Hint*: Use Corollary 2.8.)

1.3 Show that for X any recognizable subset of A^*, there exists an integer N such that, for every word w in X of length at least N, and for every factorization $w = xuy$ with $|u| \geq N$, there exists a factorization $u = pvs$ such that $0 < |v| \leq N$ and $xpv^nsy \in X$ for all $n \geq 0$. This result is known as the *pumping lemma* for recognizable (or rational) languages.

2.1 Let K be a field. The set of rational series of $K\langle\langle A\rangle\rangle$, equipped with the sum and the Hadamard product, is a K-algebra (Theorem 1.5.5). Show that the *idempotents* of this algebra are precisely the characteristic series of the rational languages.

An element S of this algebra is called *sub-invertible* if $\sum_w (S, w)^{-1}w$, where the summation is over all $w \in \text{supp}(S)$, is rational. Show that an element is sub-invertible if and only if there exists a group contained in the multiplicative monoid of this algebra and containing the given element.

2.2 Define as follows the *unambiguous rational operations* on languages:
The union $L_1 \cup L_2$ is unambiguous if the sets are disjoint. The product L_1L_2 is unambiguous if $u, u' \in L_1$, $v, v' \in L_2$, and $uv = u'v'$ imply $u = u'$, $v = v'$. The star operation $L \mapsto L^*$ is unambiguous if L is the basis of a free submonoid of A^* (that is L is a code).

A language is called *unambiguously rational* if it may be obtained from finite languages by using only unambiguous rational operations. By using Proposition 2.1 applied to $\mathbb{N}$, show that each rational language is unambiguously rational. (*Hint*: Use the part "reconizable $\Longrightarrow$ rational" in the proof of Theorem 1.7.1.)

2.3 Consider two series S and S' which differ only by values on words of length at most N. Show that they are both rational or both irrational. (*Hint*: Consider $T = S \odot \underline{A^{N+1}A^*}$, observe that $S = T + P$ and $S' = T + P'$ for some polynomials P and P', and use Corollary 2.3.)

2.4 Show how to deduce Theorem 2.10 from Corollary 2.2.3 when K is a field.

2.5 Let L be a language recognized by some finite deterministic automaton $\mathcal{A} = (Q, i, E, T)$. Let $M = (m_{p,q})$ be the matrix in $\mathbb{N}^{Q\times Q}$, where $m_{p,q}$ is the number of edges (p, a, q) in E. Let N be the inverse of the matrix $1 - xM$ over $\mathbb{Z}[[x]]$. Show that the generating function of L is equal to $\sum_{t\in\S} N_{i,t}$.

2.6 (a) Let $c(x) = \sum_{n\geq 0} c_n x^n$ be an $\mathbb{N}$-rational series without constant term. Show (without using Soittola's theorem proved in Chapter 8) that for all large enough integers $k > 0$, the series $\sum(k^n - c_n)x^n$ is $\mathbb{N}$-rational. (*Hint*: Consider a rational language C over some alphabet which has generating function $c(x)$.)

(b) Let $a(x) = \sum_{n\geq 0} a_n x^n$ be a $\mathbb{Z}$-rational series without constant term. Show, using (a), that the series $\sum(k^n + a_n)x^n$ is $\mathbb{N}$-rational for large enough integers k. (*Hint*: Write $a(x)$ as the difference $b(x) - c(x)$ of two $\mathbb{N}$-rational series and consider disjoint languages B and C with generating function $b(x)$ and $c(x)$ respectively.)

3.1 Let $L = (1 + a^3)(a^4)^*$. Show, with the notations of Proposition 3.3, that KM is not isomorphic to $\mathfrak{A}$ (show that $M = \mathbb{Z}/4\mathbb{Z}$ and $1 - a + a^2 - a^3 \in I_{\underline{L}}$).

4.1 Denote by R_K the set of supports of rational series with coefficients in the semiring K. Thus $R_{\mathbb{N}}$ is the set of rational languages (cf. Section 3.1).

(a) Show that if K and L are fields and L is an algebraic extension of K, then $R_K = R_L$.

(b) Show that if K is a finite field and t is a variable, then the support of the series over the field $K(t)$

$$\sum_{n\geq 0}((t+1)^n - t^n - 1)a^n$$

is not a rational language (use Exercise 1.1).

(c) Show that, given a field K, one has $R_K = R_{\mathbb{N}}$ if and only if K is an algebraic extension of a finite field (use Example 4.1) (see Fliess 1971).

4.2 Let $f, g : A^* \to B^*$ be two morphisms of a free monoid into another. Define the *equality set* of f and g as the language

$$E(f, g) = \{w \in A^* \mid f(w) = g(w)\}.$$

Show that the complement of $E(f, g)$ is the support of some rational series over $\mathbb{Z}$ (see Turakainen 1985).

4.3 Show that it is decidable whether the support of a rational series is empty. (*Hint*: Use Exercise 2.3.2.)

4.4 Show that it is decidable whether the support of a rational series is finite. (*Hint*: Use Exercise 2.2.3.)

4.5 Show that it is undecidable whether the support of a rational series is the whole free monoid. (*Hint*: Using Example 1.5.3, reduce this problem to the undecidability of Hilbert's tenth problem (theorem of Davis, Putnam, Robinson, Matijacevic, Cudnowski, see Manin (1977), Theorem VI.1.2 and seq.: given a polynomial $P \in \mathbb{Z}[x_1, \ldots, x_n]$, it is undecidable whether there exists $(\alpha_1, \ldots, \alpha_n) \in \mathbb{N}^n$ such that $P(\alpha_1, \ldots, \alpha_n) = 0$).)

Show that it is undecidable whether two supports are equal.

4.6 Show that the following problem is undecidable. Given a rational series $S \in \mathbb{Q}\langle\langle A \rangle\rangle$, are there infinitely many words w such that $(S, w) = 0$? Deduce that it is undecidable whether the complement of the support of a rational series is finite.

4.7 Use the undecidability of the *Post Correspondence Problem* and Exercise 4.2 to give another proof of the undecidability of the equality of two supports of rational series.

5.1 Let u_p be a quasi-power of order p, with $u_0 \neq 1$ and $u_i = u_{i-1}v_i u_{i-1}$ for $i = 1, \ldots, p$.

(a) Show that there exist words $w_1, \ldots, w_p$ such that for all $i = 1, \ldots, p$,

$$u_i = u_0 w_i w_{i-1} \cdots w_1 .$$

(b) Use question (a) to prove that for all integers n and p, there is an integer ℓ such that for every morphism

$$\mu : A^* \to K^{n \times n}$$

and for any word w of length at least ℓ, there exist nonempty words $w_1, \ldots, w_p$ such that $w_p w_{p-1} \cdots w_1$ is a factor of w and all the μw_i's have the same kernel N and the same image I with $N \cap I = 0$, and consequently belong to the same group contained in the multiplicative monoid $K^{n \times n}$ (see Jacob 1978).

Notes to Chapter 3

According to Corollary 2.2, an $\mathbb{N}$-rational series $S = \sum_{n \geq 1} a_n x^n$ with zero constant term in one variable is the generating function of a rational language L over some alphabet A. Let us say that a series S is k-realizable if the language L can be chosen over an alphabet A of k letters. In order to be k-realizable, one must have $a_n \leq k^n$ for all n. If S is k-realizable, then the series $(1 - kx)^{-1} - S$ is $\mathbb{N}$-rational since it is the generating function of the complement $A^* \setminus L$ of L. It is shown in Béal and Perrin (2003) that conversely a series S is k-realizable if both S and $(1 - kx)^{-1} - S$ are $\mathbb{N}$-rational.

Theorem 2.10 is due to Schützenberger (1961a) for fields, and to Sontag (1975) for rings.

The proof of Jacob's theorem (Theorem 5.1) is from Reutenauer (1980b); in this paper, another argument makes it possible to extend the result to infinite alphabets, and also to give a smaller bound N which depends only on the rank of the series (and not on the size of the alphabet). See also Okniński (1998, Theorem 1.12).

The *cancellation property* of Theorem 6.2 characterizes the rationality of a language: indeed, each rational language has this property, for some n, as may be easily verified.

Let us mention the following open problem (Salomaa and Soittola 1978, page 81). Does there exist a language which is the support of a $\mathbb{R}$-rational series without being the support of a $\mathbb{Q}$-rational series?

4

Rational expressions

We define rational expressions, their star height and rational identities. Section 4.1 studies the rational identity $E^* \equiv 1 + EE^* \equiv 1 + E^*E$ and its consequences and the operators $a^{-1}E$. In Section 4.2, we show that, over a commutative ring, rational identities are all consequences of the previous identities. In Section 4.3, we show that, over a field, star height may be characterized through some minimal representation, and deduce that the star height of the star of a generic matrix of order n is n. In the last section, we see that the star height may decrease under field extension and show how to compute the absolute star height, which is the star height over the algebraic closure of the ground field.

4.1 Rational expressions

Let K be a commutative semiring and let A be an alphabet. We define below the semiring of *rational expressions on A over K*. This semiring, denoted $\mathcal{E}$, is defined as the union of an increasing sequence of subsemirings $\mathcal{E}_n$ for $n \geq 0$. Each such subsemiring is of the form $\mathcal{E}_n = K\langle A_n \rangle$ for some (in general infinite) alphabet A_n; moreover, there will be a semiring morphism $E \mapsto (E, 1)$, $\mathcal{E}_n \to K$. We call $(E, 1)$ the *constant term* of the rational expression E.

Now $A_0 = A$, $\mathcal{E}_0 = K\langle A \rangle$ and the constant term is the usual constant term. Suppose that we have defined A_{n-1}, $\mathcal{E}_{n-1} = K\langle A_{n-1} \rangle$ and the constant term function on $\mathcal{E}_{n-1}$ for $n \geq 1$. We define

$$A_n = A_{n-1} \cup \{E^* \mid E \in \mathcal{E}_{n-1}, (E, 1) = 0\}.$$

Here E^* is a formal expression, obtained from E by putting $*$ as exponent. Now

$$\mathcal{E}_n = K\langle A_n \rangle$$

and the constant term function is obtained as follows: it is already defined on A_{n-1} (since $A_{n-1} \subset \mathcal{E}_{n-1}$), and we extend it to all of A_n by setting $(E^*, 1) = 1$ for $E \in \mathcal{E}_{n-1}$, $(E, 1) = 0$; now it is extended uniquely to a semiring morphism $\mathcal{E}_n = K\langle A_n \rangle \to K$ which is the identity on K.

An element of $\mathcal{E}_n \setminus \mathcal{E}_{n-1}$ is called a rational expression of *star height n*.

Example 1.1 Let $A = \{a, b\}$. Then $ab \in \mathcal{E}_0$, $(ab)^* \in A_1$ and $1 + b(ab)^*a \in \mathcal{E}_1$. Since $a \in A_0$, one gets $a^* \in A_1$, $a^*b \in \mathcal{E}_1$, $(a^*b)^* \in A_2$, $(a^*b)^*a^* \in \mathcal{E}_2$. The constant term of $1 + b(ab)^*a$ is 1, and so is also that of $(a^*b)^*a^*$.

It follows from the definitions of rational operations in Section 1.4 and of rational expressions above that there is a unique morphism $eval : \mathcal{E} \to K\langle\langle A\rangle\rangle$, extending the identity on $K \cup A$, such that the star operation is preserved. We leave the formal proof to the reader. Moreover, *eval* preserves constant terms, that is $(eval(E), 1) = (E, 1)$ for any rational expression. It follows also easily from the definitions that the image of *eval* is the semiring of all rational series on A over K. Finally, the star height of a rational series S is the least n such that $S \in eval(\mathcal{E}_n)$: this is a rephrasing of the corresponding definition in Section 1.4.

Let E, F be two rational expressions. We write $E \equiv F$ when $eval(E) = eval(F)$. We say that $E \equiv F$ is a *rational identity*. Clearly, the relation $\equiv$ is a congruence of the semiring $\mathcal{E}$. In other words, $E \equiv F$ and $E' \equiv F'$ imply $E + E' \equiv F + F'$ and $EE' \equiv FF'$.

We define another congruence on $\mathcal{E}$, denoted $\sim$. It is the least congruence of $\mathcal{E}$ such that for any $E \in \mathcal{E}$ with $(E, 1) = 0$, one has $E^* \sim 1 + EE^* \sim 1 + E^*E$.

If $E \sim F$, then $E \equiv F$ and $(E, 1) = (F, 1)$. Indeed, the first equation is true since $\equiv$ is a congruence satisfying $E \equiv 1 + EE^* \equiv 1 + E^*E$ for any E in $\mathcal{E}$ with $(E, 1) = 0$ (because for $S = eval(E)$, one has $S = 1 + SS^* = 1 + S^*S$, see Section 1.4). Thus we obtain the sequence of implications $E \sim F \implies E \equiv F \implies eval(E) = eval(F) \implies (E, 1) = (F, 1)$.

The constant term morphism $\mathcal{E} \to K$, $E \mapsto (E, 1)$ extends naturally to matrices: $\mathcal{E}^{n\times n} \to K^{n\times n}$, $M \mapsto (M, 1)$. We call M *proper* if $(M, 1) = 0$; in other words, if each entry of M has zero constant term. We write 1 for the identity matrix. The congruence $\sim$ extends naturally to matrices, too.

Proposition 1.1 *Given a proper square matrix M over $\mathcal{E}$, there exist matrices M_1, M_2 of the same size as M over $\mathcal{E}$ such that $M_1 \sim 1 + MM_1$ and $M_2 \sim 1 + M_2M$. In particular, if K is a ring, $1 - M$ is invertible modulo $\sim$.*

Proof. This is clear if M is of size 1×1. Let M be of larger size, and write $M = \begin{pmatrix} I & J \\ N & L \end{pmatrix}$ in nontrivial block form, with I, L square. By induction, since I and L are proper, there exist matrices I_1, L_1 of the same size as I, L such that $I_1 \sim 1 + II_1$, $L_1 \sim 1 + LL_1$. Let $I' = I + JL_1N$ and $L' = L + NI_1J$. By induction again, since I' and L' are proper, there exist I_1', L_1' such that $I_1' \sim 1 + I'I_1'$ and $L_1' \sim 1 + L'L_1'$. Let

$$M_1 = \begin{pmatrix} I_1' & I_1JL_1' \\ L_1NI_1' & L_1' \end{pmatrix}.$$

We verify that $M_1 \sim 1 + MM_1$ by computing the coefficients 1, 1 and 1, 2 of the right-hand side (we leave the remaining verifications to the reader). The first is

$$1 + II_1' + JL_1NI_1' = 1 + (I + JL_1N)I_1' = 1 + I'I_1' \sim I_1'.$$

The second is

$$II_1JL'_1 + JL'_1 = (II_1 + 1)JL'_1 \sim I_1JL'_1 .$$

This proves the result.

The existence of M_2 is proved symmetrically. Now, if K is a ring, then so are $\mathcal{E}$ and $\mathcal{E}/\sim$, hence $(1 - M)M_1 \sim 1 \sim M_2(1 - M)$. Consequently $M_1 \sim M_2(1 - M)M_1 \sim M_2$. Hence $1 - M$ is invertible in $\mathcal{E}/\sim$. □

We define now, for each letter a, a K-linear operator $\mathcal{E} \to \mathcal{E}$ denoted by $E \mapsto a^{-1}E$. This is done recursively on the subsemirings $\mathcal{E}_n$. For $n = 0$, it is the operator on $\mathcal{E}_0 = K\langle A\rangle$ defined in Section 1.5.

Suppose that we have defined the operator on $\mathcal{E}_{n-1}$, with $n \geq 1$. We define $a^{-1}E$ first for $E \in A_n$: if $E \in A_{n-1}$, then $a^{-1}E$ is already defined. Otherwise, $E = F^*$ for some $F \in \mathcal{E}_{n-1}$ with $(F, 1) = 0$; then $a^{-1}F$ is defined and we define $a^{-1}E = (a^{-1}F)F^*$.

Now $a^{-1}E$ is defined for $E \in A_n$, and we consider the function $\mu : A_n \to \mathcal{E}_n^{2\times 2}$ defined by

$$\mu(E) = \begin{pmatrix} E & 0 \\ a^{-1}E & (E, 1) \end{pmatrix} .$$

The function μ extends first to a monoid morphism $A_n^* \to \mathcal{E}_n^{2\times 2}$, the latter with its multiplicative structure. Then, since A_n^* is a basis of the K-module $\mathcal{E}_n$, it extends by K-linearity to $\mathcal{E}_n = K\langle A_n\rangle \to \mathcal{E}_n^{2\times 2}$. We then define the operator $a^{-1}E$, for any E in $\mathcal{E}_n$, by $a^{-1}E = \mu(E)_{2,1}$.

Thus the operator is defined on $\mathcal{E}_n$, hence recursively on all $\mathcal{E}$. Since μ is a multiplicative morphism, we have for all E, F in $\mathcal{E}$

$$\begin{pmatrix} EF & 0 \\ a^{-1}(EF) & (EF, 1) \end{pmatrix} = \begin{pmatrix} E & 0 \\ a^{-1}E & (E, 1) \end{pmatrix} \begin{pmatrix} F & 0 \\ a^{-1}F & (F, 1) \end{pmatrix} .$$

This implies

$$a^{-1}(EF) = (a^{-1}E)F + (E, 1)a^{-1}F . \tag{1.1}$$

Moreover, by construction $a^{-1}E^* = (a^{-1}E)E^*$ if $(E, 1) = 0$.

Proposition 1.2

(i) *If $a \in A$ and E is a rational expression, then $eval(a^{-1}E) = a^{-1}\, eval(E)$.*
(ii) *If E is a rational expression, then*

$$E \sim (E, 1) + \sum_{a\in A} a(a^{-1}E) .$$

Proof. (i) The formula holds by definition if $E \in \mathcal{E}_0$. We suppose that it holds for $E \in \mathcal{E}_{n-1}$, $n \geq 1$, and prove it for $E \in \mathcal{E}_n$. Define the semiring morphism $\mu' : K\langle\langle A\rangle\rangle \to K\langle\langle A\rangle\rangle^{2\times 2}$ by

$$\mu'(S) = \begin{pmatrix} S & 0 \\ a^{-1}S & (S, 1) \end{pmatrix} .$$

The fact that μ' is multiplicative follows from Lemma 1.7.2. We have for $E \in \mathcal{E}_n$

$$\mu' \circ eval(E) = \begin{pmatrix} eval(E) & 0 \\ a^{-1}\, eval(E) & (eval(E), 1) \end{pmatrix}$$

$$eval \circ \mu(E) = \begin{pmatrix} eval(E) & 0 \\ eval(a^{-1}E) & (E, 1) \end{pmatrix}.$$

Thus it is enough to show that, for $E \in \mathcal{E}_n$, $\mu' \circ eval(E) = eval \circ \mu(E)$. Since $\mu' \circ eval$ and $eval \circ \mu$ are K-linear semiring homomorphisms and since $\mathcal{E}_n = K\langle A_n \rangle$, it is enough to verify it for $E \in A_n$. It suffices to show that the 2, 1-entries of the two matrices $\mu' \circ eval(E)$ and $eval \circ \mu(E)$ coincide. Then, either $E \in A_{n-1} \subset \mathcal{E}_{n-1}$ and it holds by induction, or $E = F^*$ for some $F \in \mathcal{E}_{n-1}$ with $(F, 1) = 0$. Then we know that $a^{-1}E = (a^{-1}F)F^*$, so that

$$\begin{aligned} eval(a^{-1}E) &= eval(a^{-1}F)\, eval(F^*) = (a^{-1}\, eval(F))\, eval(F)^* \\ &= a^{-1}(eval(F)^*) = a^{-1}(eval(F^*)) = a^{-1}\, eval(E) \end{aligned}$$

using Lemma 1.7.2, and since by induction $eval(a^{-1}F) = a^{-1}\, eval(F)$.

(ii) This holds by definition and Equation (1.5.1) when $E \in \mathcal{E}_0$. We suppose it holds for $E \in \mathcal{E}_{n-1}$, $n \geq 1$, and prove it for $E \in \mathcal{E}_n$. First, let $E \in A_n$. If $E \in A_{n-1}$, we are done by induction. Otherwise $E = F^*$ for some $F \in \mathcal{E}_{n-1}$, $(F, 1) = 0$. Then by induction $F \sim \sum_{a \in A} a(a^{-1}F)$. Thus

$$\begin{aligned} E = F^* \sim 1 + FF^* &\sim 1 + \sum_{a \in A} a(a^{-1}F)F^* \\ &= 1 + \sum_{a \in A} a(a^{-1}F^*) = 1 + \sum_{a \in A} a(a^{-1}E) \end{aligned}$$

and we are done also.

Now, the formula to be proved is K-linear. Since $\mathcal{E}_n$ is a free K-module with basis A_n^*, it suffices to prove that the formula is preserved by product. Thus, suppose that it is true for E and F. We prove it for EF. We have

$$\begin{aligned} &(EF, 1) + \sum_{a \in A} a(a^{-1}(EF)) \\ &\quad = (EF, 1) + \sum_{a \in A} a\big((a^{-1}E)F + (E, 1)(a^{-1}F)\big) \qquad \text{(by (1.1))} \\ &\quad = (E, 1)(F, 1) + \sum_{a \in A} a(a^{-1}E)F + (E, 1) \sum_{a \in A} a(a^{-1}F) \\ &\quad = (E, 1)((F, 1) + \sum_{a \in A} a(a^{-1}F)) + \sum_{a \in A} a(a^{-1}E)F \\ &\quad \sim (E, 1)F + \sum_{a \in A} a(a^{-1}E)F \\ &\quad = ((E, 1) + \sum_{a \in A} a(a^{-1}E))F \sim EF. \end{aligned}$$

□

4.2 Rational identities over a ring

Our aim is to prove in this section that, if K is a commutative ring, then all rational identities over K are "trivial". This means that all rational identities are consequences of the fact that S^* is the inverse of $1 - S$, for any proper series S.

With the notations of the previous section, this means that the two congruences $\equiv$ and $\sim$ are equal. Since K is a ring, $\mathcal{E}$ is also a ring, and we may equivalently consider Ker(*eval*), called the *ideal of rational identities*. The result is as follows.

Theorem 2.1 *If K is a ring, the ideal of rational identities is generated by the rational expressions $(1 - E)E^* - 1$ and $E^*(1 - E) - 1$, with $E \in \mathcal{E}$ and $(E, 1) = 0$.*

Example 2.1 We illustrate the theorem by two examples. First, consider over $\{a, b\}$ the equality of series $(ab)^* = 1 + a(ba)^*b$. Combinatorially, it means that each word in $(ab)^*$ is either empty or of the form awb, where w is in $(ba)^*$. We show that this identity can be algebraically deduced from the identities $(1 - S)S^* = 1 = S^*(1 - S)$. We have indeed

$$\begin{aligned} 1 &= 1 - ab + ab = 1 - ab + a(1 - ba)(ba)^*b \\ &= 1 + a(ba)^*b - ab - aba(ba)^*b = (1 - ab)(1 + a(ba)^*b) \end{aligned}$$

where we use $(1 - ba)(ba)^* = 1$ in the second equality and algebraic operations in the others. Since $(ab)^*$ is the inverse of $1 - ab$, we obtain by left multiplication the identity $(ab)^* = 1 + a(ba)^*b$.

The second rational identity we consider is $(a + b)^* = (a^*b)^*a^*$. Combinatorially, it means that each word in $\{a, b\}^*$ has a unique factorization $a^{i_0}ba^{i_1}b \cdots ba^{i_n}$ with $n \geq 0$ and $i_0, \ldots, i_n \geq 0$. Algebraically, we have

$$\begin{aligned} 1 &= (a^*b)^*(1 - a^*b) = (a^*b)^* - (a^*b)^*a^*b \\ &= (a^*b)^*a^* - (a^*b)^*a^*a - (a^*b)^*a^*b = (a^*b)^*a^*(1 - a - b) \end{aligned}$$

where we use the fact that $(a^*b)^*$ (resp. a^*) is the inverse of $1 - a^*b$ (resp. of $1 - a$) in the first (resp. in the third) equality. Thus $1 = (a^*b)^*a^*(1 - a - b)$ and we obtain $(a + b)^* = (a^*b)^*a^*$ since $(a + b)^*$ is the inverse of $1 - a - b$.

Proof of Theorem 2.1.

1. Since a rational identity involves only finitely many coefficients of the ring K, it is enough to prove the theorem when K is a finitely generated ring. Then K is a Noetherian ring, hence each submodule of a finitely generated module over K is finitely generated (see Theorem 1.5.3 and the remark before it).

2. We now associate to each rational expression a finitely generated K-submodule of $\mathcal{E}$ which is stable, that is, closed under the operators $a^{-1}E$, and which contains E. This is done by lifting to rational expressions what has been done for rational series in the first part of the proof of Theorem 1.7.1.

If $E \in \mathcal{E}_0 = K\langle A\rangle$, the existence of the module is clear: we take the K-submodule spanned by the words appearing in E. For the induction step, we note that, taking the result for granted for $E \in \mathcal{E}_{n-1}$, it holds if $E \in A_{n-1}$. Now let $E \in A_n \setminus A_{n-1}$. Then $E = F^*$ for some $F \in \mathcal{E}_{n-1}$ with $(F, 1) = 0$. By induction, there is a stable finitely generated K-submodule M of $\mathcal{E}$ which contains F. Define $N = ME + KE$. Then N is a finitely generated K-submodule of $\mathcal{E}$ containing E. It is stable since $a^{-1}E = (a^{-1}F)E \in ME$ and since, for $G \in M$, $a^{-1}(GE) = (a^{-1}G)E + (G, 1)(a^{-1}E) \in ME$ because $a^{-1}G \in M$.

We prove the existence of a submodule for all elements of $\mathcal{E}_n$ by showing that if E, F possess such a submodule, so do $E + F$ and EF. Denote the corresponding submodules by M_E and M_F. It is easy to show that $M_E + M_F$ and $M_E F + M_F$ do the job. Observe that we use here only the fact that K is a commutative semiring.

3. Now let $E \equiv 0$ be some rational identity. Let M be the smallest stable K-submodule of $\mathcal{E}$ containing E. It is finitely generated by 1. and 2. Let $E_1, \ldots, E_n$ generate M. It is enough to show that $E_1, \ldots, E_n$ are in the ideal $\mathcal{J}$ of $\mathcal{E}$ generated by the elements indicated in the theorem. Note that $\sim$ is equality modulo $\mathcal{J}$. Hence we have to show that $E_i \sim 0$.

By Proposition 1.2(i), each element of M is itself a rational identity, since M is spanned by the smallest subset of $\mathcal{E}$ containing E and closed under the operations $F \mapsto a^{-1}F$, $a \in A$. In particular, $(E_i, 1) = 0$. Thus by Proposition 1.2(ii) we have

$$E_i \sim \sum_{a \in A} a(a^{-1}E_i).$$

Since M is stable, $a^{-1}E_i$ is a K-linear combination of the E_j. Thus we may find homogeneous polynomials $M_{i,j}$ of degree 1 such that $E_i \sim \sum_j M_{i,j} E_j$. In other words, if we put $M = (M_{i,j})$, we obtain

$$(1 - M) \begin{pmatrix} E_1 \\ \vdots \\ E_n \end{pmatrix} \sim 0.$$

By Proposition 1.1, $1 - M$ is invertible modulo $\mathcal{J}$. Thus $E_i \in \mathcal{J}$ for any i. □

4.3 Star height

A finite directed graph $G = (V, E)$ is *strongly connected* if there is a path between any pair of vertices. A *strongly connected component* of G is a maximal subgraph which is strongly connected. The *cycle complexity* of G is defined as follows: If G has no infinite path, its cycle complexity is 0. Otherwise, if G is strongly connected, it is 1+ the minimum of the cycle complexity of the graphs $G \setminus v$, for all vertices v in G. Finally, if G is not strongly connected, it is the maximum of the cycle complexity of the strongly connected components of G.

Lemma 3.1 *Let G be a finite directed graph. Let $\tilde{G}$ be the opposite graph, obtained by reverting the edges. Then G and $\tilde{G}$ have the same cycle complexity.*

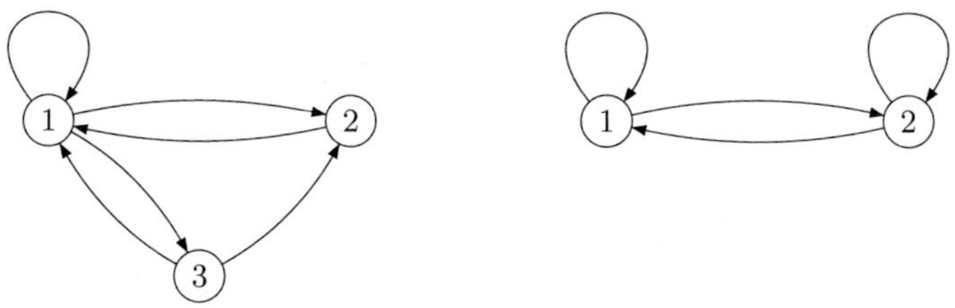

Figure 4.1 Two graphs with cycle complexity 1 and 2 respectively.

Proof. Clearly G and $\tilde{G}$ have simultaneously infinite paths or not. Moreover, the strongly connected components of G and $\tilde{G}$ are opposite graphs. Furthermore, if v is a vertex, then $\widetilde{G \setminus v} = \tilde{G} \setminus v$. From this, it is easy to verify by induction that G and $\tilde{G}$ have the same cycle complexity. Details are left to the reader. □

Let V be a totally ordered finite set and let $h : V \to \mathbb{N}$ be a function. We define another function $n : V \to V \cup \{\infty\}$, where $\infty \notin V$ and $v < \infty$ for any $v \in V$. It is called the *next* function: $n(v)$ is the smallest $v' > v$ such that $h(v') \geq h(v)$ if such a v' exists; and $n(v) = \infty$ otherwise. Note that $v < n(v)$ for any $v \in V$.

Lemma 3.2 *Let V' be an interval in V, let $h' = h|V'$ and let n' be the next function of h'. Then, for any $v \in V'$,*

$$n'(v) = \begin{cases} n(v) & \text{if } n(v) \in V', \\ \infty & \text{otherwise.} \end{cases}$$

In particular $n'(v) \geq n(v)$. If moreover V' is an upper order ideal of V, then equality $n'(v) = n(v)$ holds for any v in V'.

Proof. Since $n'(v) = \min\{u \in V' \mid h(u) \geq h(v) \text{ and } u > v\}$ and $n(v) = \min\{u \in V \mid h(u) \geq h(v) \text{ and } u > v\}$, we see that if $n(v) \in V'$ then $n(v) = n'(v)$. If $n(v) \notin V'$ then, since $v < n(v)$ and since V' is an interval, $n(v)$ is greater than each element in V'; thus for $v' \in V'$ with $v' > v$, one has $h(v') < h(v)$ and consequently $n'(v) = \infty$.

If V' is an upper order ideal, then for $v \in V'$ and $u \in V$, the relation $u > v$ implies $u \in V'$. Hence the formulas at the beginning of the proof imply $n(v) = n'(v)$. □

Let $G = (V, E)$ be a finite directed graph. We say that $h : V \to \mathbb{N}$ is a *height function* for G if there is a total order on V such that, n being the next function of h, one has:

$$\text{for any } v \in V, \text{ if } h(v) = 0 \text{ (resp. } h(v) \geq 1\text{), then for each edge } v \to v', \text{ one has } v' < v \text{ (resp. } v' < n(v)\text{).} \tag{3.1}$$

Note that in the case $h(v) \geq 1$, if $n(v) = \infty$, then the conclusion always holds.

Lemma 3.3 *Let $G = (V, E)$ be a finite directed graph with height function h, let V' be an interval of V and let G' be the graph obtained by restriction of G to V'. Then $h|V'$ is a height function for G'.*

Proof. Order V' by restriction. Let n' be the next function of h'. Let $v \in V'$ with $h(v') = 0$. Then $h(v) = 0$ and by (3.1), for each edge $v \to v'$ with $v' \in V'$, one has $v' < v$.

Now, let $v \in V'$ with $h'(v) \geq 1$. Then $h(v') \geq 1$ and by (3.1), for each edge $v \to v'$ with $v' \in V'$, one has $v' < n(v)$. By Lemma 3.2, one has $n(v) \leq n'(v)$. Thus $v' < n'(v)$.

This proves that h' is a height function for G'. □

Lemma 3.4 *Let $G = (V, E)$ be a finite directed graph, $v \in V$ and let H be a strongly connected component of G. If $v \in H$, then each strongly connected component of $H \setminus v$ is a strongly connected component of $G \setminus v$. If $v \notin H$, then H is a strongly connected component of $G \setminus v$.*

The proof is left to the reader.

Theorem 3.5 *A graph $G = (V, E)$ has cycle complexity at most m if and only if it has a height function h with* $\max(h) \leq m$.

For the graphs of Figure 4.1, one takes the natural order on the vertices, and the functions $h(1) = 1$, $h(2) = h(3) = 0$ for the first graph, and $h(1) = 2$, $h(2) = 1$ for the second.

Proof. 1. Let G have cycle complexity at most m. We may assume that G has cycle complexity exactly m. If $m = 0$, then G has no infinite path, and we may totally order V in such a way that $v \to v'$ implies $v > v'$. Hence we may take $h(v) = 0$ for all v.

Suppose now that $m \geq 1$. If G is strongly connected, there exists a vertex v such that $G \setminus v$ has cycle complexity $m - 1$. By induction, a height function $h : V \setminus v \to \mathbb{N}$ exists, and $\max(h) \leq m - 1$. We extend h to V by $h(v) = m$ and extend the order on $V \setminus v$ by $v < v'$ for all $v' \in V \setminus v$. It is readily verified that the next function of h satisfies $n(v) = \infty$ and it follows from Lemma 3.2 that $n(v') = n_{G \setminus v}(v')$ for $v' \neq v$. From this, it follows that h is a height function for G.

Suppose now that G is not strongly connected. We totally order the set of strongly connected components of G in such a way that if $H < H'$ then there is no edge from H to H'. On each strongly connected component H, there exists, by induction, a total order of its set of vertices and a height function h_H with $\max(h_H) \leq m$. We define h on V by extending these functions naturally to V, and the total order on V by gluing together all these orders in a way compatible with the total order on the strongly connected components and such that each strongly connected component is an interval of V.

Note that if v, v' are not in the same strongly connected component of G, then

$$v \to v' \quad \text{implies} \quad v' < v\,. \tag{3.2}$$

Let $v \in H$. We suppose first that $h(v) = 0$. Then $v \to v'$ implies that either $v' \in H$ and then, by (3.1) $v' < v$, or $v' \notin H$ and $v' < v$ by (3.2). Suppose now that $h(v) \geq 1$.

If $n_H(v) \in H$, then $n_H(v) = n(v)$ by Lemma 3.2; suppose that $v \to v'$: then either $v' \in H$, hence by (3.1) $v' < n_H(v) = n(v)$, or $v' \notin H$ and therefore by (3.2) $v' < v < n(v)$. If $n_H(v) = \infty$, then $n(v) \notin H$ and $v < n(v)$. Suppose that $v \to v'$: then either $v' \in H$ and $v' < n(v)$ (indeed, $v, v' \in H$, $v < n(v)$, $n(v) \notin H$ and H is an interval imply $v' < n(v)$); or $v' \notin H$ and by (3.2) $v' < v < n(v)$. Hence h is a height function for G.

2. Conversely, suppose that G has a height function h with $\max(h) \leq m$. We may assume that $\max(h) = m$. If $m = 0$, Equation (3.1) implies that there is no infinite path in G, hence G has cycle complexity 0. Assume that $m \geq 1$. Suppose first that $v = \min(V)$ is the unique vertex such that $h(v) = m$.

Consider the restriction h' of h to $V \setminus v$; since $v = \min(V)$, by Lemma 3.3, h' is a height function for $G \setminus v$ and its maximum is $\leq m - 1$. By induction, $G \setminus v$ has cycle complexity $\leq m - 1$. Let H be the strongly connected component of G containing v. Then by Lemma 3.4 $H \setminus v$ is a union of strongly connected components of $G \setminus v$, hence its cycle complexity is $\leq m - 1$, and therefore that of H is $\leq m$. If H' is another strongly connected component of G, it is by Lemma 3.4 also a strongly connected component of $G \setminus v$ and so has cycle complexity $\leq m - 1$. We conclude that G has cycle complexity at most m.

Suppose now that $h(\min(V)) \neq m$ or that $\min(V)$ is not the only vertex for which h takes the value m, and let v be the greatest vertex with $h(v) = m$ in the total order on V. Then $V_1 = \{v' \in V \mid v' < v\}$ is nonempty and distinct from V. Let $V_2 = V \setminus V_1$. Then by (3.1), there is no edge from V_1 to V_2, because $v = \min(V_2)$ and $n(v_1) \leq v$ for all $v_1 \in V_1$. Let $G_i = G|V_i$. Then by Lemma 3.3 the graphs G_i inherit a height function by restriction of h, and we conclude by induction that their cycle complexity is at most m. Now, each strongly connected component of G is contained in a strongly connected component of G_1 or G_2, which implies that G has cycle complexity at most m. □

K being a field, let E be a finite dimensional vector space over K, let B be a basis of E and let Φ be a set of endomorphisms of E. We associate to E, B, Φ a directed graph with set of vertices B, and edges $b \to b'$ whenever there is some $\phi \in \Phi$ such that $\phi(b)$ involves b' when expanded in the basis B.

The *cycle complexity* and *height functions* of E, B, Φ are defined correspondingly. We say that E, Φ has *cycle complexity* m if m is the smallest cycle complexity of triples E, B, Φ over all bases B of E.

We denote by E' the dual space of E, by B' the dual basis of B and by Φ' the set of adjoints ϕ' for $\phi \in \Phi$. Recall that the adjoint of ϕ maps the linear function λ on E onto the linear function $\lambda \circ \phi$ on E. The cycle complexity of E, B, Φ is equal to the cycle complexity of E', B', Φ'. Indeed, it is well-known that b_j appears in the B-expansion of $\phi(b_i)$ if and only if b'_i appears in the B'-expansion of $\phi'(b'_j)$. Therefore the associated graphs are opposite to each other. Since opposite graphs have the same cycle complexity, so have E, B, Φ and E', B', Φ'. Taking the minimum over the bases B, we see that E, Φ and E', Φ' have the same cycle complexity.

Observe that $h : B \to \mathbb{N}$ is a height function for E, B, Φ if and only if the following condition holds.

$$\begin{aligned}&\text{if } h(b) = 0 \text{ (resp. } h(b) \geq 1\text{), then for any } \phi \in \Phi\text{, the image } \phi(b)\\ &\text{is a linear combination of } b' < b \text{ (resp. of } b' < n(b)\text{).}\end{aligned} \tag{3.3}$$

Of course, B needs to be totally ordered, and n is the corresponding next function. We slightly generalize this notion. Let E, Φ be as before, and consider a finite totally ordered family $(b_i)_{i\in I}$ which spans E as a vector space, with a function $h : I \to \mathbb{N}$ (also called *height function*) such that the following condition holds.

$$\begin{aligned}&\text{if } h(j) = 0 \text{ (resp. } h(j) \geq 1\text{) then for any } \phi \in \Phi\text{, the image } \phi(b_j)\\ &\text{is a linear combination of } b_i \text{ with } i < j \text{ (resp. with } i < n(j)\text{).}\end{aligned} \tag{3.4}$$

Lemma 3.6 *Let E, Φ, $(b_i)_{i\in I}$, h be as previously. Then E, Φ has cycle complexity at most* $\max(h)$.

Proof. We remove successively elements of the family until we obtain a basis. This is done as follows. If (b_i) is not a basis, then for some k in I, we have a relation

$$b_k = \sum_{j<k} \alpha_j b_j$$

for some α_j in K. It is then easy to see that each linear combination of elements b_i with $i < p$ (where $p \in I \cup \infty$) is also a linear combination of elements b_i with $i < p$ and in addition with $i \neq k$. This follows from the relation above.

Consider the family $(b_i)_{i\in I\setminus k}$ and the restriction h' of h to $I \setminus k$. The next function n' of h' satisfies $n'(i) \geq n(i)$. This implies, in view of the remark above, that for $j \in I \setminus k$ such that $h(j) = 0$ (resp. $h(j) \geq 1$) the image $\phi(b_j)$ is a linear combination of elements b_i with $i \in I \setminus k$ and $i < j$ (resp. $i < n'(j)$). Thus we obtain a smaller family and conclude by induction. □

Lemma 3.7 *Let E, Φ have cycle complexity m. Let F be a subspace of E which is invariant under the action of Φ. Then E/F and F, with the set of induced endomorphisms, have cycle complexity at most m.*

Proof. 1. We know that E has a basis B with a height function h satisfying condition (3.3) above and $\max(h) = m$. Hence E/F has a spanning family and a height function h satisfying (3.3) and $\max(h) = m$. By Lemma 3.6, the cycle complexity of the set of induced endomorphisms of E/F is at most m.

2. We know that for some basis B of E, the cycle complexity of E, B, Φ is m. Hence, the dual E', B', Φ' also has cycle complexity m. Let $F^{\perp}$ be the set of linear functions in E' which are 0 on F. Then classically $F' \simeq E'/F^{\perp}$. Note that each endomorphism in Φ' maps $F^{\perp}$ into itself. Hence by the previous part, F', Φ' has cycle complexity at most m. Hence, by duality again, F, Φ has cycle complexity at most m. □

To a set $\mathcal{M}$ of square matrices of order n, we associate the graph G with set of vertices $\{1, \ldots, n\}$ and edges $i \to j$ if $M_{i,j} \neq 0$ for some matrix $M \in \mathcal{M}$. We call *cycle complexity* of $\mathcal{M}$ the cycle complexity of the graph G. Similarly, the *cycle complexity* of a representation (λ, μ, γ) is the cycle complexity of the set of matrices μa, $a \in A$.

Theorem 3.8 *A rational series in $K\langle\langle A\rangle\rangle$ has star height at most m if and only if it has a minimal representation of cycle complexity at most m.*

Note that the strength of this result resides in the condition of minimality. This is quite different from what happens for languages and automata.

A matrix $(a_{i,j})$ is called (noncommutative) *generic* if its coefficients are distinct noncommutative variables.

Corollary 3.9 *Let M be a square generic matrix of size $n \times n$. Then each entry of M^* is a rational series of star height n.*

Proof. Consider the series $S_{u,v} = (M^*)_{u,v}$. By the second part of the proof of Theorem 1.7.1, it has the representation (e_u, μ, e_v^T), where μ maps $a_{i,j}$ onto the elementary matrix $E_{i,j}$. This representation is minimal by Proposition 2.2.1. Hence $S_{u,v}$ has star height at most n, since a graph with n vertices has cycle complexity at most n. Now, it is easy to see that the complete graph on n vertices has cycle complexity exactly n. Hence, if $S_{u,v}$ has star height $< n$, the theorem shows that for some minimal representation $(\lambda', \mu', \gamma')$ of $S_{u,v}$ and some i, j, one has $(\mu' a)_{i,j} = 0$ for each letter a. Now, we have $\mu' a = P\mu a P^{-1}$ for some $P \in \mathrm{GL}_n(K)$. Hence $(PE_{k,\ell}P^{-1})_{i,j} = 0$ for each elementary matrix $E_{k,\ell}$. This is not possible, since it would imply that $(PNP^{-1})_{i,j} = 0$ for any matrix N, and so $N_{i,j} = 0$ for any N. □

One part of the theorem is a consequence of the following lemma.

Lemma 3.10 *Let (λ, μ, γ) be a representation of a series S having cycle complexity at most m. Then S has star height at most m.*

Proof. If $m = 0$, then there is no infinite path in the underlying graph. Hence S is a polynomial (by Equation (1.6.1) for example) and thus has star height 0.

We assume now that $m \geq 1$. Suppose that the associated graph G is strongly connected, of cycle complexity at most m, and that $G \setminus 1$ has cycle complexity at most $m - 1$. Then the matrix $M = \sum_{a \in A} a\mu a$ may be written as

$$M = \begin{pmatrix} M_1 & M_2 \\ M_3 & M_4 \end{pmatrix}$$

where M_1 is of size 1×1. Then M_4 has cycle complexity at most $m - 1$ and by induction, each entry of M_4^* is a series of star height at most $m - 1$. Now, setting $N = M_1 + M_2 M_4^* M_3$, one has

$$M^* = \begin{pmatrix} N^* & N^* M_2 M_4^* \\ M_4^* M_3 N^* & M_4^* + M_4^* M_3 N^* M_2 M_4^* \end{pmatrix}$$

by a variant of an identity proved in the proof of Lemma 1.7.3. Note that N is a series of star height $\leq m-1$ and consequently N^* has star height at most m. It follows that each entry of M^* has star height at most m, hence S too.

Suppose now that G is not strongly connected. Then the representation μ has a block triangular form and each diagonal block has cycle complexity at most m. We then use iteratively Lemma 9.2.2(i) to conclude. □

Proof of Theorem 3.8. It remains to show that if S has star height at most m, then S has a minimal representation of cycle complexity at most m.

1. We prove first that under this hypothesis, there exists a stable subspace E of $K\langle\langle A\rangle\rangle$ containing S, and such that the set $\Phi = \{T \mapsto a^{-1}T \mid a \in A\}$ of endomorphisms of E has cycle complexity at most m.

In view of Lemma 3.6, it suffices to show that E has a spanning family $(S_i)_{i\in I}$ with a height function $h : I \to \mathbb{N}$ satisfying (3.4) and with $\max(h) \leq m$. To do this, we argue by induction on the size of a rational expression for S. By definition of the star height, it is enough to show it when

(i) S is a polynomial and $m = 0$;
(ii) $S = T + U$ or $S = UT$, with stable subspaces F, G (for T and U respectively) and families $(T_i)_{i\in I}$, $(U_j)_{j\in J}$, and height functions k, ℓ with $\max(k)$, $\max(\ell) \leq m$;
(iii) $S = T^*$, T proper, with stable subspace F, family $(T_i)_{i\in I}$ and height function k with $\max(k) \leq m-1$.

(i) follows by taking as family the set of words appearing in S, with an order compatible with the length, with $h = 0$, noting that $a^{-1}w$ has length smaller than w or is 0 for any word w.

(ii) If $S = T + U$, assuming that I, J are disjoint, consider the stable subspace $F + G$, spanned by the union $(T_i)_{i\in I} \cup (U_j)_{j\in J}$ of the families, with a total order extending those of I and J and moreover $i < j$ for $i \in I$, $j \in J$. Furthermore, let h extend k and ℓ.

If $S = UT$, take the stable subspace $GT + F$, spanned by the family $(U_jT)_{j\in J} \cup (T_i)_{i\in I}$ with the same order and height function as before. Since we have $a^{-1}(U_jT) = (a^{-1}U_j)T + (U_j, 1)(a^{-1}T)$ and since $a^{-1}U_j$ (resp. $a^{-1}T$) is a linear combination of $U_{j'}$ (resp. T_i), we see that (3.4) is satisfied.

(iii) If $S = T^*$, take $E = KS + F$, $J = I \cup \{\omega\}$, with $\omega < i$ for $i \in I$, and let $S_i = T_iS$ for $i \in I$, $S_\omega = S$. Let h extend k by $h(\omega) = m$. We have $a^{-1}S = (a^{-1}T)S$ and for i in I, $a^{-1}(T_iS) = (a^{-1}T_i)S + (T_i, 1)S$. Since $a^{-1}T_i$ is a linear combination of elements $T_{i'}$, we see that (3.4) is satisfied.

2. By the previous part and by Lemma 3.7, we see that $S \circ K\langle A\rangle$ has cycle complexity at most m with respect to the set Φ, since $S \circ K\langle A\rangle$ is a subspace of E, invariant under the endomorphisms in Φ. This shows, by the construction of Lemma 2.1.3, that S has a representation of cycle complexity at most m and dimension $\dim(S \circ K\langle A\rangle)$. Since the latter is the rank of S, we deduce that the representation is minimal (Corollary 2.1.5 and Theorem 2.1.6). □

4.4 Absolute star height

Consider the rational series $S = \frac{1}{2}(a+ib)^* + \frac{1}{2}(a-ib)^* \in \mathbb{C}\langle\langle a,b\rangle\rangle$. Clearly, S has star height 1 over $\mathbb{C}$. But S is actually in $\mathbb{R}\langle\langle a,b\rangle\rangle$. Indeed (see also Exercise 2.2)

$$\begin{aligned} S &= \frac{1}{2}\sum_{w\in\{a,b\}^*}\left(i^{|w|_b} + (-i)^{|w|_b}\right)w \\ &= \sum_{|w|_b \text{even}} i^{|w|_b} w = \sum_{|w|_b \text{even}} (-1)^{|w|_b/2} w \\ &= \sum_{k\geq 0, u_0,\ldots,u_{2k}\in a^*} (-1)^k u_0 b u_1 \cdots b u_{2k} = (a - ba^*b)^* . \end{aligned}$$

The series S has as minimal representation (λ, μ, γ) with

$$\lambda = \gamma^T = (1,0), \quad \mu a = \begin{pmatrix} 1 & 0 \\ 0 & 1 \end{pmatrix}, \quad \mu b = \begin{pmatrix} 0 & -1 \\ 1 & 0 \end{pmatrix}$$

and associated weighted automaton

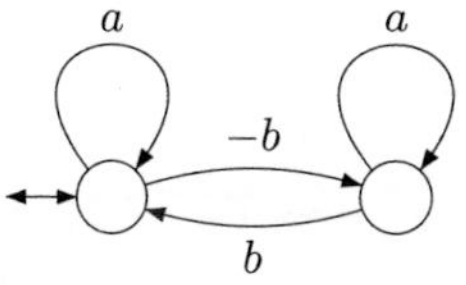

It has star height 2 over $\mathbb{R}$. Indeed, for any other minimal representation $(\lambda', \mu', \gamma')$ over $\mathbb{R}$, we have $\mu' a = \begin{pmatrix} 1 & 0 \\ 0 & 1 \end{pmatrix}$ and $\mu' b = P\begin{pmatrix} 0 & -1 \\ 1 & 0 \end{pmatrix}P^{-1}$ for some invertible matrix P over $\mathbb{R}$. Then $(\mu' b)_{1,2}, (\mu' b)_{2,1}$ are never 0, since $\begin{pmatrix} 0 & -1 \\ 1 & 0 \end{pmatrix}$ has no real eigenvalue. Thus the associated graph is complete and the representation $(\lambda', \mu', \gamma')$ has cycle complexity 2 and by Theorem 3.8, S has star height 2 over $\mathbb{R}$.

This example shows that the star height may decrease when the field of scalars is extended. If $S \in K\langle\langle A\rangle\rangle$ is rational over a field K, we call *absolute star height* the star height of S over the algebraic closure $\bar{K}$ of K.

Theorem 4.1 *The absolute star height is effectively computable.*

It is understood here that K is a field where one can compute, for example $K = \mathbb{Q}$.

Proof. 1. Given a representation $\rho = (\lambda, \mu, \gamma)$ of dimension n over K and a graph G with vertex set $\{1, \ldots, n\}$, it is decidable if ρ is conjugate over $\bar{K}$ to a representation ρ' whose associated graph G' is a subgraph (same vertices, less edges) of G. Indeed, if such a ρ' exists, then for some $P \in GL_n(\bar{K})$, G' is associated to the matrices $P\mu a P^{-1}$, $a \in A$. Note that P may be chosen of determinant 1. The existence of ρ' is therefore equivalent to the existence of a solution in $\bar{K}$ of the system of algebraic equations over K in y and $x_{i,j}$, $1 \le i, j \le n$, obtained by writing that $y \det(x_{i,j}) - 1 = 0$ and that the graph associated to the matrices $(x_{i,j})\mu a(x_{i,j})^{-1}$ is a subgraph of G (one must write that certain coefficients of these matrices are 0). The existence of a solution in $\bar{K}$ is equivalent to the fact that the ideal generated by the polynomials forming the system is not equal to $K[x_{i,j}, y]$. The latter property is decidable by Gröbner basis techniques (see Cox *et al.* (1997)).

2. Now, given a rational series over K, we may find a minimal representation ρ of it. It is then sufficient to enumerate the graphs G and to decide if ρ has a conjugate over $\bar{K}$ of a representation whose associated graph is a subgraph of G. One continues until a graph G of minimum cycle complexity is found, in view of Theorem 3.8. □

Computing the star height over $\mathbb{Q}$ of a rational series in $\mathbb{Q}\langle\langle A\rangle\rangle$ is an open problem, which may be undecidable.

Exercises for Chapter 4

1.1 Do the remaining verifications in the proof of Proposition 1.1.

1.2 For each word w, the *alphabet* of w is the set of letters that occur in w. A series S is *iso-alphabetic* if all words w in its support have the same alphabet. We consider *iso-alphabetic* rational expressions. These are expressions where the operation $E \mapsto E^*$ is restricted to expressions denoting iso-alphabetic series. For example, the expression $(a+b)^*$ is not iso-alphabetic, but is equal to the iso-alphabetic expression $(b^*a^+b)^*b^*a^*$.

Show that every rational series has an iso-alphabetic rational expression.

2.1 Improve the result obtained in the proof of Theorem 2.1 by showing that for each rational expression $E \in \mathcal{E}_n$ there exists a stable submodule of $\mathcal{E}_n$ containing E and which is generated by finitely many words on the alphabet $\bigcup_{n\ge 0} A_n$. Deduce that this module is a free K-module (K is here a commutative semiring).

2.2 Show, by using only the fact that S^* is the inverse of $1 - S$, that in $\mathbb{C}\langle\langle a, b\rangle\rangle$ one has

$$\frac{1}{2}(a+ib)^* + \frac{1}{2}(a-ib)^* = (a - ba^*b)^*$$

and

$$\frac{1}{2i}(a+ib)^* + \frac{1}{2i}(a-ib)^* = (a - ba^*b)^*ba^*.$$

2.3 Verify that the proof of Theorem 2.1 is constructive.

3.1 Show that the cycle complexity of a subgraph is less than or equal to the cycle complexity of the graph.

3.2 Show that the complete directed graph on n vertices has cycle complexity n. Give a height function for this graph.

3.3 Show that, with the notations of the proof of Corollary 3.9, $S_{u,v}$ is the sum of all paths from u to v in the complete graph with n vertices (a path is identified with the corresponding word in the $a_{i,j}$'s).

3.4 Show that if K is any commutative semiring, and if S is a rational series, then S has star height at most m if and only if S has a representation of cycle complexity at most m.

4.1 Show that the following series over $\mathbb{Q}$ has star height 2 over $\mathbb{Q}$ and star height 1 over $\mathbb{R}$: $S = \frac{1}{2}(a + b\sqrt{2})^* + \frac{1}{2}(a - b\sqrt{2})^*$. (*Hint*: Mimick the example at the beginning of Section 4.4.)

4.2 Show that if $K \subseteq L$ is an extension of algebraically closed fields, then the star height over K of a K-rational series is equal to its star height over L.

Notes to Chapter 4

The idea of lifting the operations a^{-1} to rational expressions goes back to Brzozowski (1964). The results of Section 4.2 are from Krob (1991) and those of Section 4.3 are from Reutenauer (1996). The idea of cycle complexity of a graph, Lemma 3.10, the first part of the proof of Theorem 3.8 and Exercise 3.4 go back to Eggan (1963) who introduced star height of languages. The Boolean version (for languages) of Corollary 3.9 was proved in Cohen (1970): the set of paths in a complete graph on n vertices is of star height n; however it is not clear how one could deduce one result from the other. For rational expressions and identities of languages, see Hashiguchi (1991), Kirsten (2005), Sakarovitch (2009a) and the references therein.

Part II

Arithmetic

5

Automatic sequences and algebraic series

Given an integer $k \geq 2$ and a function $f : \mathbb{N} \to K$ into some semiring K, we consider the series S defined by $(S, w) = f(n)$ whenever w is an expansion of n at base k. If S is a recognizable series, then f is called a k-regular function over K.

Section 5.1 gives a presentation of regular functions. Section 5.2 considers closure properties. In the special case of the Boolean semiring, regular functions correspond to automatic sequences. These are described in Section 5.3. In Section 5.4, we prove that q-automatic sequences over the alphabet $\mathbb{F}_q$ correspond precisely to series that are algebraic over $\mathbb{F}_q((x))$. Section 5.5 considers diagonals of rational series.

5.1 Regular functions

Let $k \geq 2$ be a fixed integer called the *base*, and let $\boldsymbol{k} = \{0, \ldots, k-1\}$. Its elements are called the *digits* in base k. Let $\nu_k : \boldsymbol{k}^* \to \mathbb{N}$ be defined for $w = d_{n-1} \cdots d_0$, with $n \geq 0$ and $d_i \in \boldsymbol{k}$, by

$$\nu_k(w) = \sum_{i=0}^{n-1} d_i k^i .$$

The number $\nu_k(w)$ is the number *represented* by w, and w is a *representation* of n at base k. In particular, $\nu_k(\varepsilon) = 0$, where ε is the empty word. Set $R = \boldsymbol{k}^* \setminus 0\boldsymbol{k}^*$. Clearly, ν_k is a bijection from R onto $\mathbb{N}$.

Conversely, the *expansion* of an integer n at base k, also called the *canonical representation* of n, is the unique word w in R such that $\nu_k(w) = n$. It is denoted by $\sigma_k(n)$. The expansion of 0 is the empty word.

To each function $f : \mathbb{N} \to K$, where K is a semiring, we associate a series S_f defined by

$$(S_f, w) = f(\nu_k(w)) \qquad w \in \boldsymbol{k}^* . \tag{1.1}$$

A function $f : \mathbb{N} \to K$ is a *k-regular function over K* (or the sequence $(f(n))_{n\geq 0}$ is a *k-regular sequence*) if the series S_f is recognizable, or equivalently rational (Theorem 1.7.1).

A subset H of $\mathbb{N}$ is called *k-recognizable* or *recognizable in base k*, if its characteristic function $H \to \mathbb{B}$ (the Boolean semiring) is k-regular.

Example 1.1 The *sum of digits function* s_k associates to each $n \in \mathbb{N}$ the sum of its digits in its expansion at base k: if

$$n = \sum c_i k^i, \qquad c_i \in \boldsymbol{k},$$

then

$$s_k(n) = \sum c_i .$$

It is k-regular over $\mathbb{N}$ because $s_k(\nu_k(w)) = \lambda\mu(w)\gamma$, where

$$\lambda = (0\ 1), \quad \mu(i) = \begin{pmatrix} 1 & 0 \\ i & 1 \end{pmatrix}, \quad i = 0, \ldots, k-1, \quad \gamma = \begin{pmatrix} 1 \\ 0 \end{pmatrix}.$$

Example 1.2 The *identity function* $\mathbb{N} \to \mathbb{N}$ is k-regular over $\mathbb{N}$ (this has been already shown in Example 1.5.2 for $k = 2$ in a different manner). The series $\sum_w \nu_k(w)\, w$ is indeed recognizable because $\nu_k(w) = \lambda\mu(w)\gamma$ with

$$\lambda = (0\ 1), \quad \mu(i) = \begin{pmatrix} k & 0 \\ i & 1 \end{pmatrix}, \quad i = 0, \ldots, k-1, \quad \gamma = \begin{pmatrix} 1 \\ 0 \end{pmatrix}.$$

Indeed, it is easily checked that

$$\mu(w) = \begin{pmatrix} k^{|w|} & 0 \\ \nu_k(w) & 1 \end{pmatrix} \qquad \text{for } w \in \boldsymbol{k}^* .$$

Proposition 1.1 *For any function $f : \mathbb{N} \to K$, the following conditions are equivalent:*

(i) *f is a k-regular function;*
(ii) *the series $S = \sum_{n \geq 0} f(n)\sigma_k(n)$ is recognizable;*
(iii) *there exists a recognizable series T which coincides with S_f on R.*

Proof. (i) $\Longleftrightarrow$ (ii). Observe that the support of the series $S = \sum_{n \geq 0} f(n)\sigma_k(n)$ is contained in R and that S coincides with S_f on R. One has $S = S_f \odot \underline{R}$. Thus if S_f is recognizable, so is S by Corollary 3.2.3. Conversely, $S_f = \underline{0}^* S$. This implies that if S is recognizable, so is S_f.

(ii) $\Longleftrightarrow$ (iii). Assume T is recognizable. Since $S = T \odot \underline{R}$, the series S is recognizable. The converse implication is clear. □

Applying this result to $\mathbb{B}$, we obtain by Corollary 3.2.5, the following result.

Corollary 1.2 *For each set H of nonnegative integers, the following conditions are equivalent:*

(i) *H is a k-recognizable subset of $\mathbb{N}$;*
(ii) *$\nu_k^{-1}(H)$ is a rational subset of $\boldsymbol{k}^*$;*

(iii) *$\sigma_k(H)$ is a rational subset of $\boldsymbol{k}^*$;*
(iv) *there exists a rational subset X of $\boldsymbol{k}^*$ such that $H = \nu_k(X)$.* □

Corollary 1.3 *If f is a k-regular function over $\mathbb{N}$, and a is in $\mathbb{N}$, then the sets $f^{-1}(a)$, $f^{-1}(\{n \mid n \leq a\})$ and $f^{-1}(\{n \mid n > a\})$ are k-recognizable subsets of $\mathbb{N}$.*

Proof. This follows from Proposition 1.1, Corollary 1.2 and Corollary 3.2.7. □

Example 1.3 The set of powers of 2 is 2-recognizable since the set of its canonical representations is the rational language 10^*.

Example 1.4 The set of squares is not 2-recognizable. Indeed, let L be the language of canonical representations of squares at base 2, and consider the language $L' = L \cap (11)^+(00)^+01$. This is the language of canonical representations of squares of the form $(2^{2m} - 1)2^{2n+2} + 1$ for some integers $n, m \geq 1$. We claim that a positive integer y satisfies $y^2 = (2^{2m} - 1)2^{2n+2} + 1$ with $n, m \geq 1$ if and only if $m = n$ and $y = 2^{2n+1} - 1$. Assume this claim for a moment. Then $L' = \{1^{2n}0^{2n+1}1 \mid n \geq 1\}$ and this is not a rational language, by the pumping lemma for rational languages (Exercise 3.1.3).

To prove the claim, let y be such that $y^2 = (2^{2m} - 1)2^{2n+2} + 1$ for some $m, n \geq 1$. Then $y^2 - 1$ is divisible by 2^{2n+2}. Since y is odd, only one of the numbers $y - 1$ and $y + 1$ is divisible by 4. Let $y - e$ be this number, with $e = \pm 1$; then $y + e$ is not divisible by 4 and since $(2^{2m} - 1)2^{2n+2} = y^2 - 1 = (y - e)(y + e)$, we can write $y = 2^{2n+1}z + e$, where z is odd. Since $(2^{2m} - 1)2^{2n+2} = y^2 - 1 = 2^{4n+2}z^2 + 2^{2n+2}ze$, we get $2^{2n}z^2 + ze = 2^{2m} - 1$. This shows that $2^{2n} - 1 \leq 2^{2m} - 1$ and therefore $n \leq m$. Assume now $n < m$ and set $p = m + n + 1$. Then $p > 2n + 1$, whence $(2^p - 1)^2 = 2^{2p} - 2^{p+1} + 1 < 2^{2p} - 2^{2n+2} + 1 = y^2 < (2^p)^2$. This shows that y^2 lies between two consecutive squares, a contradiction. Consequently $m = n$ and $y = 2^{2n+1} - 1$.

Proposition 1.4 *If $f, g : \mathbb{N} \to K$ are k-regular, then the functions $f + g$ and λf, $f\lambda$ for $\lambda \in K$ are k-regular. If K is commutative, then $f \odot g$ defined by $f \odot g(n) = f(n)g(n)$ is k-regular.*

Proof. The first assertions follow from Corollary 1.5.2. For the last assertion, it suffices to observe that $S_{f \odot g} = S_f \odot S_g$ and to apply Theorem 1.5.5. □

5.2 Stable submodules and operations on k-regular functions

The set $K^{\mathbb{N}}$ of functions $\mathbb{N} \to K$ is a right K-module for addition and multiplication by a constant defined in the usual way. We define a left action of $\boldsymbol{k}^*$ on $K^{\mathbb{N}}$ by setting, for $j \in \boldsymbol{k}$ and $f \in K^{\mathbb{N}}$,

$$(j \circ f)(n) = f(nk + j).$$

This action is extended to $\boldsymbol{k}^*$ by $u \circ (v \circ f) = uv \circ f$ for $u, v \in \boldsymbol{k}^*$. It follows that for $w \in \boldsymbol{k}^*$

$$(w \circ f)(n) = f(nk^{|w|} + \nu_k(w)).$$

Indeed, by induction, for $j \in \boldsymbol{k}$,

$$\begin{aligned}(jw \circ f)(n) &= (j \circ (w \circ f))(n) = (w \circ f)(nk + j)\\ &= f((nk + j)k^{|w|} + \nu_k(w))\\ &= f(nk^{1+|w|} + jk^{|w|} + \nu_k(w)) = f(nk^{|jw|} + \nu_k(jw)).\end{aligned}$$

A K-submodule V of $K^{\mathbb{N}}$ is *stable* if V is closed by the operations $f \mapsto w \circ f$ for $w \in \boldsymbol{k}^*$. This is equivalent to saying that if V contains f, then V contains all functions $n \mapsto f(nk^e + s)$, for $e \geq 0$ and $0 \leq s < k^e$.

We define, for $u \in \boldsymbol{k}^*$ and $S \in K\langle\langle \boldsymbol{k} \rangle\rangle$, the series Su^{-1} by $(Su^{-1}, v) = (S, vu)$. This is a left-right symmetric version of the operation defined in Section 1.5. We write also $Su^{-1} = u \circ S$, since it is a left action of $\boldsymbol{k}^*$ on $K\langle\langle \boldsymbol{k} \rangle\rangle$.

Consider the subset

$$E = \{S \in K\langle\langle \boldsymbol{k} \rangle\rangle \mid \forall w \in \boldsymbol{k}^*, (S, 0w) = 0\}$$

of $K\langle\langle \boldsymbol{k} \rangle\rangle$. It is a right K-submodule which is closed under the left action of $\boldsymbol{k}^*$ on $K\langle\langle \boldsymbol{k} \rangle\rangle$. Indeed, if $S \in E$ and $u, w \in \boldsymbol{k}^*$, then $(u \circ S, 0w) = (S, 0wu) = 0$.

Lemma 2.1 *The function $h : f \mapsto S_f \odot \underline{R}$ is a right K-linear isomorphism from $K^{\mathbb{N}}$ onto E which commutes with the left actions of $\boldsymbol{k}^*$ on $K^{\mathbb{N}}$ and on E.*

Proof. One has $h(f) = \sum_{n \in \mathbb{N}} f(n)\sigma_k(n)$. This shows that h is a right K-linear isomorphism. Concerning the left action, let $u \in \boldsymbol{k}^*$. We prove that $u \circ h(f) = h(u \circ f)$. Let $w \in \boldsymbol{k}^*$. If $w \in 0\boldsymbol{k}^*$, then $(u \circ h(f), w) = (h(f), wu) = 0 = (h(u \circ f), w)$, and if $w \notin 0\boldsymbol{k}^*$, then

$$\begin{aligned}(u \circ h(f), w) &= (h(f), wu) = f(k^{|u|}\nu_k(w) + \nu_k(u)) = (u \circ f, \nu_k(w))\\ &= (h(u \circ f), w).\end{aligned}$$

□

Proposition 2.2 *A function $f : \mathbb{N} \to K$ is k-regular over K if and only if there exists a stable finitely generated right K-submodule of $K^{\mathbb{N}}$ which contains f.*

Proof. If such a submodule exists, then there is a stable finitely generated right K-submodule of $K\langle\langle \boldsymbol{k} \rangle\rangle$ which contains $S = \sum_{n \geq 0} f(n)\sigma_k(n)$, by Lemma 2.1. Hence S is recognizable by the left-right symmetric statement of Proposition 1.5.1. Thus f is k-regular by Proposition 1.1(ii).

Conversely, suppose that S is recognizable. Then, by Proposition 1.5.1, there exists a finitely generated right K-submodule of $K\langle\langle \boldsymbol{k} \rangle\rangle$, containing S, and which is closed under the left actions $T \mapsto u \circ T$ for $u \in \boldsymbol{k}^*$. Let $S_1, \ldots, S_n$ be generators of this right K-module. For any $a \in \boldsymbol{k}$, we have for $j = 1, \ldots, n$

$$a \circ S_j = \sum_{i=1}^{n} S_i \alpha_{i,j} \,,$$

where $\alpha_{i,j} \in K$. Thus for any word $w \in \boldsymbol{k}^*$

$$(S_j, wa) = \sum_{i=1}^{n} (S_i, w)\alpha_{i,j} \,. \tag{2.1}$$

Define $U = 1 \in K\langle\langle \boldsymbol{k} \rangle\rangle$ and $T_j = S_j \odot \underline{\boldsymbol{k}^+ \setminus 0\boldsymbol{k}^*}$ for $j = 1, \ldots, n$. Observe that each T_j has constant term 0. We show that the right K-submodule M spanned by $U, T_1, \ldots, T_n$ contains S and is stable. This will imply the proposition by Lemma 2.1, since the series $U, T_1, \ldots, T_n$ are in E.

The submodule M contains S since $\operatorname{supp}(S) \subset R$ and since S is a right K-linear combination of $S_1, \ldots, S_n$. We verify that $a \circ T_j = \sum_{i=1}^{n} T_i \alpha_{i,j} + U(T_j, a)$. This will imply stability. Let $w \in \boldsymbol{k}^*$. We have to show that, for $a \in \boldsymbol{k}$,

$$(T_j, wa) = \sum_{i=1}^{n} (T_i, w)\alpha_{i,j} + (U, w)(T_j, a) \,.$$

This is clear if $w \in 0k^*$. If $w \notin 0k^*$, suppose first that w is the empty word. Then both sides of the equation are equal to (T_j, a). Suppose now that the word w is not empty. Then $(T_j, wa) = (S_j, wa)$. Moreover, $(U, w) = 0$ and $(T_i, w) = (S_i, w)$. Thus the equality follows from (2.1). □

Example 2.1 We show by using Proposition 2.2 that the sum of digits function s_k already considered in Example 1.1 is k-regular.

For this, observe first that the constant functions $c_i : \mathbb{N} \to \mathbb{N}$, defined for $i \in \boldsymbol{k}$ by $c_i(n) = i$ for all n, are k-regular since $j \circ c_i = c_i$. Next, $j \circ s_k = s_k + c_j$ because $(j \circ s_k)(n) = s_k(nk + j) = s_k(n) + j$. Thus s_k together with the constant functions $c_0, \ldots, c_{k-1}$ span a stable finitely generated submodule of $K^{\mathbb{N}}$.

Corollary 2.3 *Let K be a finite semiring or a commutative ring. Then a function $f : \mathbb{N} \to K$ is k-regular over K if and only if the right K-submodule of $K^{\mathbb{N}}$ generated by the functions $w \circ f$, for $w \in \boldsymbol{k}^*$, is finitely generated. In this case, it is generated by finitely many of these functions.*

Proof. This follows from Proposition 2.2 and Corollary 1.5.4. □

An interesting property of k-regular functions is closure by extraction of an arithmetic progression on the argument. We start with a lemma.

Lemma 2.4 *If $f : \mathbb{N} \to K$ is k-regular, then the functions g and g' defined by $g(n) = f(n+1)$ for $n \geq 0$, and $g'(n) = f(n-1)$ for $n \geq 1$, and $g'(0) = 0$ are k-regular.*

The exact value of $g'(0)$ in the previous statement has no importance because two series which differ only by a finite number of values are both rational or both irrational, by Exercise 3.2.3.

Proof. We start with g. Let M be a finitely generated stable K-submodule of $K^{\mathbb{N}}$ containing f, and let N be the K-submodule generated by the functions in M and the functions $n \mapsto h(n+1)$ for $h \in M$. Clearly N is finitely generated and contains g. It remains to show that N is stable. For this, consider a function $h \in M$, and set $u(n) = h(n+1)$. Let j be an integer with $0 \leq j < k$. If $j < k-1$,

$$(j \circ u)(n) = u(kn + j) = h(kn + j + 1) = ((j+1) \circ h)(n)$$

and thus $j \circ u \in M$, and if $j = k - 1$,

$$((k-1) \circ u)(n) = u(kn + k - 1) = h(kn + k) = h(k(n+1)) = (0 \circ h)(n+1)\,.$$

Since $0 \circ h \in M$, the function $n \mapsto (0 \circ h)(n+1)$ is in N. This shows that $j \circ u \in N$ for $0 \leq j < k$ and that N is stable.

A similar argument holds for the function g'. Here, the case distinction is between $j > 0$ and $j = 0$. □

Proposition 2.5 *Let $a \geq 1, b \geq 0$ be integers. If $f : \mathbb{N} \to K$ is k-regular, then the function g defined by $g(n) = f(an + b)$ is k-regular.*

Proof. Assume first $b < a$. Let M be a finitely generated stable K-submodule of $K^{\mathbb{N}}$ containing f, and let N be the K-submodule generated by the functions in M and by all functions $n \mapsto h(an + c)$, for $0 \leq c < a$ and $h \in M$. Clearly N is finitely generated and contains g. It remains to show that N is stable. For this, observe that for $0 \leq j < k$, one has $aj + c \leq a(k-1) + a - 1 = (a-1)k + k - 1$. Euclidean division of $aj + c$ by k therefore gives

$$aj + c = c'k + \ell, \quad \text{with } 0 \leq c' < a\,,\ 0 \leq \ell < k\,.$$

Let now $h \in M$ and define $p \in N$ by $p(n) = h(an + c)$. Then

$$\begin{aligned}(j \circ p)(n) &= p(kn + j) = h(a(kn + j) + c) = h(kan + aj + c) \\ &= h(k(an + c') + \ell) = (\ell \circ h)(an + c')\,.\end{aligned}$$

The function $h' = \ell \circ h$ is in M because M is stable, and by construction, the function $n \mapsto h'(an + c')$ is in N. This shows that $j \circ p$ is in N and thus that N is stable.

This proves the claim if $b < a$. If $b \geq a$, we argue by induction on b. Assuming that the function $n \mapsto f(an + b - 1)$ is k-regular, it follows by Lemma 2.4 that the function $n \mapsto f(an + b)$ is k-regular. □

Proposition 2.5 is used in the proof of the following property.

Proposition 2.6 *Assume that K is a finite semiring or a commutative ring. Let $k, \ell \geq 2$ be integers. If $f : \mathbb{N} \to K$ is both k-regular and ℓ-regular, then f is $k\ell$-regular.*

Proof. In this proof, we use both the left action of $\boldsymbol{k}^*$ and the left action of $\boldsymbol{\ell}^*$ on $K^{\mathbb{N}}$. Although it follows from the context which of the actions is meant, it is better to use the notation $\circ_k$ (resp. $\circ_\ell$) for the left action of $\boldsymbol{k}^*$ (resp. of $\boldsymbol{\ell}^*$) on $K^{\mathbb{N}}$. Similarly, a submodule of $K^{\mathbb{N}}$ will be called k-stable (resp. ℓ-stable) if it is stable under the action of $\boldsymbol{k}^*$ (resp. of $\boldsymbol{\ell}^*$).

Let $f : \mathbb{N} \to K$. We first prove that, for $u \in \boldsymbol{k}^*$ and $v \in \boldsymbol{\ell}^*$, there exist $u' \in \boldsymbol{k}^*$, $v' \in \boldsymbol{\ell}^*$, of the same length as u and v respectively, such that

$$u \circ_k (v \circ_\ell f) = v' \circ_\ell (u' \circ_k f). \tag{2.2}$$

Indeed, set $\alpha = |u|$, $\beta = |v|$, $r = \nu_k(u)$, $s = \nu_\ell(v)$. Then for $n \geq 0$,

$$u \circ_k (v \circ_\ell f)(n) = f(k^\alpha(\ell^\beta n + s) + r),$$

and since $k^\alpha s + r \leq k^\alpha(\ell^\beta - 1) + r \leq k^\alpha(\ell^\beta - 1) + (k^\alpha - 1) = k^\alpha \ell^\beta - 1$, there exist integers $q < k^\alpha, t < \ell^\beta$ such that $k^\alpha s + r = \ell^\beta q + t$ (Euclidean division of $k^\alpha s + r$ by ℓ^β). Let $u' \in \boldsymbol{k}^*$ and $v' \in \boldsymbol{\ell}^*$ be the words such that $|u'| = \alpha$, $\nu_k(u') = q$, $|v'| = \beta$, $\nu_\ell(v') = t$. Then

$$u \circ_k (v \circ_\ell f)(n) = f(\ell^\beta(k^\alpha n + q) + t) = v' \circ_\ell (u' \circ_k f)(n).$$

Now, let M be the K-submodule of $K^{\mathbb{N}}$ spanned by the functions $u \circ_k f$ for $u \in \boldsymbol{k}^*$. By Corollary 2.3, it is spanned by a finite number $f_1, \ldots, f_d$ of functions with $f_i = u_i \circ_k f$ for some $u_i \in \boldsymbol{k}^*$.

Next, since the function f is ℓ-regular, Proposition 2.5 implies that each f_i is ℓ-regular. Let M_i be the K-submodule of $K^{\mathbb{N}}$ spanned by the functions $v \circ_\ell f_i$ for $v \in \boldsymbol{\ell}^*$. By Proposition 2.2 again, each M_i is spanned by a finite number of functions $f_{i,j}$, for $j = 1, \ldots, d_i$, with $f_{i,j} = v_{i,j} \circ_\ell f_i$ for some $v_{i,j} \in \boldsymbol{\ell}^*$. Let N be the K-submodule spanned by the $f_{i,j}$. It is ℓ-stable by definition. It is also k-stable since for $r \in \boldsymbol{k}$, and in view of Equation (2.2)

$$r \circ_k f_{i,j} = r \circ_k (v_{i,j} \circ_\ell f_i) = v' \circ_\ell (r' \circ_k f_i) = v' \circ_\ell (r' u_i \circ_k f),$$

for some $r' \in \boldsymbol{k}$ and $v' \in \boldsymbol{\ell}^*$. Now $r' u_i \circ_k f$ is in M and thus is a linear combination of the f_i and moreover each $v' \circ_\ell f_i$ is in N.

It follows that N contains all functions $u \circ_k (v \circ_\ell f)$ and all functions $v \circ_\ell (u \circ_k f)$ for $u \in \boldsymbol{k}^*$ and $v \in \boldsymbol{\ell}^*$. It remains to show that N is $k\ell$-stable, but this follows from the fact that for $0 \leq j < k\ell$, and setting $j = kq + r$ with $0 \leq r < k$,

$$(j \circ_{k\ell} f)(n) = f(k\ell n + j) = f(k(\ell n + q) + r) = r \circ_k (q \circ_\ell f)(n). \qquad \square$$

Given two functions $f, g : \mathbb{N} \to K$, define their *Cauchy product* $f * g$ by

$$f * g(n) = \sum_{i+j=n} f(i)g(j).$$

This is just another way to consider the product of one variable series.

Proposition 2.7 *The Cauchy product of two k-regular functions is again k-regular.*

Proof. Let $u, v : \mathbb{N} \to K$ be two k-regular functions, and let $w = u * v$. Let M and N be stable finitely generated submodules of $K^{\mathbb{N}}$ containing u and v respectively, and let L be the submodule spanned by the functions $f * g$ for $f \in M$, $g \in N$ and the functions $n \mapsto (f * g)(n-1)$ for $f \in M$, $g \in N$ (with the convention that $(f * g)(-1) = 0$). Clearly, L is finitely generated and contains w. It suffices to show that L is stable. It will be more readable to write f_i instead of $i \circ f$ for $i \in \boldsymbol{k}$.

Let $f \in M$, $g \in N$, and define functions h and h' by $h(n) = f * g(n)$ and $h'(n) = f * g(n-1)$. We show that $h_d, h'_d \in L$ for each $d \in \boldsymbol{k}$. This will show that L is stable. We start with h_d. By definition

$$h_d(n) = h(nk + d) = \sum_{r+s=kn+d} f(r)g(s)\,. \tag{2.3}$$

Consider a pair (r, s) with $r + s = kn + d$ and consider the Euclidean division of r by k. This gives $r = ki + e$ for some $0 \le i \le n$ and $0 \le e < k$. It follows that $s = kn + d - r = kn + d - ki - e = k(n - i) + d - e$. We write this as

$$s = \begin{cases} kj + d - e & \text{with } j = n - i, \text{ if } 0 \le e \le d\,, \\ kj + (k + d - e) & \text{with } j = n - 1 - i, \text{ if } d < e < k\,. \end{cases}$$

This ensures that the rest $d - e$ (resp. $k + d - e$) is always nonnegative. Note that $j \ge 0$ in both cases. Accordingly, the sum (2.3) is split into two parts:

$$\begin{aligned} h(nk + d) &= \sum_{0 \le e \le d} \sum_{i+j=n} f(ik + e)g(jk + d - e) \\ &\quad + \sum_{d<e<k} \sum_{i+j=n-1} f(ik + e)g(jk + k + d - e) \\ &= \sum_{0 \le e \le d} (f_e * g_{d-e})(n) + \sum_{d<e<k} (f_e * g_{k+d-e})(n - 1)\,. \end{aligned}$$

This shows that $d \circ h$ is in L.

Next $h'_d(n) = h_{d-1}(n)$ for $d > 0$, so $d \circ h'$ is in L for $d > 0$. In the case $d = 0$, one has $h'_d(n) = h'(nk) = h(nk - 1) = h((n - 1)k + k - 1)$. This yields the decomposition

$$\begin{aligned} h'(nk) &= h((n - 1)k + k - 1) \\ &= \sum_{0 \le e \le k-1} \sum_{i+j=n-1} f(ik + e)g(jk + k - 1 - e) \\ &= \sum_{0 \le e \le k-1} (f_e * g_{k-1-e})(n - 1)\,. \end{aligned}$$

This shows that $0 \circ h'$ is in L and concludes the proof that L is stable. □

Proposition 2.8 *For any k-regular function $f : \mathbb{N} \to K$, where K is a field equipped with an absolute value $|\;|$, there is a constant c such that $|f(n)| = O(n^c)$.*

Proof. The series S_f is recognizable. By Exercise 1.5.1(a), there is a constant C such that $|(S_f, w)| \le C^{1+|w|}$ for all words w. If $w = \sigma_k(n)$, then $|w| \le 1 + \log_k n$, and consequently $|f(n)| = |(S_f, \sigma_k(n))| \le C^{2+\log_k n}$. Since $C^{\log_k n} = n^{\log_k C}$, it follows that $|f(n)| \le C^2 n^{\log_k C}$ and $|f(n)| = O(n^c)$ with $c = \log_k C$. □

5.3 Automatic sequences

We consider now partitions of the set $\mathbb{N}$ of integers into a finite number of k-recognizable sets. We assign, to each integer, a symbol denoting its class in the partition. When these symbols are enumerated as a sequence, one gets an infinite sequence called k-automatic.

More precisely, a sequence or *infinite word* u over the finite alphabet A is a mapping $u : \mathbb{N} \to A$. It is usual to write u as the sequence of its symbols $u = u(0)u(1)\cdots u(n)\cdots$. For instance, the sequence $u : \mathbb{N} \to \{0, 1\}$ defined by $u(n) = 1$ if n is a square and $u(n) = 0$ otherwise is displayed as $11001000010000001\cdots$. If the infinite word u is periodic and has period p, then we write $u = v^\omega$, where $v = u(0)\cdots u(p-1)$. For instance, $u = (012)^\omega$ denotes the infinite word over $\{0, 1, 2\}$ where $u(n)$ is the remainder of the division of n by 3. We write $u = vw^\omega$ to denote an eventually periodic infinite word which has a periodic suffix w^ω.

Let $k \ge 2$ be an integer. An infinite sequence u over the alphabet A is *k-automatic* if for each letter $a \in A$, the set $u^{-1}(a) = \{n \in \mathbb{N} \mid u(n) = a\}$ is recognizable in base k. Equivalently, consider the mapping

$$\boldsymbol{k}^* \xrightarrow{\nu_k} \mathbb{N} \xrightarrow{u} A\,.$$

Then u is k-automatic if the languages $\nu_k^{-1}(u^{-1}(a))$ or, what is equivalent by Corollary 1.2, if the languages $\sigma_k(u^{-1}(a))$, are recognizable for all letters $a \in A$.

It is useful to consider a left action of $\boldsymbol{k}$ on u defined for r in $\boldsymbol{k}$ by

$$(r \circ u)(n) = u(nk + r)\,.$$

This operation extracts from u the sequence composed of the letters appearing at the positions $\equiv r \bmod k$. The action extends to words on $\boldsymbol{k}$ by

$$rs \circ u = r \circ (s \circ u)\,.$$

It follows that, for a word $r \in \boldsymbol{k}^*$,

$$(r \circ u)(n) = u(nk^{|r|} + \nu_k(r))\,. \tag{3.1}$$

The set of sequences $r \circ u$ for $r \in \boldsymbol{k}^*$ is sometimes called the *k-kernel* of u. By Equation (3.1), it is the set of infinite sequences

$$n \mapsto u(nk^e + j)\,, \quad e \ge 0\,,\ 0 \le j < k^e\,.$$

Proposition 3.1 *A sequence u is k-automatic if and only if the set of sequences $r \circ u$, for $r \in \boldsymbol{k}^*$, is finite.*

Proof. We may assume that A is a semiring, since there exist semirings of any finite cardinality. Then the proposition is a consequence of Proposition 2.2. Indeed, a finitely generated module over a finite semiring is always finite. □

Corollary 3.2 *A subset H of $\mathbb{N}$ is k-recognizable if and only if the set of subsets $r \circ H = \{n \in \mathbb{N} \mid nk^{|r|} + \nu_k(r) \in H\}$, for $r \in \mathbf{k}^*$, is finite.* □

Example 3.1 The *Thue–Morse* sequence is the infinite binary sequence t over the letters a and b defined by $t(0) = a$, and $t(2m) = t(m)$, $t(2m+1) = \overline{t(m)}$, where $\bar{a} = b$ and $\bar{b} = a$. Thus

$$t = abbabaabbaababba \cdots$$

To see that it is 2-automatic, we consider the sequence $\bar{t}$ defined by $\bar{t}(n) = \overline{t(n)}$. Then $0 \circ t = t$, $1 \circ t = \bar{t}$, $0 \circ \bar{t} = \bar{t}$, $1 \circ \bar{t} = t$. Thus the 2-kernel of t is composed of t and $\bar{t}$. It is easily checked on the definition that $t(n) = a$ if and only if $s_2(n)$ is even (see Example 1.1 for the notation $s_2(n)$).

Example 3.2 We consider the so-called *paper-folding* sequence. This is the infinite binary sequence p over the letters a and b defined for $m \geq 0$ by

$$\begin{aligned} p(4m) &= a\,, \\ p(4m+2) &= b\,, \\ p(2m+1) &= p(m)\,. \end{aligned} \tag{3.2}$$

Thus

$$p = aabaabbaaabbabb \cdots$$

To see that it is 2-automatic, we observe that by definition, symbols in even positions are alternatively a and b, so that $0 \circ p = (ab)^\omega$. Moreover

$$1 \circ p = p\,, \quad 0 \circ (ab)^\omega = a^\omega\,, \quad 1 \circ (ab)^\omega = b^\omega\,,$$
$$0 \circ a^\omega = 1 \circ a^\omega = a^\omega\,, \quad 0 \circ b^\omega = 1 \circ b^\omega = b^\omega\,.$$

This shows that p is 2-automatic. Moreover, $p(n) = a$ if and only if $n = (4m+1)2^\ell - 1$ for some $m, \ell \geq 0$. Indeed, assume first $n = (4m+1)2^\ell - 1$. If $\ell = 0$, then $n = 4m$ and $p(n) = a$. If $\ell > 0$, then $n = 2^\ell 4m + 1 + 2 + \cdots + 2^{\ell-1}$, and by iterating (3.2) ℓ times, one gets $p(n) = p(4m) = a$. Conversely, assume $p(n) = a$. If n is even, then by (3.2) $n = 4m$ for some m. If n is odd, define ℓ by $n = 1 + 2 + \cdots + 2^{\ell-1} + 2^\ell m'$ with $m' \geq 0$ even. Then by iterating (3.2) ℓ times, $p(m') = p(n) = a$. Thus m' is a multiple of 4 and therefore $n = (4m+1)2^\ell - 1$.

The first numbers in the set $p^{-1}(a)$ are $0, 1, 3, 4, 7, 8, 9, 12, \ldots$

The next proposition describes how k-regular functions and k-automatic sequences are related.

Proposition 3.3 *Any k-automatic sequence with values in a semiring is k-regular. Conversely, a k-regular function with values in a commutative ring that takes only finitely many values is k-automatic. Similarly, a k-regular function over a finite semiring is k-automatic.*

Proof. Let $f : \mathbb{N} \to A$ be a k-automatic sequence, and assume A is a subset of a semiring K. For each $a \in A$, the language $Z_a = \nu_k^{-1}(f^{-1}(a)) \subset \boldsymbol{k}^*$ is rational, and consequently $S_f = \sum_{a \in A} a\underline{Z_a}$ is a rational series over the semiring K. Thus f is a k-regular function.

Conversely, let $f : \mathbb{N} \to K$ be a k-regular function, where K is a commutative ring, that takes only finitely many values, and set $A = f(\mathbb{N})$. Then for each $a \in A$, the set $H_a = \{n \in \mathbb{N} \mid f(n) = a\}$ is k-recognizable by Theorem 3.2.10. Thus f, viewed as a sequence with values in A, is k-automatic. The proof of the last assertion is similar, using Proposition 3.2.4. □

5.4 Automatic sequences and algebraic series

In this section, q denotes a positive power of some prime, and $\mathbb{F}_q$ is the field with q elements. To each infinite sequence u over the the field $\mathbb{F}_q$, we associate the formal series

$$u(x) = \sum_{n \geq 0} u_n x^n ,$$

where u_n is the element at position n in u. Series over $\mathbb{F}_q$ have some properties which are useful in computations. In particular, $u(x^q) = u(x)^q$, as it is easily checked. As usual, we denote by $\mathbb{F}_q(x)$ the field of rational fractions with coefficients in $\mathbb{F}_q$, by $\mathbb{F}_q[[x]]$ the ring of formal series with coefficients in $\mathbb{F}_q$, and by $\mathbb{F}_q((x))$ its quotient field, the field of Laurent series.

A series f is *algebraic* over the field $\mathbb{F}_q(x)$ of rational fractions with coefficients in $\mathbb{F}_q$ if there exist polynomials $a_0, \ldots, a_n \in \mathbb{F}_q[x]$ with $n \geq 1$ and $a_n \neq 0$ such that

$$a_0 + a_1 f + \cdots + a_n f^n = 0 .$$

Later we will use the observation that if f is algebraic, then the powers f^i are linearly dependent elements of $\mathbb{F}_q((x))$ viewed as a vector space over the field $\mathbb{F}_q(x)$.

The aim of this section is to prove the following result.

Theorem 4.1 (Christol 1979, Christol *et al.* 1980) *An infinite sequence u over the alphabet $\mathbb{F}_q$ is q-automatic if and only if its associated series $u(x)$ is algebraic over $\mathbb{F}_q(x)$.*

Example 4.1 Consider the Thue–Morse sequence t. This infinite sequence satisfies the relations $t_0 = 0$, $t_{2n} = t_n$ and $t_{2n+1} = 1 + t_n$, over $\mathbb{F}_2$. It follows that

$$\begin{aligned} t(x) &= \sum_{n=0}^{\infty} t_n x^n = \sum_{n=0}^{\infty} t_{2n} x^{2n} + \sum_{n=0}^{\infty} t_{2n+1} x^{2n+1} \\ &= \sum_{n=0}^{\infty} t_n x^{2n} + \sum_{n=0}^{\infty} (1 + t_n) x^{2n+1} = t(x^2) + \sum x^{2n+1} + x t(x^2) \\ &= (1+x) t(x^2) + \frac{x}{1+x^2} = (1+x) t(x)^2 + \frac{x}{(1+x)^2} . \end{aligned}$$

Thus

$$(1+x)^3 t^2 + (1+x)^2 t + x = 0 ,$$

showing that $t(x)$ is algebraic over $\mathbb{F}_2(x)$.

We define a left action of the set $\boldsymbol{q} = \{0, \ldots, q-1\}$ on series by setting, for $u = u(x)$ and $0 \le r < q$,

$$(r \circ u)(x) = \sum_{n=0}^{\infty} u_{nq+r} x^n .$$

With this notation, one gets

$$u(x) = \sum_{r=0}^{q-1} x^r ((r \circ u)(x))^q = \sum_{r=0}^{q-1} x^r (r \circ u)(x^q) , \tag{4.1}$$

since indeed

$$u(x) = \sum_{r=0}^{q-1} x^r \sum_{n=0}^{\infty} u_{nq+r} x^{nq} .$$

We start with the following lemma.

Lemma 4.2 *Let $u(x)$ and $v(x)$ be two series over $\mathbb{F}_q$. For each $r \in \boldsymbol{q}$,*

$$r \circ (u v^q) = (r \circ u) v .$$

Proof. Set $w = u v^q$. Since $v(x)^q = v(x^q)$,

$$w(x) = \sum_{n=0}^{\infty} w_n x^n = \sum_{m,\ell \ge 0} u_m v_\ell x^{\ell q + m} ,$$

with

$$w_n = \sum_{n = \ell q + m} u_m v_\ell .$$

By definition $(r \circ w)(x) = \sum_{n=0}^{\infty} w_{nq+r}x^n$ and

$$w_{nq+r} = \sum_{\substack{m,\ell \geq 0 \\ nq+r=\ell q+m}} u_m v_\ell \,.$$

In this sum, the equality $nq + r = \ell q + m$ shows that $m \equiv r \mod q$, and therefore $m = m'q + r$ for some $m' \geq 0$. Thus

$$w_{nq+r} = \sum_{\substack{m',\ell \geq 0 \\ m'+\ell=n}} u_{m'q+r} v_\ell \,.$$

On the other hand,

$$((r \circ u)v)(x) = \sum_{n=0}^{\infty} \sum_{m+\ell=n} u_{mq+r} v_\ell x^n \,.$$

The coefficient of x^n is $\sum_{m+\ell=n} u_{mq+r} v_\ell$. This implies the equality. □

Corollary 4.3 *Let u and v be two series over $\mathbb{F}_q$. For each $0 \leq r < q$ and $i \geq 1$*

$$r \circ (uv^{q^i}) = (r \circ u)v^{q^{i-1}} \,.$$

We use the corollary in the proof of the following statement.

Lemma 4.4 *A series f is algebraic over $\mathbb{F}_q(x)$ if and only if there exist polynomials $c_0, \ldots, c_d$ in $\mathbb{F}_q[x]$, with $c_0 \neq 0$, such that*

$$c_0 f = \sum_{i=1}^{d} c_i f^{q^i} \,.$$

Proof. If such a relation exists, then f is algebraic. Conversely, if f is algebraic, then the vector space spanned by the powers of f has finite dimension. Consequently, there exist an integer d and polynomials $c_0, \ldots, c_d$, not all 0, such that

$$\sum_{i=0}^{d} c_i f^{q^i} = 0 \,. \tag{4.2}$$

Let j be the smallest integer for which there is such a relation with $c_j \neq 0$. It is enough to show that $j = 0$. For this, observe that since $c_j \neq 0$, in view of (4.1), there exists r such that $r \circ c_j \neq 0$. Assume now $j \geq 1$. Then for this r, the relation (4.2) implies, with the use of Corollary 4.3, the relation

$$r \circ \Big(\sum_{i=j}^{d} c_i f^{q^i}\Big) = \sum_{i=j}^{d} (r \circ c_i) f^{q^{i-1}} = 0 \,,$$

and this contradicts the minimality of j. □

Proof of Theorem 4.1. Let u be a q-automatic sequence. By Proposition 3.1, the set W of sequences of the form $s \circ u$ where s is a word over the alphabet $\boldsymbol{q}$, is finite. Let

d be their number. Let U_0 be the set of series $v(x)$ associated to the sequences v in W, and for $h \geq 1$, let U_h be the set of series $v(x^{q^h})$ with $v(x) \in U_0$. Finally, denote by V_h the vector space over $\mathbb{F}_q(x)$ spanned by U_h for $h \geq 0$. Each of these vector spaces has dimension at most d.

Recall that by (4.1), one has

$$v(x) = \sum_{r=0}^{q-1} x^r (r \circ v)(x^q).$$

This shows that U_0 is contained in the vector space V_1, and, more generally, using the formula

$$v(x^{q^h}) = \sum_{r=0}^{q-1} (x^{q^h})^r (r \circ v)(x^{q^{h+1}})$$

one gets the inclusions $V_0 \subset V_1 \subset \cdots \subset V_d$.

The $d+1$ series $u(x), u(x^q), \ldots, u(x^{q^d})$ are in the spaces $V_0, V_1 \ldots, V_d$ respectively, hence are all in V_d. They are therefore linearly dependent over $F(x)$, and, using the identity $u(x^{q^h}) = u(x)^{q^h}$, there exist polynomials a_h, not all 0, such that

$$\sum_{h=0}^{d} a_h u(x)^{q^h} = 0.$$

This proves that u is algebraic.

Conversely, if u is algebraic, then in view of Lemma 4.4, there is a relation

$$c_0 u = \sum_{i=1}^{d} c_i u^{q^i}$$

with $c_0 \neq 0$. Set $v = u/c_0$. Then

$$c_0(c_0 v) = \sum_{i=1}^{d} c_i c_0^{q^i} v^{q^i},$$

and consequently

$$v = \sum_{i=1}^{d} b_i v^{q^i}$$

where each $b_i = c_i c_0^{q^i - 2}$ is a polynomial with coefficients in $\mathbb{F}_q$. Let $N = \deg c_0 + \max\{\deg b_i \mid i = 1, \ldots, d,\ b_i \neq 0\}$, and let F be the finite set of series over $\mathbb{F}_q$ of the form

$$f = \sum_{i=0}^{d} a_i v^{q^i} \qquad a_i \in \mathbb{F}_q[x], \deg(a_i) \leq N.$$

The series $u(x) = c_0 v(x)$ is in F. In order to prove that the infinite sequence u corresponding to $u(x)$ is q-automatic, it suffices to show that the set F is closed under the operation $\circ$. Let $f \in F$. Then using Corollary 4.3

$$r \circ f = r \circ \left(a_0 v + \sum_{i=1}^{d} a_i v^{q^i}\right) = r \circ \left(a_0 \sum_{i=1}^{d} b_i v^{q^i} + \sum_{i=1}^{d} a_i v^{q^i}\right)$$

$$= r \circ \left(\sum_{i=1}^{d} (a_0 b_i + a_i) v^{q^i}\right) = \sum_{i=1}^{d} (r \circ (a_0 b_i + a_i)) v^{q^{i-1}}.$$

Next, for any polynomial $h(x) = \sum_{n=0}^{M} h_n x^n$ of degree at most M, the polynomial $r \circ h(x) = \sum_{0 \le nq+r \le M} h_{nq+r} x^n$ has degree at most $(M-r)/q \le M/q$. In our case, since $\deg(a_0 b_i + a_i) \le 2N$, one has $\deg(r \circ (a_0 b_i + a_i)) \le 2N/q \le N$. This proves that $r \circ f$ is in F. □

5.5 Algebraic series and diagonals of rational series

A series

$$f(x_1, \ldots, x_m) = \sum_{n_1, \ldots, n_m} (f, x_1^{n_1} \cdots x_m^{n_m}) x_1^{n_1} \cdots x_m^{n_m}$$

in the commuting variables $x_1, \ldots, x_m$ over a field K is a Laurent series if, up to a finite number of them, all nonzero coefficients $(f, x_1^{n_1} \cdots x_m^{n_m})$ satisfy $n_i \ge 0$ for $i = 1, \ldots, m$. We denote by $K((x_1, \ldots, x_m))$ the K-algebra of Laurent series in the commuting variables $x_1, \ldots, x_m$ over K. The *diagonal* of the Laurent series f is the series $\mathcal{D}f(t)$ in one variable t defined by

$$\mathcal{D}f(t) = \sum_n (f, x_1^n \cdots x_m^n) t^n .$$

For example, the diagonal of the rational series $(1-x-y)^{-1} = \sum \binom{n+m}{m} x^n y^m$ is the algebraic series $\sum \binom{2n}{n} t^n$. The purpose of this section is to prove the following theorems.

Theorem 5.1 (Furstenberg 1967) *Any algebraic series $f(t)$ over a finite field is the diagonal of a rational series $r(x, y)$ in two variables: $f(t) = \mathcal{D}r(t)$.*

We call *rational Laurent series* a Laurent series of the form P/Q, where P, Q are polynomials, with Q of the form zQ_1, z a Laurent monomial, Q_1 a polynomial with $Q_1(0) = 1$.

Theorem 5.2 (Furstenberg 1967) *The diagonal of a rational series in m variables over a field of positive characteristic is an algebraic series.*

We first consider Theorem 5.1. Observe that $f(x) = \mathcal{D}f(xy)$, so the result holds if f has finite support. This means that it suffices to prove the theorem up to a finite number of monomials, and therefore we may assume that $f(x)$ has only positive exponents in its expansion.

The theorem will be a consequence of two lemmas. As before, denote by $\mathbb{F}_q$ the finite field with q elements. First, we prove the following lemma.

Lemma 5.3 *Let $f(x)$ be an algebraic Laurent series over the field $\mathbb{F}_q$. Then $f(x) = r(x) + x^h g(x)$, where $r(x)$ is a Laurent polynomial, and $g(x)$ is algebraic, satisfying an equation of the form*

$$b_0 g = b_1 g^q + \cdots + b_d g^{q^d} + s\,, \tag{5.1}$$

where $b_0(x), \ldots, b_d(x)$ and $s(x)$ are polynomials, and $b_0(x)$ is not divisible by x.

Proof. We may suppose that the function $f(x)$ has no negative exponents in its expansion. By Lemma 4.4 there exist polynomials $c_0(x), \ldots, c_d(x)$ such that

$$c_0 f = c_1 f^q + \cdots + c_d f^{q^d}\,, \quad c_0 \neq 0\,. \tag{5.2}$$

If $c_0(x)$ is not divisible by x, we are through. Otherwise, we write $f(x) = f(0) + xg(x)$. We then find

$$c_0 x g = c_1 x^q g^q + \cdots + c_d x^{q^d} g^{q^d} + s \tag{5.3}$$

for an appropriate polynomial $s(x)$. Let m be the exponent of the highest power of x that divides $c_0(x)$, and set $n = \min(m+1, q)$. Each term $c_j(x)x^{q^j}$ is divisible by x^n, and so $s(x)$ is also divisible by x^n. We may therefore divide (5.3) by x^n which yields an equation of the same form as (5.2) for which the power dividing $c_0(x)$ is at most $m-1$. Iterating this procedure we obtain the lemma. □

The next lemma holds in an arbitrary field. We denote by P'_y the partial derivative of $P(x, y)$ with respect to the variable y.

Lemma 5.4 *Let $P(x, y) \in K[x, y]$ be a polynomial with coefficients in a field K and assume that $P'_y(0, 0) \neq 0$. Let $f(x) \in K[[x]]$ be a series satisfying $f(0) = 0$ and $P(x, f(x)) = 0$. Then*

$$f = \mathcal{D}\Big(\frac{y^2}{P(xy, y)} P'_y(xy, y)\Big)\,. \tag{5.4}$$

Note that $P(0, 0) = 0$ and therefore the hypothesis $P'_y(0, 0) \neq 0$ implies that $P(xy, y) = yP_1(x, y)$ with $P_1(x, y) \in K[x, y]$ and $P_1(0, 0) \neq 0$. Thus $P(xy, y)$ is invertible as Laurent series and $y/P(xy, y) \in K[[x, y]]$.

Proof. Considering $P(x, y)$ as a polynomial in y over the ring $K[[x]]$, we see that $P(x, y)$ is divisible by $y - f(x)$. Set

$$P(x, y) = (y - f(x))Q(x, y)\,, \tag{5.5}$$

with $Q(x, y) \in K[[x]][y]$. This equality holds in $K[[x, y]]$. Next, we have

$$P'_y(x, y) = Q(x, y) + (y - f(x))Q'_y(x, y)\,. \tag{5.6}$$

Since $f(0) = 0$, we get $P'_y(0, 0) = Q(0, 0)$, and since $P'_y(0, 0) \neq 0$, it follows that $Q(0, 0) \neq 0$. Hence $Q(xy, y)$ is invertible in $K[[x, y]]$ and a fortiori in $K((x, y))$. Moreover, $y^{-1} f(xy) \in K[[x, y]]$ has constant term 0, so $1 - y^{-1} f(xy)$ is invertible in $K[[x, y]]$ and $y - f(xy)$ is invertible in $K((x, y))$. In Equations (5.5) and (5.6), replace x by xy. We obtain

$$P(xy, y) = (y - f(xy))Q(xy, y), \tag{5.7}$$

and

$$P'_y(xy, y) = Q(xy, y) + (y - f(xy))Q'_y(xy, y). \tag{5.8}$$

Since the three series appearing in Equation (5.7) are invertible in $K((x, y))$, we may divide Equation (5.8) by Equation (5.7) and then multiply by y^2. We obtain

$$\frac{y^2 P'_y(xy, y)}{P(xy, y)} = \frac{y^2}{y - f(xy)} + \frac{y^2 Q'_y(xy, y)}{Q(xy, y)}. \tag{5.9}$$

The diagonal of the first term in the right-hand side is

$$\mathcal{D}\Big(\frac{y^2}{y - f(xy)}\Big) = \mathcal{D}\Big(\frac{y}{1 - y^{-1} f(xy)}\Big) = \mathcal{D}\Big(\sum_{n\geq 0} y^{-n+1} f(xy)^n\Big).$$

Each of the series $y^{-n+1} f(xy)^n$ has only terms in the powers $x^{nk} y^{nk-n+1}$ with $k \geq 1$ (since $f(0) = 0$), and these contribute to the diagonal only when $n = 1$. Thus

$$\mathcal{D}\Big(\frac{y^2}{y - f(xy)}\Big) = \mathcal{D}(f(xy)) = f.$$

Since $Q(0, 0) \neq 0$, we have

$$\frac{1}{Q} = \frac{1}{Q(0, 0)} \frac{1}{1 - R} = \frac{1}{Q(0, 0)} \sum_{n\geq 0} R^n(x, y)$$

for some series R. In the series $\frac{1}{Q} Q'_y$, each monomial $x^m y^n$ gives, after replacing x by xy and multiplying by y^2, a monomial of the form $x^m y^{m+n+2}$, for $m, n \geq 2$. Such a monomial does not contribute to the diagonal, and therefore the diagonal of the second term in the right-hand side of Equation (5.9) is null. This shows (5.4). □

Proof of Theorem 5.1. We may suppose that the series f has only positive exponents in its expansion. By Lemma 5.3, $f(x) = r(x) + x^h g(x)$, where $g(x)$ satisfies (5.1). We may assume that $g(0) = 0$. By Lemma 5.4, the series $g(x)$ is the diagonal of some rational series, and the same is true for f, since r is a Laurent polynomial. □

Example 5.1 We have seen in Example 4.1 that the Thue–Morse sequence satisfies the equation

$$(1 + x)^3 t^2 + (1 + x)^2 t + x = 0,$$

showing that $t(x)$ is algebraic over $\mathbb{F}_2(x)$. Write this equation as $P(x, t(x)) = 0$ with

$$P(x, y) = x + (1 + x)^2 y + (1 + x)^3 y^2.$$

Next, $P'_y(x, y) = (1 + x)^2$, so the hypotheses of Lemma 5.4 are satisfied since $t_0 = 0$. By (5.4), the algebraic function $t(x)$ is the diagonal of the rational function

$$y^2 \frac{1}{P(xy, y)} P'_y(xy, y) = y^2 \frac{(1 + xy)^2}{xy + (1 + xy)^2 y + (1 + xy)^3 y^2}$$
$$= \frac{y}{1 + (1 + xy)y + \dfrac{x}{(1 + xy)^2}}.$$

In the proof of Theorem 5.2, we use the following result.

Theorem 5.5 (see (Lang 1984), Proposition VIII.5.3) *Let F be a field and let $\alpha_1, \ldots, \alpha_n$ be elements in some extension field of F. Suppose that for some polynomials $P_i(x_1, \ldots, x_n)$ in $F[x_1, \ldots, x_n]$ for $i = 1, \ldots, n$, one has*

(i) $P_i(\alpha_1, \ldots, \alpha_n) = 0$ *for* $i = 1, \ldots, n$;
(ii) *the determinant* $\left(\frac{\partial P_i}{\partial x_j}(\alpha_1, \ldots, \alpha_n)\right)_{1 \le i, j \le n}$ *is nonzero.*

Then $\alpha_1, \ldots, \alpha_n$ are algebraic over F.

Proof of Theorem 5.2. Let $f(x_1, \ldots, x_m)$ be a rational series in the commuting variables $x_1, \ldots, x_m$. A monomial $w = x_1^{n_1} \cdots x_m^{n_m}$ is called diagonal if $n_1 = \cdots = n_m$. Notice that if w is diagonal, then $\mathcal{D}(wf)(t) = t^{|w|/m} \mathcal{D}f(t)$ where $|w|$ denotes the degree of w. Let $P(x_1, \ldots, x_m)$ and $Q(x_1, \ldots, x_m)$ be polynomials such that $f = P/Q$, and $Q = zQ_1$, where z is a Laurent polynomial and Q_1 is a polynomial in $K[x_1, \ldots, x_m]$ with $Q_1(0) = 1$. Set $\varphi = \mathcal{D}f$. Because of the special form of the denominator Q, for any Laurent monomial w, the quotient w/Q is a rational Laurent series. Therefore, it will suffice, for the proof that φ is algebraic, to consider the case where P reduces to a monomial. Now, we have, for monomials z, w, $w/Q(x_1, \ldots, x_m)) = wz/zQ(x_1, \ldots, x_m)$ so if we choose z such that wz is diagonal, we reduce the general case to the case

$$f(x_1, \ldots, x_m) = 1/Q(x_1, \ldots, x_m). \tag{5.10}$$

Let K be the underlying field and p be the characteristic of K. Define two functions σ and ρ for any monomial $w = x_1^{n_1} \cdots x_m^{n_m}$ by considering the Euclidean divisions $n_i = pq_i + r_i$ with $0 \le r_i < p$ for $1 \le i \le m$ and by setting

$$\sigma(w) = x_1^{q_1} \cdots x_m^{q_m}, \quad \rho(w) = x_1^{r_1} \cdots x_m^{r_m},$$

so that $w = (\sigma(w))^p \rho(w)$. Observe that for monomials x, y with $x = \rho(x)$, xy^p is diagonal if and only if x and y are diagonal; hence $\mathcal{D}(xf^p) \neq 0$ implies x diagonal, and in this case $\mathcal{D}(xf^p) = \mathcal{D}(x)\mathcal{D}(f^p)$.

Let now ℓ be the degree of Q, and set $R = Q^{p-1}$. Since $Qf = 1$ by (5.10), one has $Rf^p = f$. Also $\deg(R) = (p-1)\ell$. For any monomial v, one gets

$$vf = vRf^p = \sum_{|u| \le \ell(p-1)} v(R, u)uf^p = \sum_{|u| \le \ell(p-1)} (R, u)\rho(vu)(\sigma(vu)f)^p. \tag{5.11}$$

Assume $|v| \le \ell$. Then $|vu| \le p\ell$ for all u with $|u| \le \ell(p-1)$, and since $vu = \sigma(vu)^p \rho(vu)$, one gets $p\ell \ge |vu| \ge p|\sigma(vu)|$. Therefore $|\sigma(vu)| \le \ell$. The development of $(\sigma(vu)f)^p$ has the form $(\sigma(vu)f)^p = \sum_w (f, w)^p \sigma(vu)^p w^p$. Thus, by a previous observation, $\mathcal{D}(\rho(vu)(\sigma(vu)f)^p)$ is nonzero only if $\rho(vu)$ is diagonal, and in this case it is equal to $\mathcal{D}(\rho(vu))\mathcal{D}((\sigma(vu)f)^p)$. Denote by H the set of monomials h of length at most $p\ell$ such that $\rho(h)$ is diagonal. Then the diagonal of vf, as computed in (5.11), can be written as

$$\mathcal{D}(vf) = \sum_{vu \in H} (R, u)\mathcal{D}(\rho(vu))\mathcal{D}((\sigma(vu)f)^p). \tag{5.12}$$

We group the terms on the right-hand side according to the value of $\sigma(vu)$. This gives

$$\mathcal{D}(vf) = \sum_w \sum_{\substack{u: vu \in H \\ \sigma(vu) = w}} (R, u)\mathcal{D}(\rho(vu))\mathcal{D}((wf)^p).$$

Denoting by $c_{w,v}(t)$ the polynomial given by

$$c_{w,v} = \sum_{\substack{u: vu \in H \\ \sigma(vu) = w}} (R, u)\mathcal{D}(\rho(vu)),$$

and setting $\varphi_v(t) = \mathcal{D}(vf)(t)$, we get

$$\varphi_v(t) = \sum_{|w| \le \ell} c_{w,v}(t)(\varphi_w(t))^p, \qquad |v| \le \ell. \tag{5.13}$$

Let Y be the set of commuting variables y_v, for v a monomial in $x_1, \ldots, x_m$ of degree $\le \ell$. Define polynomials P_v in these variables over $K[t]$, by

$$P_v = y_v - \sum_{|w| \le \ell} c_{w,v}(t) y_w^p, \qquad |v| \le \ell.$$

Then the series $\varphi_v(t)$ are solutions of the algebraic system

$$P_v = 0, \qquad |v| \le \ell.$$

Now the Jacobian matrix $\left(\dfrac{\partial P_v}{\partial y_w}\right)_{|v|,|w| \le \ell}$ is equal to the identity matrix, since $p = 0$ in K. Thus, by Theorem 5.5, all $\varphi_v(t)$ are algebraic over $K(t)$. In particular, $\varphi_1 = \mathcal{D}(f)$ is algebraic. □

Example 5.2 Consider the series $f = 1/Q$ with $Q = 1 - x - y$, for $p = 3$. Here $\ell = 1, R = Q^2 = 1 + x + y + x^2 - xy + y^2$ and $H = \{1, xy, x^3, y^3\}$. There are two equations (5.12), for $v = 1$ and $v = x$ (the case $v = y$ is symmetric).

For $v = 1$, since the set of words u in H with $(R, u) \ne 0$ is $\{1, xy\}$, one gets

$$\begin{aligned} \mathcal{D}f &= \mathcal{D}(\rho(1))\mathcal{D}((\sigma(1)f)^3 - \mathcal{D}(\rho(xy))\mathcal{D}((\sigma(xy)f)^3) \\ &= \mathcal{D}(1)\mathcal{D}(f^3) - \mathcal{D}(xy)\mathcal{D}(f^3) = (1-t)(\mathcal{D}f)^3. \end{aligned}$$

For $v = x$, the set of words u such that $ux \in H$ and $(R, u) \neq 0$ is $\{y, x^2\}$. One gets

$$\begin{aligned}\mathcal{D}(xf) &= \mathcal{D}(\rho(xy))\mathcal{D}((\sigma(xy)f)^3 + \mathcal{D}(\rho(x^3))\mathcal{D}((\sigma(x^3)f)^3) \\ &= \mathcal{D}(xy)\mathcal{D}(f^3) + \mathcal{D}(1)\mathcal{D}((xf)^3) = t(\mathcal{D}f)^3 + \mathcal{D}(xf)^3 .\end{aligned}$$

The first of these equations gives, for $\varphi = \mathcal{D}f$, the equation $\varphi = (1-t)\varphi^3$ which has the solutions $\varphi = 0$ and $\varphi = \pm(1-t)^{-1/2}$. One of these solutions is the series $\sum \binom{2n}{n} t^n$, with binomial coefficients taken modulo 3.

Exercises for Chapter 5

1.1 Show that if f is k-regular, then the function F defined by $F(n) = \sum_{0 \le i \le n} f(i)$ is k-regular. (*Hint*: Use the product with the sequence composed of 1 only.)

1.2 The *Kimberling* function $c : \mathbb{N} \to \mathbb{N}$ is defined by $c(n) = k(n+1)$, where $k(n) = \frac{1}{2}\left(\frac{n}{2^{v_2(n)}} + 1\right)$ for $n \ge 1$. Here $v_2(n)$ is the 2-adic valuation of n, that is the exponent of the highest power of 2 dividing n. The first values of the Kimberling sequence are

n	0	1	2	3	4	5	6	7	8	9	10
$c(n)$	1	1	2	1	3	2	4	1	5	3	6

Show that the Kimberling function is 2-regular. (*Hint*: Show that $c(2n) = n + 1$ and $c(2n + 1) = c(n)$ for $n \ge 0$.)

Check the following scheme for the construction of the sequence: write down integers in increasing order, leaving one place free at each step, and iterate this. Here is the beginning of the process:

n	0	1	2	3	4	5	6	7	8	9	10	11	12	13	14
$c(n)$	1	.	2	.	3	.	4	.	5	.	6	.	7	.	8
		1		.		2		.		3		.		4	
				1				.				2			
								1							

Show that the Kimberling sequence has the property that deleting the first occurrence of each positive integer in it leaves the sequence unchanged.

1.3 It is known that an integer $n \ge 0$ is the sum of three integer squares if and only if it is not of the form $n = 4^a(8r + 7)$ for integers $a, r \ge 0$. Denote by $f(n)$ the number of integers $\le n$ which are the sum of three squares. Show that the function f is 2-regular.

1.4 Let $\ell = k^p$ with $k \ge 2$, $p > 1$. Show that a subset H of $\mathbb{N}$ is k-recognizable if and only if it is ℓ-recognizable. (*Hint*: Consider the morphism α from $\{0, 1, \ldots, \ell - 1\}^*$ into $\{0, 1, \ldots, k - 1\}^*$ that maps a digit d of $\{0, 1, \ldots, \ell - 1\}$ onto the unique word

u of length p over $\{0, 1, \ldots, k-1\}$ such that $\nu_\ell(d) = \nu_k(u)$. Show that $\nu_\ell^{-1}(H) = \alpha^{-1}\nu_k^{-1}(H)$ and that $H = \nu_k(\alpha(\sigma_\ell(H)))$.)

1.5 If $a_0, a_1, \ldots, a_n \in \boldsymbol{k}$, denote by $\widetilde{\nu}_k(a_0 a_1 \cdots a_n)$ the number $n = a_0 + a_1 k + \cdots a_n k^n$. The word $a_0 a_1 \cdots a_n$ is a *reverse representation* of n. Show that H is k-recognizable if and only if $\widetilde{\nu}_k^{-1}(H)$ is a recognizable subset of $\boldsymbol{k}^*$.

1.6 Let a and b be positive integers. Show that the arithmetic progression $a\mathbb{N} + b$ is k-recognizable for every $k \geq 2$.

1.7 Show that if H, H' are k-recognizable sets, then so is $H + H' = \{h + h' \mid h \in H, h' \in H'\}$. (*Hint*: Consider automata $\mathcal{A}$ and $\mathcal{A}'$ with sets of states Q and Q' and recognizing $L = \nu_k^{-1}(H)$ and $L' = \nu_k^{-1}(H')$ respectively. Build an automaton $\mathcal{B}$ which has as set of states the disjoint union of two copies of the product $Q \times Q'$, according to the value of a carry, and edges $(p, q, c) \xrightarrow{\ell} (p', q', c')$ if and only if $p \xrightarrow{i} p'$ in $\mathcal{A}$, $q \xrightarrow{j} q'$ in $\mathcal{A}'$, and $i + j + c = \ell + c'$. Here c, c' are carries, and $i, j, \ell \in \boldsymbol{k}$.)

2.1 Show that the mapping $f \mapsto S_f$ is a right K-linear isomorphism which commutes with the left actions of $\boldsymbol{k}^*$, from $K^{\mathbb{N}}$ onto the submodule $\{S \in K\langle\langle \boldsymbol{k} \rangle\rangle \mid \forall w \in \boldsymbol{k}^*, (S, 0w) = (S, w)\}$. Show the same statement for the mapping $S \mapsto S \odot \underline{R}$ from the latter submodule onto the submodule E of Lemma 2.1.

3.1 A morphism $\alpha : A^* \to B^*$ is *k-uniform* if all words $\alpha(a)$, for $a \in A$, have length k. An infinite sequence w over A is *purely k-morphic* if there exists a k-uniform endomorphism $\alpha : A^* \to A^*$ such that $w = \alpha(w)$. A sequence is *k-morphic* if it is the image of a pure k-morphic sequence by a 1-uniform morphism.

Show that a sequence w is k-automatic if and only if w is k-morphic.

3.2 Show that if u is a k-automatic sequence, then the sequence u' defined by $u'(n) = u(k^n)$ is eventually periodic. (For the Thue–Morse sequence $t = abbabaab\cdots$, one gets $t' = (ba)^\omega$.)

Conversely, given an eventually periodic sequence u', define u by $u(k^n) = u'(n)$, and $u'(i) = 0$ if i is not a power of k. Show that u is k-automatic.

3.3 Show that the sequence starting with 0 and consisting of the *first* digit in the canonical representation of $n > 0$ in base k is k-automatic. (For $k = 2$, this is 01^ω, for $k = 3$, it is $012111222111111111\cdots$)

4.1 Give a polynomial equation for the series associated to the paper-folding sequence.

4.2 The set of powers of 2 is 2-recognizable. Give the polynomial equation for the series associated to the characteristic sequence of this set.

4.3 What are the polynomial equations for the arithmetic progressions?

5.1 Prove directly that the Thue–Morse series $t(x)$ is the diagonal of the series

$$y^2 \frac{1}{P} \frac{\partial P}{\partial y}(xy, y) = \frac{y}{1 + (1 + xy)y + \frac{x}{(1+xy)^2}}$$

without using Furstenberg's theorem. (*Hint*: Show that the diagonal has the form $\sum_{k\geq 0} \binom{2k+1}{k} x^{k+1}(1+x)^{-k-2}$ and use the identities $\binom{4k+2}{2k+1} \equiv \binom{2k+1}{k}$ modulo 2 for $k \geq 0$ and $\binom{4k+1}{2k} \equiv \binom{2k}{k}$ modulo 2 for $k > 0$ to prove that the diagonal satisfies the same equation as $t(x)$.)

Notes to Chapter 5

Recognizable sets of integers have been considered since the very beginning of the theory of automata. A fundamental and difficult result, not included here, is the so-called base dependence and is due to Cobham (1969). It states that if k and ℓ are multiplicatively independent, that is if there are no positive integers such that $k^n = \ell^m$, then the only sets of integers that are both k-recognizable and ℓ-recognizable are finite unions of arithmetic progressions. See Allouche and Shallit (2003a).

The description of recognizable sets of integers by automatic sequences starts with Cobham (1972). It is used in Eilenberg (1974). It is one of the main themes of the book of Allouche and Shallit (2003a). The paper-folding sequence takes its name from the following method that can be used to build it (full details are in Allouche and Shallit (2003a)): take a strip of paper, fold it in the middle, then fold it again in the middle, and iterate. When the paper is unfolded, a sequence of peaks and valleys appear. Coding these with the letters a and b yields the sequence.

k-regular functions were introduced by Allouche and Shallit (1992). Their paper contains about thirty examples of k-regular sequences from the literature of number theory. Other results appear in Allouche and Shallit (2003b).

Theorem 4.1 was first proved by Christol (1979) for series with values 0 and 1, then completed by Christol *et al.* (1980).

The definition of rational Laurent series in Section 5.5 is justified by a result of Gessel (1981): he shows that if P, Q are polynomials in several commuting variables, relatively prime, and S a formal series in these variables with $QS = P$, then $Q(0) \neq 0$.

Theorem 5.2 holds without restriction on the characteristic of the field for rational functions in two variables (Furstenberg (1967)). Exercise 5.1 is from Allouche (1987), see also Allouche and Shallit (2003a).

There have been extensions to algebraic (or context-free) series in noncommuting variables by Fliess (1974b) and Haiman (1993), see also Fagnot (1996). The last three authors give a proof of Theorem 5.2 based on formal languages.

6

Rational series in one variable

This chapter gives a short introduction to some striking arithmetic properties of the expansion of rational functions.

In the first section, the notions of rational series, Hankel matrix and rank are shown to coincide, in the case of series in one variable, with the classical definitions. The exponential polynomial is defined in Section 6.2, with emphasis on its algebraic aspects. As an application, we obtain Benzaghou's theorem on the invertible series in the Hadamard algebra (Theorem 2.3).

Section 6.3 is devoted to a theorem of George Pólya concerning arithmetic properties of the coefficients of a rational series.

In the final section, we give an elementary proof, due to Georges Hansel, of the famous Skolem–Mahler–Lech theorem on the positions of vanishing coefficients of a rational series.

6.1 Rational functions

We consider a commutative ring K and an alphabet consisting of a single letter x. We write, as usual, $K[x]$ and $K[[x]]$ instead of $K\langle x\rangle$ and $K\langle\langle x\rangle\rangle$. An element S of $K[[x]]$ is written as

$$S = \sum_{n\geq 0} a_n x^n .$$

Recall from Section 1.4 that S is called rational if S belongs to the smallest subalgebra of $K[[x]]$ which contains $K[x]$ and which is closed under inversion.

Proposition 1.1 *A series S is rational if and only if there exist polynomials P and Q in $K[x]$ with $Q(0) = 1$ such that S is the power series expansion of the rational function P/Q.*

Note that $Q(0)$ is the constant term of the denominator Q of P/Q. Note also that since $Q(0) = 1$, the polynomial Q is invertible in $K[[x]]$ so that P/Q makes sense.

Proof. Let $\mathbf{E}$ be the set of series which are the power series expansion of the form described. Then, since a series with constant term 1 is invertible in $K[[x]]$, $\mathbf{E}$ is contained in the algebra of rational series. Moreover, $\mathbf{E}$ is a subalgebra of $K[[x]]$ closed under inversion, since if $S \in \mathbf{E}$, and $S = P(x)/Q(x)$ is invertible in $K[[x]]$, then its constant term is invertible in K. This constant term is $P(0)/Q(0) = P(0) = \lambda$. Thus

$$S^{-1} = \frac{\lambda^{-1}Q(x)}{\lambda^{-1}P(x)} \in \mathbf{E}\,,$$

since the constant term of the denominator is 1. This shows that any rational series is in $\mathbf{E}$. □

From now on, we assume that K is a field. Let S be a rational series which corresponds to the rational function $P(x)/Q(x)$. The quotient is called *normalized* if P and Q have no common factor in $K[x]$ and if $Q(0) = 1$. In this case, Q is called the *minimal denominator* of S. The roots of Q, which are the poles of the rational function, are called the *poles* of S.

What about the syntactic ideal of S? Set $S = \sum_{n\geq 0} a_n x^n$.

Recall from Section 2.1 that the function $x^n \mapsto a_n$ is extended to a linear function $K[x] \to K$, and that the syntactic ideal (resp. right ideal) of S is the greatest ideal (resp. right ideal) of $K[x]$ contained in the kernel of this linear function.

Let

$$R = x^k - \alpha_1 x^{k-1} - \cdots - \alpha_k \in K[x]$$

be a polynomial. Since K is commutative, the syntactic ideal I of S and the syntactic right ideal coincide. Thus $R \in I$ if and only if $S \circ R = 0$ by Proposition 2.1.4. Since

$$S \circ x^i = \sum_{n\geq 0} a_{n+i} x^n$$

this gives the equivalence

$$R \in I \iff \text{for all } n \in \mathbb{N},\ a_{n+k} - \alpha_1 a_{n+k-1} - \cdots - \alpha_k a_n = 0\,.$$

Observe that in view of Theorem 2.1.2, the series S is rational if and only if its syntactic ideal is not null, since a nonnull ideal in $K[x]$ always has a finite codimension, and the latter is equal to the degree of any generator of this ideal. Recall that a sequence (a_n) over K satisfies a *linear recurrence relation* if, for some $k \geq 0$ and some elements $\alpha_1, \ldots, \alpha_k$ in K, one has

$$\forall n \in \mathbb{N}\ a_{n+k} = \alpha_1 a_{n+k-1} + \cdots + \alpha_k a_n\,. \tag{1.1}$$

This yields the classical result stating that *a series is rational if and only if the sequence of its coefficients satisfies a linear recurrence relation.* The syntactic ideal of S is thus precisely the ideal of polynomials associated with the linear recurrence relations satisfied by S. More precisely, the linear recurrence relations satisfied by (a_n) correspond bijectively to the elements in the syntactic ideal of S whose leading coefficient is 1. We refer to the generator of the syntactic ideal of S having leading coefficient equal to 1 as the *minimal polynomial* of S. It is the polynomial associated with the shortest linear recurrence relation.

By Theorem 2.1.6, the rank of S is equal to the codimension of I; thus it is equal to the length of the shortest linear recurrence relation satisfied by the sequence (a_n). The *eigenvalues* of S are the roots of its minimal polynomial, and their *multiplicities* are defined similarly.

We say that the linear recurrence relation (1.1) is *strict* if $\alpha_k \neq 0$. A rational series $S = \sum_{n\geq 0} a_n x^n$ is *strict* if it satisfies some strict linear recurrence relation. The interest of such series is that one may compute backwards: knowing k consecutive coefficients (not only the k first ones) determines all of them.

Proposition 1.2 *The following conditions are equivalent:*

(i) *S is strict;*
(ii) *the shortest linear recurrence relation satisfied by S is strict;*
(iii) *there exists a polynomial P such that $S \circ P = 0$ and $P(0) \neq 0$;*
(iv) *the minimal polynomial of S has a nonzero constant term;*
(v) *the eigenvalues of S are nonzero;*
(vi) *S has a linear representation (λ, μ, γ) with μx invertible;*
(vii) *for the minimal linear representation (λ, μ, γ) of S, the matrix μx is invertible;*
(viii) *$S = P/Q$ with $P, Q \in K[x]$ and $\deg P < \deg Q$;*
(ix) *for $S = P/Q$ written as an irreducible fraction, one has $\deg P < \deg Q$.*

Proof. Let P_0 be the minimal polynomial of S. Then P_0 divides P (defined in (iii)) by the discussion above; thus (iii), (iv) and (v) are equivalent. By the same discussion, (i) and (ii) are equivalent, and so are (ii) and (iv).

Exercise 1.1 shows that (i) implies (vi). The implication (vi) $\Longrightarrow$ (vii) follows from Corollary 2.2.5, and the implication (vii) $\Longrightarrow$ (i) follows from Exercise 1.2. Now suppose that Equation (1.1) holds, with $\alpha_k \neq 0$. Then, for any $n \geq k$, one has $a_n - \alpha_1 a_{n-1} - \cdots - \alpha_k a_{n-k} = 0$, showing that $(1 - \alpha_1 x - \cdots - \alpha_k x^k)S$ is a polynomial of degree $< k$; thus (i) implies (viii). Simplifying numerator and denominator, we see that (viii) implies (ix). Now, the previous equality also shows that (ix) implies (i). □

The last part of the proof also shows the following result.

Corollary 1.3 *Let $S = P/Q$ be a strict rational series written as an irreducible fraction with $\deg P < \deg Q$. Then the rank of S is equal to the degree of Q. The minimal polynomial is equal to the reciprocal polynomial of the minimal denominator Q.* □

Recall that the *reciprocal polynomial* of $1 - \alpha_1 x - \cdots - \alpha_\ell x^\ell$, with $\alpha_\ell \neq 0$, is $x^\ell - \alpha_1 x^{\ell-1} - \cdots - \alpha_\ell$.

Proposition 1.4 *The rank of a rational series $S = \sum_{n\geq 0} a_n x^n = P(x)/Q(x)$, where $P, Q \in K[x]$ are relatively prime and $Q \neq 0$, is equal to $\max(1 + \deg P, \deg Q)$ and also to the size of the greatest nonzero principal minor of its Hankel matrix.*

Recall that a *principal minor* of the Hankel matrix $(a_{i+j})_{i,j\geq 0}$ is a determinant of the form $|a_{i+j}|_{0\leq i,j\leq n}$.

Proof. Note that a finite prefix-closed (or suffix-closed) subset of x^* is necessarily of the form $\{1, x, \ldots, x^n\}$. Thus the last statement follows from Corollary 2.3.7: indeed, if the rank of S is r, then this result shows that the principal minor of size $r \times r$ of the Hankel matrix is nonzero; since r is the rank of this matrix, all greater minors are 0.

Observe that if S is strict, then $\deg P < \deg Q$ and the first assertion is contained in Corollary 1.3.

Suppose now that S is not strict, so $\deg P \geq \deg Q$. Then by Euclidean division, $P = A + BQ$ with $\deg(A) < \deg(Q)$. Thus $S = P/Q = A/Q + B$, and $\deg B = \deg P - \deg Q$. Let $d = \deg B + 1$. We have $B \circ x^d = 0$. Moreover, the rank of A/Q is $\deg Q$ by Corollary 1.3, since A/Q is strict. Let $\overline{Q}$ be the reciprocal polynomial of Q. Then $(A/Q) \circ \overline{Q} = 0$ by Corollary 1.3, and therefore $S \circ (\overline{Q}x^d) = 0$. Thus the rank of S is $\leq \deg Q + d = \deg P + 1$.

Conversely, let k be the rank of S. Then we have Equation (1.1) and $\alpha_k = 0$, since S is not strict, and by Proposition 1.2. Let ℓ be such that $\alpha_\ell \neq 0$, with ℓ maximum. Then $\ell < k$. By using the recurrence relation backwards, we may find a sequence (a'_n) which coincides with (a_n) for $n \geq k - \ell$ and which satisfies the strict linear recurrence $a'_{n+\ell} = \alpha_1 a'_{n+\ell-1} + \cdots + \alpha_\ell a'_n$ for all n in $\mathbb{N}$. Indeed, for $n \geq k - \ell$ we have $n + \ell = (n + \ell - k) + k$ and $n + \ell - k \geq 0$, hence $a_{n+\ell} = \alpha_1 a_{n+\ell-1} + \cdots + \alpha_\ell a_n$ by using Equation (1.1).

The series $S' = \sum a'_n x^n$ is strict, of the form A/B with $\deg A < \deg B \leq \ell$ by Corollary 1.3. Moreover $S = S' + C$, where C is a nonzero polynomial with $\deg C \leq k - \ell - 1$. Thus $S = (A + BC)/B$ and we conclude that $1 + \deg P \leq 1 + \deg BC = 1 + \deg B + \deg C \leq 1 + \ell + k - \ell - 1 = k$.

Finally, $k = 1 + \deg P$, what was to be shown. □

Corollary 1.5 *With $S = P/Q$ as in the proposition, let $Q = 1 - \alpha_1 x - \cdots - \alpha_\ell x^\ell$, $\alpha_\ell \neq 0$, and let $d = 0$ if $\deg P < \deg Q$, $d = \deg P - \deg Q + 1$ if $\deg P \geq \deg Q$. Then the minimal polynomial of S is equal to $x^d(x^\ell - \alpha_1 x^{\ell-1} - \cdots - \alpha_\ell)$ and also to the characteristic polynomial of μx, for any minimal linear representation (λ, μ, γ) of S.*

Proof. The first equality has been obtained in the proof of Proposition 1.4, since the minimal polynomial of S corresponds to the shortest linear recurrence satisfied by (a_n). The second equality follows from the fact that the minimal polynomial is the generator, with leading coefficient 1, and of the smallest degree, of the syntactic ideal of S, and from Corollary 2.2.2. Alternatively, it is a consequence of Exercises 1.1 and 1.2. □

Proposition 1.6 *For every rational series S, there exists a unique pair (T, P), where T is a strict series and P is a polynomial, such that $S = P + T$.*

Proof. The result is equivalent to: each rational function is, in a unique way, the sum of a polynomial and of a rational function P/Q with $\deg P < \deg Q$. The details are left to the reader. □

In view of Proposition 1.6, it suffices for many purposes to study strict rational series.

Proposition 1.7 *The subset of strict rational series of $K[[x]]$ is closed under linear combination, product and Hadamard product.*

Observe that this set does not contain any nonvanishing polynomials by Proposition 1.6.

Proof. Let $S_1 = P_1/Q_1$ and $S_2 = P_2/Q_2$ be strict series with $\deg(P_1) < \deg(Q_1)$ and $\deg(P_2) < \deg(Q_2)$. Then $S_1 + S_2 = (P_1Q_2 + P_2Q_1)/Q_1Q_2$ and $S_1S_2 = P_1P_2/Q_1Q_2$. Since clearly $\deg(P_1Q_2 + P_2Q_1) < \deg(Q_1Q_2)$ and $\deg(P_1P_2) < \deg(Q_1Q_2)$, the series $S_1 + S_2$ and S_1S_2 are strict. Moreover, if $(S_1, x^n) = \lambda_1\mu_1x^n\gamma_1$ and $(S_2, x^n) = \lambda_2\mu_2x^n\gamma_2$, where μ_1x and μ_2x are invertible matrices, then

$$(S_1 \odot S_2, x^n) = (S_1, x^n)(S_2, x^n) = (\lambda_1 \otimes \lambda_2)(\mu_1 \otimes \mu_2)(x^n)(\gamma_1 \otimes \gamma_2),$$

and since $(\mu_1 \otimes \mu_2)(x)$ is invertible, this shows that $S_1 \odot S_2$ is strict. □

The set of strict rational series equipped with the structure of vector space and with the Hadamard product is the *Hadamard algebra of strict rational series*. Its neutral element is the series $\sum x^n = 1/(1-x)$.

6.2 The exponential polynomial

We assume from now on that K has *characteristic zero*. Let Λ be the multiplicative group $K \setminus 0$, and let t be an indeterminate. We consider the algebra

$$K[t][\Lambda]$$

of the group Λ over the ring $K[t]$. It is in particular an algebra over K. An element of $K[t][\Lambda]$ is called an *exponential polynomial.*

Theorem 2.1 *Let K be algebraically closed. The function which associates to an exponential polynomial*

$$\sum_{\lambda\in\Lambda} P_\lambda(t)\lambda$$

of $K[t][\Lambda]$ the strict rational series

$$\sum_{n\geq 0} a_n x^n$$

defined by

$$a_n = \sum_{\lambda \in \Lambda} P_\lambda(n)\lambda^n$$

(with the sum computed in K) is an isomorphism of K-algebras from $K[t][\Lambda]$ onto the Hadamard algebra of strict rational series.

Proof. Let ϕ be the function of the statement. Let $E = \sum P_\lambda(t)\lambda$ and $F = \sum Q_\lambda(t)\lambda$ be two exponential polynomials, and let $G = E + F = \sum R_\lambda(t)\lambda$, $H = EF = \sum S_\lambda(t)\lambda \in K[t][\Lambda]$. Then

$$R_\lambda = P_\lambda + Q_\lambda, \quad S_\lambda = \sum_{\mu\nu=\lambda} P_\mu Q_\nu .$$

Consequently

$$\begin{aligned}(\phi(G), x^n) &= \sum R_\lambda(n)\lambda^n = \sum P_\lambda(n)\lambda^n + \sum Q_\lambda(n)\lambda^n \\ &= (\phi(E), x^n) + (\phi(F), x^n), \\ (\phi(H), x^n) &= \sum S_\lambda(n)\lambda^n = \sum_\lambda \lambda^n \sum_{\mu\nu=\lambda} P_\mu(n)Q_\nu(n) \\ &= \Big(\sum_\mu P_\mu(n)\mu^n\Big)\Big(\sum_\nu Q_\nu(n)\nu^n\Big) \\ &= (\phi(E), x^n)(\phi(F), x^n).\end{aligned}$$

Thus

$$\phi(E + F) = \phi(E) + \phi(F), \quad \phi(EF) = \phi(E) \odot \phi(F).$$

Let us now verify that ϕ is a bijection. Let $\alpha_1, \ldots, \alpha_k$ be elements of K with $\alpha_k \neq 0$, and let V be the set of all rational series $S = \sum a_n x^n$ satisfying the relation

$$a_{n+k} = \alpha_1 a_{n+k-1} + \cdots + \alpha_k a_n, \quad (n \geq 0).$$

These series are necessarily strict. Clearly, V is a vector space of dimension k over K. Let $\lambda_1, \ldots, \lambda_p$ be the roots of the polynomial

$$R(x) = x^k - \alpha_1 x^{k-1} - \cdots - \alpha_k$$

with multiplicities $n_1, \ldots, n_p$ respectively. Consider the subspace V' of $K[t][\Lambda]$ of dimension k

$$V' = \Big\{ \sum_{1 \leq i \leq p} P_i(t)\lambda_i \;\Big|\; \deg(P_i) \leq n_i - 1 \Big\}.$$

We show that ϕ induces a surjection $V' \to V$ (and consequently an injection) and this will prove the theorem.

Any $S = \sum a_n x^n$ in V can be written as $P(x)/Q(x)$, with $\deg(P) < \deg(Q)$ and $Q = 1 - \alpha_1 x - \cdots - \alpha_k x^k$. Decomposing P/Q into simple elements shows that S

is a linear combination over K of series

$$\frac{1}{(1-\lambda_i x)^j}, \quad 1 \leq i \leq p,\ 1 \leq j \leq n_j .$$

Next, it is well-known (see Exercise 2.7) that

$$\frac{1}{(1-\lambda x)^j} = \sum_{n \geq 0} \binom{n+j-1}{j-1} \lambda^n x^n .$$

Since $\binom{n+j-1}{j-1}$ is a polynomial of degree $j-1$ in the variable n, the surjectivity of $\phi : V' \to V$ is proved. □

Observe that in the bijection described in the theorem and its proof, the *support* of an exponential polynomial $E = \sum P_\lambda(t)\lambda$ (that is the set of $\lambda \in \Lambda$ such that $P_\lambda \neq 0$) is exactly the set of nonzero eigenvalues (that is, inverses of poles) of S, and that the multiplicity of an eigenvalue λ is equal to $1 + \deg(P_\lambda)$. Furthermore, if the coefficients and the eigenvalues of S are in some subfield K_1 of K, then the corresponding exponential polynomial is in $K_1[t][\Lambda_1]$, with $\Lambda_1 = K_1 \setminus 0$.

Corollary 2.2 *Let $S = \sum a_n x^n$ be a rational series over an algebraically closed field K of characteristic* 0.

(i) *The coefficients a_n are given, for large enough n, by*

$$a_n = \sum_{1 \leq i \leq p} \lambda_i^n P_i(n) , \tag{2.1}$$

where $\lambda_1, \ldots, \lambda_p \in K \setminus 0$ and $P_i(t) \in K[t]$.

(ii) *The expression* (2.1) *is unique if the λ_i's are distinct; in particular, the nonzero eigenvalues of S are the λ_i's with $P_i \neq 0$.*

Proof. (i) By Proposition 1.4, $S = P + T$ for some polynomial P and some rational strict series T. Thus, it suffices to use Theorem 2.1.

(ii) Let

$$T = \sum_{n \geq 0} \Big(\sum_{1 \leq i \leq p} \lambda_i^n P_i(n) \Big) x^n .$$

Then, in view of Theorem 2.1, T is rational strict. Moreover $S = P + T$ for some polynomial P (because S and T have by assumption the same coefficients for large enough n). By Proposition 1.4, T depends only on S, and by Theorem 2.1, the exponential polynomial of T is unique. This proves the first assertion. By the remark following the proof of Theorem 2.1, the λ_i's with $P_i \neq 0$ are exactly the eigenvalues of T. Now, it is clear that T and S have the same poles, so they have the same nonzero eigenvalues. □

Definition Let $S_0, \ldots, S_{p-1}$ be formal series in $K[[x]]$. The *merge* of these series is the formal series defined for $m \in \mathbb{N}$ and $i \in \{0, \ldots, p-1\}$ by

$$(S, x^{mp+i}) = (S_i, x^m) .$$

In other words, if $n = mp + i$ (Euclidean division of n by p), then $(S, x^n) = (S_i, x^m)$. This can also be written as

$$S(x) = \sum_{0 \le i < p} x^i S_i(x^p)$$

with self-evident notation.

An example. If $p = 2$ and $S_0 = \sum a_n x^n$ and $S_1 = \sum b_n x^n$, then the *merge* of S_0 and S_1 is the series $\sum c_n x^n$ where the sequence (c_n) is

$$a_0, b_0, a_1, b_1, a_2, b_2, a_3, \ldots$$

Observe that for any series $S = \sum a_n x^n \in K[[x]]$ and any p, there is a unique p-tuple of series $(S_0, \ldots, S_{p-1})$ whose merge is S. These series are indeed

$$S_i = \sum_{n \ge 0} a_{i+np} x^n .$$

Definition A series $\sum a_n x^n$ is *geometric* if there exist b, c in K with $c \neq 0$ such that $a_n = bc^n$.

Theorem 2.3 (Benzaghou 1970) *If a strict rational series is invertible in the Hadamard algebra of strict rational series, then it is a merge of geometric series.*

The conclusion can also be formulated as follows: there exist an integer p and elements $a_0, \ldots, a_{p-1}, b_0, \ldots, b_{p-1}$ in K with $b_0, \ldots, b_{p-1} \neq 0$ and such that the series is

$$\sum_{0 \le i \le p-1} \frac{a_i x^i}{1 - b_i x^p} .$$

Proof. (i) Let i and p be natural numbers and consider the K-linear function $\psi : K[t][\Lambda] \to K[t][\Lambda]$ defined on monomials by

$$\psi(P(t)\lambda) = (\lambda^i P(i + pt))\lambda^p ,$$

where $P(t) \in K[t]$, $\lambda \in \Lambda$ and where $\lambda^i P(i + pt)$ is viewed as an element of $K[t]$. The function ψ is a morphism of K-algebra. To see this, it suffices to compute ψ on products of monomials, and indeed

$$\begin{aligned} \psi(P(t)Q(t)\lambda\mu) &= (\lambda^i \mu^i P(i + pt)Q(i + pt))\lambda^p \mu^p \\ &= \psi(P(t)\lambda)\psi(Q(t)\mu) . \end{aligned}$$

(ii) Consider now two exponential polynomials $E, F \in K[t][\Lambda]$ and let Λ_1 be the subgroup of Λ generated by $\mathrm{supp}(E) \cup \mathrm{supp}(F)$. The group Λ_1 is a finitely generated Abelian group, thus is isomorphic to the product of a finite group (of p elements, say) and of a finitely generated free Abelian group. Consequently, the subgroup Λ_2 of Λ_1 generated by the λ^p, for $\lambda \in \Lambda_1$, is free.

By construction, the supports of $\psi(E)$ and $\psi(F)$ are in Λ_2 (for any i, and for the fixed p), and $\psi(E), \psi(F) \in K[t][\Lambda_2]$. Assume now $EF = 1$. Then $\psi(E)\psi(F) = 1$. Since Λ_2 is free, the only invertible elements of $K[t][\Lambda_2]$ have the form $a\lambda$, with $a \in K, \lambda \in \Lambda_2$. See Exercise 2.3.

(iii) Consider now two strict rational series S and T such that $S \odot T = \sum_{n\geq 0} x^n$ (the neutral element of the Hadamard algebra). Let $E, F \in K[t][\Lambda]$ be such that $\phi(E) = S$, $\phi(F) = T$, where ϕ is the isomorphism of Theorem 2.1. Then $EF = 1$. Thus $\psi(E)\psi(F) = 1$.

Set $S = \sum a_n x^n$. If $E = \sum P_\lambda(t)\lambda$, then $\psi(E) = \sum \lambda^i P_\lambda(i + tp)\lambda^p$ and

$$\phi(\psi(E)) = \sum_{n\geq 0}\Big(\sum_{\lambda} \lambda^i P_\lambda(i + pn)\lambda^{pn}\Big)x^n = S_i\,,$$

where

$$S_i = \sum_{n\geq 0} a_{i+pn}x^n\,.$$

In view of the conclusion of (ii), $\psi(E) = a\lambda$ for some $a \in K, \lambda \in \Lambda$. Consequently,

$$S_i = \sum_{n\geq 0} a\lambda^n x^n\,.$$

This proves the theorem because S is the merge of the S_i's, $i = 0, \ldots, p-1$. □

The proof of the theorem suggests the following definition and proposition which will be of use later.

Definition A strict rational series is *simple* if the Abelian multiplicative subgroup of $K \setminus 0$ generated by its eigenvalues is a free Abelian group. Similarly, a set of strict rational series is *simple* if the set of all its eigenvalues generates a free Abelian group.

Proposition 2.4 *Let* $\mathbf{S}$ *be a finite set of strict rational series. There exists an integer* $p \geq 1$ *such that the set of series of the form*

$$\sum_{n\geq 0} a_{i+pn}x^n$$

for $i \in \mathbb{N}$ *and for* $\sum a_n x^n \in \mathbf{S}$ *is simple.*

Proof. Since $\mathbf{S}$ is finite, there exists an invertible matrix $m \in K^{q\times q}$ such that each $S \in \mathbf{S}$ can be written as

$$S = \sum_{n\geq 0} \phi_S(m^n)x^n$$

for some linear function ϕ_S on $K^{q\times q}$. Let Λ_1 be the set of eigenvalues of m. The group generated by Λ_1 in $K \setminus 0$ is finitely generated, and consequently there is an integer $p \geq 1$ such that the group G generated by the λ^p, for $\lambda \in \Lambda_1$, is free Abelian. Let P be the characteristic polynomial of m^p. The roots of P are in G. For each

$i \in \mathbb{N}$ and $S = \sum a_n x^n \in \mathbf{S}$, the series $S_i = \sum a_{i+pn} x^n$ has the form

$$S_i = \sum_n \phi_S(m^i (m^p)^n) x^n ,$$

showing that $S_i \circ P = 0$ (see Exercise 1.2). Consequently, the eigenvalues of S_i are in G. □

6.3 A theorem of Pólya

In this section, we consider series with coefficients in $\mathbb{Q}$. Recall that for any prime number p, the *p-adic valuation* v_p over $\mathbb{Q}$ is defined by $v_p(0) = \infty$ and $v_p(p^n a/b) = n$ for $n, a, b \in \mathbb{Z}$, and p dividing neither a nor b.

Definition Let $S = \sum a_n x^n \in \mathbb{Q}[[x]]$. The set of *prime factors* of S is the set of prime numbers

$$P(S) = \{p \mid \exists n \in \mathbb{N},\, v_p(a_n) \neq 0, \infty\} .$$

Theorem 3.1 (Pólya 1921) *The set of prime factors of a rational series S is finite if and only if S is the sum of a polynomial and of a merge of geometric series.*

We start with a result of independent interest.

Theorem 3.2 (Benzaghou 1970) *Let $S = \sum a_n x^n$ be a rational series which is not a polynomial, and let p be a prime number. There exist integers $n_0 \geq 0$ and $q \geq 1$ such that the function $n \mapsto v_p(a_{n_0+qn})$ is affine.*

Proof. (i) We start by proving a preliminary result. Let K be a field with a discrete valuation $v : K \to \mathbb{N} \cup \{\infty\}$. Let A be its valuation ring, $A = \{z \in K \mid v(z) \geq 0\}$, let I be the maximal ideal of A, $I = \{z \in K \mid v(z) \geq 1\}$ and let $U = A \setminus I = \{z \in K \mid v(z) = 0\}$ be the group of invertible elements of A. Suppose further that the residual field $F = A/I$ is finite. Since v is discrete, I is a principal ideal, and consequently $I = \pi A$ for some $\pi \in A$ with $v(\pi) = 1$. (For an exposition of these concepts, see e.g. Amice (1975), Koblitz (1984).) Let $\lambda_1, \ldots, \lambda_k$ be elements of $A \setminus 0$, let $P_1, \ldots, P_k \in K[t]$ be polynomials and let (a_n) be a sequence of elements in A defined by

$$a_n = \sum_{1 \leq i \leq k} P_i(n) \lambda_i^n . \tag{3.1}$$

Then we claim that there exist integers n_0 and q such that the function $n \mapsto v(a_{n_0+qn})$ is affine.

The proof is in three steps.

1. One may assume that all the P_i are in $A[t]$ (by multiplying the polynomials by a common denominator, if necessary).

2. Assuming that $\lambda_i \in I$ for all $i = 1, \dots, k$, set

$$r = \inf\{v(\lambda_i) \mid i = 1, \dots, k\}.$$

Then $r \geq 1$. Since each P_i has coefficients in A and $v(\lambda_i) \geq r$ for all i, it follows that

$$a'_n = \frac{a_n}{\pi^{rn}} = \sum_{1 \leq i \leq k} P_i(n)\left(\frac{\lambda_i}{\pi^r}\right)^n \in A.$$

Since $v(a_n) = v(a'_n) + rn$, we may assume in addition that $\lambda_i \in U$ for at least one index i.

3. Let $\ell \geq 1$ be such that $\lambda_1, \dots, \lambda_\ell \in U$ and $\lambda_{\ell+1}, \dots, \lambda_k \in I$ (possibly $\ell = k$). Set

$$b_n = \sum_{i=1}^{\ell} P_i(n)\lambda_i^n, \quad c_n = \sum_{i=\ell+1}^{k} P_i(n)\lambda_i^n$$

($c_n = 0$ if $\ell = k$). We prove that there is an arithmetic progression of integers n where $v(b_n)$ is constant. For this, observe that the minimal polynomial of the strict series $\sum b_n x^n$ is

$$P(x) = \prod_{i=1}^{\ell}(x - \lambda_i)^{\deg(P_i)+1}$$

(cf. Theorem 2.1 and the observation following its proof). By setting

$$P(x) = x^h - \alpha_1 x^{h-1} - \cdots - \alpha_h,$$

one has $\alpha_h \in U$. Let

$$s = \inf\{v(b_0), \dots, v(b_{h-1})\}.$$

Since the sequence (b_n) satisfies the recurrence relation associated with P, and since the coefficients of P are in A, it follows that $v(b_n) \geq s$ for all n. Consequently, the sequence (b'_n) defined by

$$b'_n = b_n/\pi^s$$

is also in A. It has the same minimal polynomial as (b_n) and there is an integer j such that

$$v(b'_j) = 0,$$

that is $b'_j \in U$. Next, by Exercise 1.1,

$$b'_n = \lambda m^n \gamma,$$

where

$$\lambda = (1, 0, \dots, 0), \quad m = \begin{pmatrix} 0 & 1 & 0 & \cdots & 0 \\ \vdots & & \ddots & \ddots & \vdots \\ \vdots & & & \ddots & 0 \\ 0 & \cdots & & 0 & 1 \\ \alpha_h & \cdots & & & \alpha_1 \end{pmatrix}, \quad \gamma = \begin{pmatrix} b'_0 \\ b'_1 \\ \vdots \\ b'_{h-1} \end{pmatrix}.$$

Since the determinant of the matrix m is $\pm\alpha_h \in U$, and since $F = A/I$ is finite, there is an integer q such that $m^q \equiv 1 \mod I$ (with I the identity matrix). This shows that the sequence (b'_n) has period q modulo I and in particular for all $n \geq 0$,

$$b'_{j+qn} \equiv b'_j \mod I.$$

Thus, $v(b'_{j+qn}) = v(b'_j) = 0$, and consequently

$$v(b_{j+qn}) = s \quad \text{for } n \geq 0.$$

Finally, observe that $v(c_n) \geq n$. This implies that, for large n (more precisely for $j + qn > s$),

$$v(a_{j+qn}) = v(b_{j+qn}) = s.$$

This proves the preliminary result.

(ii) The series S is rational over $\mathbb{Q}$. We may assume that it is strict by Proposition 1.6. By Exercise 1.5.1.b, we may assume that it is rational over $\mathbb{Z}$ and has a linear representation (λ, μ, γ) with μx over $\mathbb{Z}$ and of nonzero determinant. Let $P(x) = x^r - \alpha_1 x^{r-1} - \cdots - \alpha_r$ be its characteristic polynomial. Then (a_n) satisfies the linear recurrence relation associated to P (see Exercise 1.2). The roots $\lambda_1, \dots, \lambda_k$ of P are algebraic integers. Let K be the number field $K = \mathbb{Q}[\lambda_1, \dots, \lambda_k]$. By Theorem 2.1, the a_n admit the expression given by Equation (3.1). Moreover, for any prime ideal $\mathfrak{p}$ of K, the α_i and a_n are in the valuation ring of K for the valuation $v_\mathfrak{p}$ and by our preliminary result (i), there exist integers j and ℓ such that

$$n \mapsto v_\mathfrak{p}(a_{j+\ell n})$$

is an affine function.

(iii) Let B be the ring of algebraic integers of K, and let p be a prime number. The ideal pB of B decomposes as

$$pB = \mathfrak{p}_1^{m_1} \cdots \mathfrak{p}_s^{m_s},$$

where $\mathfrak{p}_1, \dots, \mathfrak{p}_s$ are distinct prime ideals of K. By applying the preceding argument for $\mathfrak{p} = \mathfrak{p}_1$ one obtains integers j, ℓ such that the function

$$n \mapsto v_{\mathfrak{p}_1}(a_{j+\ell n})$$

is affine. By iteration of this computation for $\mathfrak{p}_2, \dots, \mathfrak{p}_s$, one gets successive subsequences and finally one obtains an arithmetic progression $n'_0 + q'\mathbb{N}$ such that for each $i = 1, \ldots, s$, the function

$$n \mapsto v_{\mathfrak{p}_i}(a_{n_0'+q'n})$$

is affine. Thus there exist integers x_i and y_i such that

$$v_{\mathfrak{p}_i}(a_{n_0'+q'n}) = x_i + y_i n .$$

Note that x_i, y_i are integers, since $x_i + y_i n$ is an integer for n in $\mathbb{N}$. Now observe that for all $a \in \mathbb{Z}$,

$$v_p(a) = \inf\left\{\left\lfloor \frac{v_{\mathfrak{p}_i}(a)}{m_i} \right\rfloor ; i = 1, \ldots, s\right\}$$

where $\lfloor z \rfloor$ denotes the integral part of z. Since the functions

$$n \mapsto \frac{v_{\mathfrak{p}_i}(a_{n_0'+q'n})}{m_i} = \frac{x_i + y_i n}{m_i}$$

also are affine, there exists an integer $i_0 \in \{1, \ldots, s\}$ such that for all $i = 1, \ldots, s$ and all sufficiently large n,

$$\frac{1}{m_i}(x_i + y_i n) \geq \frac{1}{m_{i_0}}(x_{i_0} + y_{i_0} n),$$

showing that

$$v_p(a_{n_0'+q'n}) = \left\lfloor \frac{x_{i_0} + y_{i_0} n}{m_{i_0}} \right\rfloor$$

for sufficiently large n. Since the function

$$n \mapsto \left\lfloor \frac{x_{i_0} + y_{i_0} m_{i_0} n}{m_{i_0}} \right\rfloor = \left\lfloor \frac{x_{i_0}}{m_{i_0}} \right\rfloor + y_{i_0} n$$

also is affine, the lemma follows. □

Proof of Theorem 3.1. Let S be a rational series having a finite set of prime factors. Clearly we may assume that S is strict (Proposition 1.6). In view of Proposition 2.4, we may even assume that S is simple.

Let $S = \sum a_n x^n$ and let $p_1, \ldots, p_\ell$ be the prime factors of S. Applying Lemma 3.2 successively to $p_1, \ldots, p_\ell$, one obtains integers n_0 and q such that, for every $i = 1, \ldots, \ell$, the function

$$n \mapsto v_{p_i}(a_{n_0+qn})$$

is affine. Set $\epsilon_k = -1, 0, 1$ according to $a_n < 0$, $a_n = 0$, $a_n > 0$. Then for $n \geq 0$, one has

$$a_{n_0+qn} = \theta_n b c^n$$

with $\theta_n = \epsilon_{n_0+qn}$.

Now let $\lambda_1, \ldots, \lambda_k$, with $k \geq 1$, be the distinct eigenvalues of S. In view of Theorem 2.1, there are nonvanishing polynomials $P_1, \ldots, P_k$ such that

$$a_n = \sum_{i=1}^{k} P_i(n)\lambda_i^n \,. \tag{3.2}$$

Thus, setting

$$b_n = a_{n_0+qn}\,, \quad Q_i(t) = P_i(n_0 + qt)\lambda_i^{n_0}\,, \quad \mu_i = \lambda_i^q\,,$$

one has

$$b_n = \theta_n bc^n = \sum_{i=1}^{k} Q_i(n)\mu_i^n \,.$$

Since the group generated by the λ_i's is free, all the μ_i are distinct. Moreover, the polynomials $Q_i(t)$ do not vanish, and consequently $\sum b_n x^n$ is not a polynomial. Therefore $\theta_n \neq 0$ for infinitely many n, and we may suppose that $\theta_n = 1$ for infinitely many n. The series

$$\sum \frac{b_n}{c^n} x^n$$

has finite image. By Theorem 3.2.10 and Exercise 3.1.1, there exists an arithmetic progression $n_1 + r\mathbb{N}$ such that $\theta_n = 1$ for $n \in n_1 + r\mathbb{N}$. It follows that

$$b_{n_1+rn} = bc^{n_1}(c^r)^n = \sum_{i=1}^{k} Q_i(n_1 + rn)\mu_i^{n_1}(\mu_i^r)^n \,.$$

As before, the μ_i^r are pairwise distinct. In view of the uniqueness of the exponential polynomial, one has $k = 1$ and $Q_1(n_1 + rt) = C$, for some constant C. Thus Q_1 is a constant and also P_1. By Equation (3.2), $a_n = P_1\lambda_1^n$. This completes the proof. □

6.4 A theorem of Skolem, Mahler, Lech

The following result describes completely the supports of rational series in one variable with coefficients in a field of characteristic zero. They are exactly the rational one-letter languages. This does not hold for more than one variable (see Example 3.4.1).

Theorem 4.1 (Skolem 1934, Mahler 1935, Lech 1953) *Let K be a field of characteristic 0, and let $S = \sum a_n x^n$ be a rational series with coefficients in K. The set*

$$\{n \in \mathbb{N} \mid a_n = 0\}$$

is the union of a finite set and of a finite number of arithmetic progressions.

In fact, this result has been proved for $K = \mathbb{Z}$ by Skolem, it has been extended to algebraic number fields by Mahler and to fields of characteristic 0 by Lech. The latter also gives the following example showing that the theorem does not hold in characteristic $p \neq 0$. Indeed, let θ be transcendent over the field $\mathbf{F}_p$ with p elements. Then the series $\sum a_n x^n$ with

$$a_n = (\theta + 1)^n - \theta^n - 1$$

is rational over $\mathbf{F}_p(\theta)$ and, however, $\{n \mid a_n = 0\} = \{p^r \mid r \in \mathbb{N}\}$ is not a rational subset of the monoid $\mathbb{N}$ (see Exercise 3.4.1.b).

The proof given here is elementary and does not use p-adic analysis. It requires several definitions and lemmas, and goes through three steps. First, the result is proved for series with integral coefficients. Then it is extended to transcendental extensions and finally to the general case.

Definition A set A of nonnegative integers is called *purely periodic* if there exist an integer $N \geq 0$ and integers $k_1, k_2, \ldots, k_r \in \{0, 1, \ldots, N-1\}$ such that

$$A = \{k_i + nN \mid n \in \mathbb{N}, 1 \leq i \leq r\}.$$

The integer N is *a period* of A. A *quasi-periodic* set (of period N) is a subset of $\mathbb{N}$ which is the union of a finite set and of a purely periodic set (of period N).

Lemma 4.2 *The intersection of a family of quasi-periodic sets of period N is quasi-periodic of period N.*

Proof. Let $(A_i)_{i \in I}$ be a family of quasi-periodic sets, all having period N. Given a $j \in \{0, 1, \ldots, N-1\}$, for any $i \in I$, the set $(j + N\mathbb{N}) \cap A_i$ is either finite or equal to $j + N\mathbb{N}$. Thus the same holds for $(j + N\mathbb{N}) \cap (\cap A_i)$. □

Definition Given a series $S = \sum a_n x^n$ with coefficients in a semiring K, the *annihilator* of S is the set

$$\operatorname{ann}(S) = \{n \in \mathbb{N} \mid a_n = 0\}.$$

Thus the annihilator is the complement of the support.

With these definitions, the first (and most difficult) step in the proof of Theorem 4.1 can be formulated as follows.

Proposition 4.3 *Let $S = \sum a_n x^n \in \mathbb{Q}[[x]]$ be a strict rational series with rational coefficients. Then the annihilator of S is quasi-periodic.*

Let p be a fixed prime number. The p-adic valuation v_p is defined at the beginning of Section 6.3. Observe that

$$v_p(q_1 \cdots q_n) = \sum_{1 \leq i \leq n} v_p(q_i),$$

$$v_p(q_1 + \cdots + q_n) \geq \inf\{v_p(q_1), \ldots, v_p(q_n)\}.$$

Observe also that for $n \in \mathbb{N}$

$$v_p(n!) \le n/(p-1) \tag{4.1}$$

since indeed (Exercise!)

$$\begin{aligned} v_p(n!) &= \lfloor n/p \rfloor + \lfloor n/p^2 \rfloor + \cdots + \lfloor n/p^k \rfloor + \cdots \\ &\le n/p + n/p^2 + \cdots + n/p^k + \cdots \\ &\le n \sum_{k \ge 1} \frac{1}{p^k} = n\frac{1/p}{1 - 1/p} = n/(p-1)\,. \end{aligned}$$

From Equation (4.1), we deduce

$$v_p\left(\frac{p^n}{n!}\right) = v_p(p^n) - v_p(n!) \ge n - \frac{n}{p-1}\,,$$

thus

$$v_p\left(\frac{p^n}{n!}\right) \ge n\frac{p-2}{p-1}\,. \tag{4.2}$$

Next, consider an arbitrary polynomial

$$P(x) = a_0 + a_1 x + \cdots + a_n x^n$$

with integral coefficients. For any integer $k \ge 0$, let

$$\omega_k(P) = \inf\{v_p(a_j) \mid j \ge k\}\,.$$

Clearly

$$\omega_0(P) \le \omega_1(P) \le \cdots \le \omega_k(P) \le \cdots$$

and

$$\omega_k(P) = \infty \qquad \text{for } k > n\,.$$

Observe also that $v_p(P(t)) \ge \inf\{a_0, a_1 t, \ldots, a_n t^n\}$ for any integer $t \in \mathbb{Z}$, and consequently

$$v_p(P(t)) \ge \inf\{v_p(a_0), v_p(a_1), \ldots, v_p(a_n)\} = \omega_0(P)\,. \tag{4.3}$$

Lemma 4.4 *Let P and Q be two polynomials with rational coefficients such that*

$$P(x) = (x - t)Q(x)$$

for some $t \in \mathbb{Z}$. Then for all $k \in \mathbb{N}$

$$\omega_{k+1}(P) \le \omega_k(Q)\,.$$

Proof. Set

$$Q(x) = a_0 + a_1 x + \cdots + a_n x^n\,, \qquad P(x) = b_0 + b_1 x + \cdots + b_{n+1} x^{n+1}\,.$$

Then $b_{j+1} = a_j - ta_{j+1}$ for $0 \le j \le n-1$, $b_{n+1} = a_n$, whence for $j = 0, \dots, n$,

$$a_j = b_{j+1} + tb_{j+2} + \cdots + t^{n-j} b_{n+1} .$$

This shows that $v_p(a_j) \ge \omega_{j+1}(P)$ for any $j \in \mathbb{N}$. Thus, given any $k \in \mathbb{N}$, one has for $j \ge k$

$$v_p(a_j) \ge \omega_{j+1}(P) \ge \omega_{k+1}(P)$$

and consequently

$$\omega_k(Q) \ge \omega_{k+1}(P) . \qquad \square$$

Corollary 4.5 *Let Q be a polynomial with rational coefficients, let $t_1, t_2, \dots, t_k \in \mathbb{Z}$, and let*

$$P(x) = (x - t_1)(x - t_2) \cdots (x - t_k) Q(x) .$$

Then

$$\omega_k(P) \le \omega_0(Q) .$$

The main argument is the following lemma.

Lemma 4.6 *Let $(d_n)_{n \in \mathbb{N}}$ be any sequence of integers and let $(b_n)_{n \in \mathbb{N}}$ be the sequence defined by*

$$b_n = \sum_{i=0}^{n} \binom{n}{i} p^i d_i ,$$

where p is an odd prime number. If $b_n = 0$ for infinitely many indices n, then the sequence $(b_n)_{n \in \mathbb{N}}$ vanishes.

Proof. For $n \in \mathbb{N}$, let

$$R_n(x) = \sum_{i=0}^{n} d_i p^i \frac{x(x-1)\cdots(x-i+1)}{i!} .$$

Then for $t \in \mathbb{N}$,

$$R_n(t) = \sum_{i=0}^{n} \binom{t}{i} p^i d_i$$

and since $\binom{t}{i} = 0$ for $i > t$, it follows that

$$b_t = R_t(t) = R_n(t) \quad (n \ge t) . \tag{4.4}$$

Next, we show that for all $k, n \ge 0$,

$$\omega_k(R_n) \ge k \frac{p-2}{p-1} .$$

For this, let

$$R_n(x) = \sum_{k=0}^{n} c_k^{(n)} x^k .$$

Each $c_k^{(n)} x^k$ is a linear combination, with integral coefficients, of numbers $d_i \dfrac{p^i}{i!}$, for indices i with $k \leq i \leq n$. Consequently,

$$v_p(c_k^{(n)}) \geq \inf_{k \leq i \leq n} \left(v_p\left(d_i \frac{p^i}{i!} \right) \right).$$

In view of Equation (4.2), this implies

$$v_p(c_k^{(n)}) \geq \inf\left(i \, \frac{p-2}{p-1} \right) \geq k \, \frac{p-2}{p-1}$$

which in turn shows that

$$\omega_k(R_n) \geq k \, \frac{p-2}{p-1} . \tag{4.5}$$

Consider now any coefficient b_t of the sequence $(b_n)_{n \in \mathbb{N}}$. We shall see that

$$v_p(b_t) \geq k \, \frac{p-2}{p-1}$$

for any integer k, which of course shows that $b_t = 0$. For this, let $t_1 < t_2 < \cdots < t_k$ be the first k indices with $b_{t_1} = \cdots = b_{t_k} = 0$, and let $n \geq \sup(t, t_k)$. By Equation (4.4), $R_n(t_i) = b_{t_i} = 0$ for $i = 1, \ldots, k$. Thus

$$R_n(x) = (x - t_1)(x - t_2) \cdots (x - t_k) Q(x) \tag{4.6}$$

for some polynomial $Q(x)$ with integral coefficients. By Corollary 4.5, one has

$$\omega_k(R_n) \leq \omega_0(Q) . \tag{4.7}$$

Next, by Equation (4.4), $v_p(b_t) = v_p(R_n(t))$ and by Equations (4.6), (4.3) and (4.7),

$$v_p(R_n(t)) \geq v_p(Q(t)) \geq \omega_0(Q) \geq \omega_k(R_n) .$$

Thus, in view of Equation (4.5),

$$v_p(b_t) \geq k \, \frac{p-2}{p-1}$$

for all $k \geq 0$. □

Lemma 4.7 *Let $S = \sum a_n x^n \in \mathbb{Z}[[x]]$ be a strict rational series and let (λ, μ, γ) be a linear representation of S of dimension k with integral coefficients. For any odd prime p not dividing* $\det(\mu(x))$*, the annihilator* $\operatorname{ann}(S)$ *is quasi-periodic of period* $N = \operatorname{Card}(\mathrm{GL}_k(\mathbb{Z}/p\mathbb{Z}))$.

Proof. Let p be an odd prime that does not divide $\det(\mu(x))$. Let

$$n \mapsto \overline{n}$$

be the canonical morphism from $\mathbb{Z}$ onto $\mathbb{Z}/p\mathbb{Z}$. Since $\det(\overline{\mu(x)}) = \overline{\det(\mu(x))} \neq 0$, the matrix $\overline{\mu(x)}$ is invertible over $\mathbb{Z}/p\mathbb{Z}$, and setting $N = \mathrm{Card}(\mathrm{GL}_k(\mathbb{Z}/p\mathbb{Z}))$, one has

$$\overline{\mu(x^N)} = \overline{I}\,.$$

Reverting to the original matrix, this means that

$$\mu(x^N) = I + pM$$

for some matrix M with integral coefficients.

Consider now a fixed integer $j \in \{0, \ldots, N-1\}$ and set for $n \geq 0$

$$b_n = a_{j+nN}\,.$$

Then

$$b_n = \lambda\mu(x^{j+nN})\gamma = \lambda\mu(x^j)(I + pM)^n\gamma = \sum_{i=0}^{n} \binom{n}{i} p^i \lambda\mu(x^j)M^i\gamma\,.$$

Thus, setting $d_i = \lambda\mu(x^j)M^i\gamma$, one obtains

$$b_n = \sum_{i=0}^{n} \binom{n}{i} p^i d_i\,.$$

In view of Lemma 4.6, the sequence $(b_n)_{n\geq 0}$ either vanishes or contains only finitely many vanishing terms. This shows that the annihilator of S is quasi-periodic with period N. □

Proof of Proposition 4.3. Let (λ, μ, γ) be a linear representation of S with $\mu(x)$ invertible, and let q be a common multiple of the denominators of the coefficients in λ, μ and γ. Then $(q\lambda, q\mu, q\gamma)$ is a linear representation of the strict series $S' = \sum q^{n+2}a_n x^n$. Clearly $\mathrm{ann}(S) = \mathrm{ann}(S')$. By Lemma 4.7, the set $\mathrm{ann}(S')$ is quasi-periodic. Thus $\mathrm{ann}(S)$ is quasi-periodic. □

We now turn to the second part of the proof. For this, we consider the ring $\mathbb{Z}[y_1, \ldots, y_m]$ of polynomials over $\mathbb{Z}$ in commutative variables $y_1, \ldots, y_m$ and the quotient field $\mathbb{Q}(y_1, \ldots, y_m)$ of rational functions. As usual, if $P \in \mathbb{Q}(y_1, \ldots, y_m)$ and $a_1, \ldots, a_m \in \mathbb{Q}$, then $P(a_1, \ldots, a_m)$ is the value of P at that point. The result to be proved is the following.

Proposition 4.8 *Let $S = \sum a_n x^n$ be a strict rational series with coefficients in the field $\mathbb{Q}(y_1, \ldots, y_m)$. Then* $\mathrm{ann}(S)$ *is quasi-periodic.*

We start with the following well-known property of polynomials.

Lemma 4.9 *Let K be a field, and let $P \in K[y_1, \ldots, y_m]$. Let δ_i be the degree of P in the variable y_i. Assume that there exist subsets $A_1, \ldots, A_m$ of K with* $\mathrm{Card}(A_i) > \delta_i$ *for $i = 1, \ldots, m$ such that $P(a_1, \ldots, a_m) = 0$ for all $(a_1, \ldots, a_m) \in A_1 \times \cdots \times A_m$. Then $P = 0$.* □

Corollary 4.10 *Let $S = \sum a_n x^n$ be any series with coefficients in $K[y_1, \ldots, y_m]$ and let $H_1, \ldots, H_m$ be arbitrary infinite subsets of K. For each $(h_1, \ldots, h_m) \in K^m$, let*

$$S_{h_1,\ldots,h_m} = \sum a_n(h_1, \ldots, h_m)x^n .$$

Then

$$\operatorname{ann}(S) = \bigcap_{(h_1,\ldots,h_m)\in H_1\times\cdots\times H_m} \operatorname{ann}(S_{h_1,\ldots,h_m}) .$$

Proof. It follows immediately from Lemma 4.9 that $a_n = 0$ if and only if $a_n(h_1, \ldots, h_m) = 0$ for all $(h_1, \ldots, h_m) \in H_1 \times \cdots \times H_m$. □

Lemma 4.11 *Let $P \in \mathbb{Z}[y_1, \ldots, y_m]$, $P \neq 0$. For all but a finite number of prime numbers p, there exists a subset $H \subset \mathbb{Z}^m$ of the form*

$$H = (k_1, \ldots, k_m) + p\mathbb{Z}^m \tag{4.8}$$

such that for all $(h_1, \ldots, h_m) \in H$,

$$P(h_1, \ldots, h_m) \not\equiv 0 \mod p .$$

Proof. Let

$$P = \sum c_{i_1,i_2,\ldots,i_m} y_1^{i_1} y_2^{i_2} \cdots y_m^{i_m} .$$

Let δ_i be the degree of P in the variable y_i, and let p be any prime number strictly greater than the δ_i's and not dividing all the coefficients $c_{i_1,i_2,\ldots,i_m}$. Again let $n \mapsto \overline{n}$ be the morphism from $\mathbb{Z}$ onto $\mathbb{Z}/p\mathbb{Z}$. The polynomial

$$\overline{P} = \sum \overline{c}_{i_1,i_2,\ldots,i_m} y_1^{i_1} y_2^{i_2} \cdots y_m^{i_m}$$

is a nonvanishing polynomial with coefficients in $\mathbb{Z}/p\mathbb{Z}$. Since $p > \delta_i$ for $i = 1, \ldots, m$, it follows from Lemma 4.9 that there exists $(k_1, \ldots, k_m) \in \mathbb{Z}^m$ such that $\overline{P}(\overline{k}_1, \ldots, \overline{k}_m) \neq 0$. This proves the lemma. □

Proof of Proposition 4.8. Let (λ, μ, γ) be a linear representation of S of dimension k with μx invertible. As in the proof of Proposition 4.3, consider a common multiple $q \in \mathbb{Z}[y_1, \ldots, y_m]$ of the denominators of the coefficients of λ, μ and γ. Then $(q\lambda, q\mu, q\gamma)$ is a linear representation of the series $S' = \sum q^{n+2} a_n x^n$ and $\operatorname{ann}(S') = \operatorname{ann}(S)$. Thus we may suppose that the coefficients of λ, μ and γ are in $\mathbb{Z}[y_1, \ldots, y_m]$.

Let $P = \det(\mu(x)) \in \mathbb{Z}[y_1, \ldots, y_m]$. Then $P \neq 0$ and by Lemma 4.11, there exist a prime number p and an infinite set $H \subset \mathbb{Z}^n$ of the form (4.8) such that

$$\det\big(\mu(x)(h_1, \ldots, h_m)\big) \not\equiv 0 \mod p$$

for all $(h_1, \ldots, h_m) \in H$. Setting

$$S_{h_1,\ldots,h_m} = \sum_n a_n(h_1, \ldots, h_m)x^n$$

this implies, in view of Lemma 4.7, that the set $\mathrm{ann}(S_{h_1,\dots,h_m})$ is quasi-periodic, for all $(h_1,\dots,h_m) \in H$, with a period at most p^{k^2}. Thus $r = (p^{k^2})!$ is a period for all these annihilators. In view of Lemma 4.2, the set

$$\bigcap_{(h_1,\dots,h_m)\in H} \mathrm{ann}(S_{h_1,\dots,h_m})$$

is quasi-periodic. By Corollary 4.10, this intersection is the set $\mathrm{ann}(S)$. The proof is complete. □

It is convenient to introduce the following

Definition A field K is an *SML field* (Skolem–Mahler–Lech field) if K satisfies Theorem 4.1.

We have seen already that the field $\mathbb{Q}$ of rational numbers, and the field $\mathbb{Q}(y_1,\dots,y_m)$ are *SML* fields.

Proposition 4.12 *Let K and L be fields. If L is an SML field and K is a finite algebraic extension of L, then K is an SML field.*

Proof. Let $S = \sum a_n x^n$ be a rational series over K. Let k be the dimension of K over L, and let $\phi_1,\dots,\phi_k$ be L-linear functions $K \to L$ such that, for any $h \in K$

$$h = 0 \iff \phi_i(h) = 0, \quad \forall\, i = 1,\dots,k\,.$$

Define

$$S_i = \sum_n \phi_i(a_n)x^n \in L[[x]]\,.$$

Then, by the choice of the function ϕ_i, one has

$$\mathrm{ann}(S) = \bigcap_{1\le i\le k} \mathrm{ann}(S_i)\,. \tag{4.9}$$

Thus, it suffices by Lemma 4.2 to prove that the series S_i are rational over L. By Proposition 1.5.1, there exists a finite dimensional subvector space M of $K[[x]]$, containing S and which is stable, that is closed for the operation $T \mapsto T \circ x$. Since K has finite dimension over L, the space M also has finite dimension over L.

The functions ϕ_i, extended to series

$$\phi_i : K[[x]] \to L[[x]]$$

by

$$\phi_i\Big(\sum_n b_n x^n\Big) = \sum_n \phi_i(b_n)x^n$$

are L-linear. Consequently, $\phi_i(M)$ is a finite dimensional vector space over L. Since $\phi_i(T \circ x) = \phi_i(T) \circ x$, the space $\phi_i(M)$ is stable. Moreover, it contains the series $S_i = \phi_i(S)$. Thus, again by Proposition 1.5.1, each series S_i is rational over L. □

Proof of Theorem 4.1. Let S be a rational series with coefficients in K. Then by Proposition 1.6, there is a polynomial P such that $S - P$ is strict. Since $\mathrm{ann}(S - P)$ and $\mathrm{ann}(S)$ differ only by a finite set, it suffices to prove the result for $S - P$. Thus we may assume that S is strict.

Let (λ, μ, γ) be a linear representation of S, and let K' be the subfield of K generated by the set Z of coefficients of λ, $\mu(x)$, γ. Then S has coefficients in K' and we may assume that K is a finite extension of $\mathbb{Q}$, since $K' = \mathbb{Q}(Z)$.

Let Y be a maximal subset of Z that is algebraically independent over $\mathbb{Q}$. The field $\mathbb{Q}(Y)$ is isomorphic to the field of rational functions $\mathbb{Q}(y_1, \ldots, y_m)$ with $Y = \{y_1, \ldots, y_m\}$. In view of Proposition 4.8, the field $\mathbb{Q}(Y)$ is an *SML* field. Next, K is a finite algebraic extension of $\mathbb{Q}(Y)$. By Proposition 4.12, the field K is an *SML* field. This concludes the proof. □

Exercises for Chapter 6

1.1 Let $P(x) = x^d - g_1 x^{d-1} - \cdots - g_d$ be a polynomial over some commutative ring K. Its *companion matrix* is the matrix

$$M = \begin{pmatrix} 0 & 1 & 0 & \cdots & 0 \\ \vdots & & \ddots & \ddots & \vdots \\ \vdots & & & \ddots & 0 \\ 0 & \cdots & \cdots & 0 & 1 \\ g_d & g_{d-1} & \cdots & g_2 & g_1 \end{pmatrix}.$$

Show that the characteristic and minimal polynomials of M are both equal to $P(x)$. Show that if a sequence (a_n) satisfies the linear recurrence relation $a_{n+d} = g_1 a_{n+d-1} + \cdots + g_d a_n$ for all $n \geq 0$, then $a_n = \lambda M^n \gamma$, where $\lambda = (1, 0, \ldots, 0)$ and $\gamma = (a_0, \ldots, a_{d-1})^T$. (*Hint*: Let e_i be the i-th canonical basis row vector. Show that $e_1 M^{i-1} = e_i$ for $i = 1, \ldots, d$. Show that $e_1 P(M) = 0$ and then $vP(M) = 0$ for any v in K^n, knowing that e_1 generates K^n under the action of M.)

1.2 Let $M \in K^{d \times d}$ and φ be a linear function on $K^{d \times d}$. Let $a_n = \varphi(M^n)$. Show that (a_n) satisfies the linear recurrence relation associated to the characteristic polynomial of M.

1.3 Show that $\sum_{n \geq 0} a_n x^n$ is a strict rational series if and only if it is of the following form: there exists a finite dimensional algebra $\mathfrak{M}$ over K, a linear function $\varphi : \mathfrak{M} \to K$ and a homomorphism μ of algebras from the algebra of Laurent polynomials $K[x, x^{-1}]$ into $\mathfrak{M}$ such that $a_n = \varphi \circ \mu(x^n)$ for all $n \in \mathbb{N}$.

1.4 Show that the set of sequences $(a_n)_{n \geq 0}$ over a field K satisfying a given recurrence relation of length n is a vector space of dimension n closed under the shift operation $(a_n) \mapsto (a_{n+1})$. Show that the converse holds (Lidl and Niederreiter (1983), Theorem 8.5.6).

2.1 Consider the *Fibonacci sequence* (see Example 3.2.1). Find the exponential polynomial for F_n.

2.2 Consider the *Lucas sequence* defined by $L_0 = 2$, $L_1 = 1$, $L_{n+2} = L_{n+1} + L_n$. Find the exponential polynomial for L_n.

2.3 Show that if L is a commutative ring without zero divisors, the only invertible elements in the ring of Laurent polynomials $L[x_1, \ldots, x_n, x_1^{-1}, \ldots, x_n^{-1}]$ are the Laurent monomials $\alpha x_1^{d_1} \cdots x_n^{d_n}$ with α invertible in L.

2.4 Show that if K is an algebraically closed field of positive characteristic and S is a strict rational series over K, then S is the merge of series, each of which is a K-linear combination of geometric series.

2.5 Let K be a semiring and $S = \sum_{n\geq 0} a_n x^n \in K[[x]]$. Show that the following conditions are equivalent:

(i) S is rational;
(ii) S is the merge of rational series;
(iii) for some $h \in \mathbb{N}$, $\sum_{n\geq 0} a_{n+h} x^n$ is rational;
(iv) for some $h \in \mathbb{N}$, $\sum_{n\geq h} a_n x^n$ is rational.

2.6 Let $S = \sum_{n\geq 0} a_n x^n$ be a rational series in $\mathbb{C}[[x]]$ with denominator $P(x)$. Let $P = \prod_{i=1}^{d}(1 - \lambda_i x)$ and $P_k(x) = \prod_{i=1}^{d}(1 - \lambda_i^k x)$. Let $S_0, \ldots, S_{k-1}$ be the k series whose merge is S. Show that each S_i has denominator $P_k(x)$.

2.7 (a) Show that $\dfrac{1}{(1-x)^d} = \sum_{n\geq 0} \binom{n+d-1}{d-1} x^n$. (*Hint*: Use induction and derivation, starting with $\frac{1}{1-x} = \sum_{n\geq 0} x^n$.)

(b) Deduce that, for any λ, one has $\dfrac{1}{(1-\lambda x)^d} = \sum_{n\geq 0} \binom{n+d-1}{d-1} \lambda^n x^n$.

(c) Show that $\dfrac{x^d}{(1-x)^{d+1}} = \sum_{n\geq 0} \binom{n}{d} x^n$.

3.1 A *Pólya series* in $\mathbb{Q}\langle\langle A\rangle\rangle$ is a series which has only a finitely number of prime factors in the numerators and denominators of its coefficients (this extends the definition of Section 6.3 to several variables).

The *unambiguous rational operations* on series are defined as follows. A rational operation (sum, product, star) on series is unambiguous if the corresponding operation on the support (union, product, star) is unambiguous. More formally: the sum $S + T$ (resp. product ST) is unambiguous if $\mathrm{supp}(S) \cap \mathrm{supp}(T) = \emptyset$ (resp. the product $\mathrm{supp}(S)\,\mathrm{supp}(T)$ is an unambiguous product of languages); the star S^* is unambiguous if the star $\mathrm{supp}(S)^*$ is unambiguous. A rational series $S \in \mathbb{Q}\langle\langle A\rangle\rangle$ is *unambiguous* if it is obtained from polynomials using only unambiguous rational operations. (For unambiguous rational operations of languages, see Exercise 3.2.2.)

(a) Show that each unambiguous rational series is Hadamard sub-invertible (see Exercise 3.2.1 of Chapter 3).

(b) Show that each rational series in $\mathbb{Q}\langle\langle A\rangle\rangle$ which is Hadamard sub-invertible is a Pólya series.

(c) Show that a Pólya series in one variable is unambiguously rational (use Theorem 4.1).

3.2 Show that if $S \in \mathbb{Q}[[x]]$ has only finitely many prime factors, and if S is neither a polynomial nor a geometric series, then for some eigenvalues λ, μ of S, the quotient λ/μ is a root of unity $\neq 1$.

4.1 Set $B(x) = \sum_{n=0}^{\infty} b_n x^n$, $D(x) = \sum_{n=0}^{\infty} d_n x^n$ with integers b_n, d_n related as in Lemma 4.6. Show that $B(x) = \sum_{n=0}^{\infty} d_n \frac{p^n x^n}{(1-x)^{n+1}}$.

4.2 Let $S \in K[[x]]$ be a rational series, where K is a field of characteristic 0. Suppose that S is not a polynomial and has infinitely many vanishing coefficients. Show that for some eigenvalues λ, μ of S, the quotient λ/μ is a root of unity $\neq 1$.

Notes to Chapter 6

The notion of an exponential polynomial is a classical one. The formalism we use here is from Reutenauer (1982). It allows us to give an algebraic proof of Benzaghou's theorem. His proof was based on analytic techniques. The algebraic method makes it possible to prove Benzaghou's theorem in positive characteristic. Some modifications are necessary, since in that case, the exponential polynomial may not exist nor be unique. Pólya's theorem is extended to general fields by Bézivin (1984).

There are a great number of arithmetic and combinatorial properties of linear recurrence sequences. A recent book is Everest *et al.* (2003). See also Chapter 8 of Lidl and Niederreiter (1983) for linear recurrence sequences over finite fields. The use of symmetric functions to derive divisibility properties is illustrated by Duboué (1983). Lascoux (1986) gives numerous applications of expressions of the exponential polynomial by means of symmetric functions.

The proof of the Skolem–Mahler–Lech theorem given here is due to Hansel (1986). The original proofs, by Skolem (1934), Mahler (1935) and Lech (1953) depend on p-adic analysis. The specialists will recognize, in Lemma 4.6, a key property of p-adic analysis. Recently, Derksen (2007) has shown that the set of zeroes of a rational series over a field of characteristic p is a p-automatic set.

An open problem, stated by C. Pisot, is the following. Is it decidable, for a rational series $\sum a_n x^n$, whether there exists an n such that $a_n = 0$? It is decidable whether there exist infinitely many n with $a_n = 0$, see Berstel and Mignotte (1976).

The notion of Pólya series may be extended to noncommuting variables, see Exercise 3.1. The following problem remains open.

Conjecture Each rational Pólya series over $\mathbb{Q}$ is an unambiguous rational series.

7

Changing the semiring

If K is a subsemiring of a semiring L, each K-rational series is clearly L-rational. The main problem considered in this chapter is the converse: how to determine which of the L-rational series are rational over K. This leads to the study of semirings of a special type, and also shows the existence of remarkable families of rational series.

In the first section, we examine principal rings from this aspect. Fatou's Lemma is proved and the rings satisfying this lemma are characterized (Chabert's Theorem 1.5).

In the second section, Fatou extensions are introduced. We show in particular that $\mathbb{Q}_+$ is a Fatou extension of $\mathbb{N}$ (Theorem 2.2 due to Fliess).

In the third section, we apply Shirshov's theorem on rings with polynomial identities to prove criteria for rationality of series and languages. This is then applied, in the last section, to Fatou ring extensions.

7.1 Rational series over a principal ring

Let K be a commutative principal ring and let F be its quotient field. Let $S \in K\langle\langle A\rangle\rangle$ be a formal series over A with coefficients in K. If S is a rational series over F, is it also rational over K? This question admits a positive answer, and there is even a stronger result, namely that S has a minimal linear representation with coefficients in K.

Theorem 1.1 (Fliess 1974a) *Let $S \in K\langle\langle A\rangle\rangle$ be a series which is rational of rank n over F. Then S is rational over K and has a linear representation over K of dimension n. In other words, S has a minimal representation with coefficients in K.*

Proof. Let (λ, μ, γ) be a minimal linear representation of S over F. According to Corollary 2.2.3, there exist polynomials $P_1, \ldots, P_n, Q_1, \ldots, Q_n \in F\langle A\rangle$ such that for $w \in A^*$

$$\mu w = ((S, P_i w Q_j))_{1\le i,j\le n}\,.$$

Let d be an element in $K \setminus 0$ such that $dP_i, dQ_j \in K\langle A\rangle$ and $d\lambda \in K^{1\times n}$. Then for any polynomial $P \in K\langle A\rangle$

$$d^3\lambda\mu P = (d\lambda)((S, dP_i\, P\, dQ_j))_{i,j} \in K^{1\times n}\,,$$

since $(S, R) \in K$ whenever $R \in K\langle A\rangle$. Consequently,

$$\lambda\mu(K\langle A\rangle) \subset \frac{1}{d^3} K^{1\times n}.$$

This shows that $\lambda\mu(K\langle A\rangle)$ is a submodule of a free K-module of rank n, hence is also free and has rank $\leq n$. It suffices now to apply Lemma 2.1.3: we obtain a representation of S over K of dimension $\leq n$, thus of dimension n by Theorem 2.1.6. □

In particular, a series which is rational over $\mathbb{Q}$ and with coefficients in $\mathbb{Z}$ has a minimal representation with coefficients in $\mathbb{Z}$. The theorem admits the following corollary, known as *Fatou's Lemma*.

Corollary 1.2 (Fatou 1904) *Let $P(x)/Q(x) \in \mathbb{Q}(x)$ be an irreducible rational function such that the constant term of Q is 1. If the coefficients of its series expansion are integers, then P and Q have integral coefficients.*

Proof. We have $Q(0) = 1$. Then $S = \sum a_n x^n = P(x)/Q(x)$ is a rational series. Let (λ, μ, γ) be a minimal linear representation of S. Since $\mathbb{Z}$ is principal, this representation is similar, by Theorem 1.1 and Theorem 2.2.4, to a representation over $\mathbb{Z}$. In particular, the characteristic polynomial of $\mu(x)$ has integral coefficients. Now, $Q(x)$ is the reciprocal polynomial of this polynomial (Corollary 6.1.5). Thus $Q(x)$ has integral coefficients, and so does $P = SQ$. □

The previous result holds for rings other than the ring $\mathbb{Z}$ of integers. We shall characterize these rings completely.

Let K be a *commutative integral domain* and let F be its quotient field. Let $\mathfrak{M}$ be an F-algebra. An element $m \in \mathfrak{M}$ is *quasi-integral* over K if $K[m]$ is contained in some finitely generated K-submodule of $\mathfrak{M}$. It is easy to see that, in this case, the K-submodule may be chosen to lie in $F[m]$, see Exercise 1.3. Hence this definition is intrinsic.

Proposition 1.3 *An element $m \in F$ is quasi-integral over K if and only if there exists $d \in K \setminus 0$ such that $dm^n \in K$ for all $n \in \mathbb{N}$.*

Proof. If the last condition holds, then $m^n \in d^{-1}K$ for all $n \in \mathbb{N}$ and therefore $K[m] \subset d^{-1}K$, which is the K-module spanned by d^{-1}. Conversely, if M is a finitely generated K-submodule of F containing $K[m]$, then $dM \subset K$ for some $d \in K \setminus 0$. Thus $dK[m] \subset K$, which implies the last condition. □

Corollary 1.4 *If $\mathfrak{M}$ is a commutative F-algebra, then the set of elements of $\mathfrak{M}$ which are quasi-integral over K is a subring of $\mathfrak{M}$.*

Proof. This follows from the fact that if M_1, M_2 are finitely generated K-submodules of $\mathfrak{M}$, then so are $M_1 + M_2$ and $M_1 M_2$. □

Definition The domain K is called *completely integrally closed* if any m in F which is quasi-integral over K is already in K.

Recall that an element m of $\mathfrak{M}$ is called *integral* if there are elements $a_1, \ldots, a_k$ in K such that

$$m^k = a_1 m^{k-1} + \cdots + a_{k-1} m + a_k \,.$$

In other words, the K-subalgebra of $\mathfrak{M}$ generated by m is a finitely generated K-module. Observe that an element in F which is integral over K is also quasi-integral over K. Thus, if K is completely integrally closed, it is integrally closed.

Theorem 1.5 (Chabert 1972) *The following conditions are equivalent:*

(i) *the domain K is completely integrally closed;*
(ii) *for any irreducible rational function $P(x)/Q(x) \in F(x)$ whose series expansion has coefficients in K, and such that the constant term of Q is* 1*, both P and Q have coefficients in K.*

We use the following lemma.

Lemma 1.6 *Let m be a matrix in $F^{n\times n}$ which is quasi-integral over K. Then the coefficients of the characteristic and of the minimal polynomials of m are quasi-integral over K.*

Proof. Let $P(t) = t^n + a_1 t^{n-1} + \cdots + a_n \in F[t]$ be the characteristic polynomial of m. Since m is quasi-integral over K, there exists a finitely generated K-submodule of $F^{n\times n}$ containing all powers of m. Thus there exists some $d \in K \setminus 0$ such that

$$dm^k \in K^{n\times n}$$

for all $k \in \mathbb{N}$. Consequently, since a_i is a $\mathbb{Z}$-linear combination of products of i entries of m,

$$da_1, d^2 a_2, \ldots, d^n a_n \in K \,.$$

Let λ be an eigenvalue of m. Then $d\lambda$ is integral over K. Indeed, $0 = d^n P(\lambda) = (d\lambda)^n + da_1 (d\lambda)^{n-1} + \cdots + d^n a_n$. Consequently, the K-algebra $L = K[d\lambda]$ is a finitely generated K-module. The element λ is in the quotient field E of L, and there exists $q \in \mathrm{GL}_n(E)$ such that

$$m' = q^{-1} m q = \begin{pmatrix} \lambda & * & \cdots & * \\ 0 & * & \cdots & * \\ \vdots & & & \vdots \\ 0 & * & \cdots & * \end{pmatrix}.$$

Let $d' \in L \setminus 0$ be a common denominator of the coefficients of q and q^{-1}, that is, such that $d'q$ and $d'q^{-1}$ have coefficients in L. Then for all $k \in \mathbb{N}$

$$(d'^2 d) m'^k = (d' q^{-1}) dm^k (d' q) \in L^{n\times n} \,.$$

Thus $(d'^2 d)\lambda^k \in L$, whence $K[\lambda] \subset (d'^2 d)^{-1} L$. This shows that λ is quasi-integral over K.

Since all eigenvalues of m are quasi-integral, the same holds for the coefficients a_i by Corollary 1.4. Similarily, it holds for the coefficients of the minimal polynomial of m. □

Proof of Theorem 1.5. Assume that K is completely integrally closed. Let $P(x)/Q(x)$ be a function satisfying the hypotheses of (ii). We have $Q(0) = 1$. The series

$$S = \sum a_n x^n = P(x)/Q(x)$$

is F-rational and has coefficients in K. Let (λ, μ, γ) be a minimal linear representation of S. By Corollary 2.2.3, the matrix $\mu(x)$ is quasi-integral over K. In view of Lemma 1.6 and (i), the characteristic polynomial of $\mu(x)$ has coefficients in K, and since Q is its reciprocal polynomial (Corollary 6.1.5), the polynomial Q has coefficients in K, and the same holds for $P = SQ$.

Assume conversely that (ii) holds. Let $m \in F$ be quasi-integral over K. Then there exists $d \in K \setminus 0$ such that

$$dm^n \in K$$

for all $n \in \mathbb{N}$. Set $P(x) = d$, $Q(x) = 1 - mx$. Then

$$P(x)/Q(x) = d \sum m^n x^n \in K[[x]] .$$

Thus by hypothesis $Q(x) \in K[x]$, whence $m \in K$. This shows that K is completely integrally closed. □

7.2 Fatou extensions

According to Fatou's Lemma (Corollary 1.2) any rational series in $\mathbb{Q}[[x]]$ with integral coefficients is rational in $\mathbb{Z}[[x]]$. The same result holds for an arbitrary alphabet A, by Theorem 1.1. This leads to the following definition.

Definition Let $K \subset L$ be two semirings. Then L is a *Fatou extension* of K if every L-rational series with coefficients in K is K-rational.

Theorem 2.1 (Fliess 1974a) *If $K \subset L$ are fields, then L is a Fatou extension of K.*

Proof. This follows immediately from the expression of rationality by means of the rank of the Hankel matrix (Theorem 2.1.6). □

Theorem 2.2 (Fliess 1975) *The semiring $\mathbb{Q}_+$ is a Fatou extension of $\mathbb{N}$.*

We need some preliminary lemmas.

Lemma 2.3 (Eilenberg and Schützenberger 1969) *The intersection of two finitely generated submonoids of an Abelian group is still a finitely generated submonoid.*

Proof. Let M_1 and M_2 be two finitely generated submonoids of an Abelian group G, with law denoted by $+$. There exist integers k_1, k_2 and surjective monoid morphisms $\phi_i : \mathbb{N}^{k_i} \to M_i, i = 1, 2$. Let $k = k_1 + k_2$ and let S be the submonoid of $\mathbb{N}^k = \mathbb{N}^{k_1} \times \mathbb{N}^{k_2}$ defined by

$$S = \{x = (x_1, x_2) \in \mathbb{N}^k \mid \phi_1 x_1 = \phi_2 x_2\}\,.$$

Let $p_1 : \mathbb{N}^k \to \mathbb{N}^{k_1}$ be the projection. Then

$$M_1 \cap M_2 = \phi_1 \circ p_1(S)\,.$$

Thus it suffices to prove that S is finitely generated. Observe that S satisfies the following condition

$$x, x + y \in S \implies y \in S\,. \tag{2.1}$$

Indeed, since $\phi_1 x_1 = \phi_2 x_2$ and $\phi_1 x_1 + \phi_1 y_1 = \phi_2 x_2 + \phi_2 y_2$ and since all these elements are in G, it follows that $\phi_1 y_1 = \phi_2 y_2$, whence $y \in S$.

Let X be the set of minimal elements of $S \setminus 0$ (for the natural ordering of $\mathbb{N}^k$). For all $z \in S$, there is $x \in X$ such that $x \leq z$. Thus $z = x + y$ for some $y \in \mathbb{N}^k$ and by Equation (2.1), $y \in S$. This shows by induction that X generates S. In view of the following well-known lemma, the set X is finite, since the elements in X are mutually incomparable. □

Lemma 2.4 *Every infinite sequence in $\mathbb{N}^k$ contains an infinite increasing subsequence.*

Proof. By induction on k. Let (u_n) be a sequence of elements of $\mathbb{N}^k$. If $k = 1$, either the sequence is bounded, and one can extract a constant sequence, or it is unbounded, and one can extract a strictly increasing subsequence. For $k > 1$, one first extracts a sequence that is increasing in the first coordinate, and then uses induction for this subsequence. □

Lemma 2.5 (Eilenberg and Schützenberger 1969) *Let I be a set and let M be a finitely generated submonoid of $\mathbb{N}^I$. Then the submonoid M' of $\mathbb{N}^I$ given by*

$$M' = \{x \in \mathbb{N}^I \mid \exists n \geq 1, nx \in M\}$$

is finitely generated.

Proof. Let $x_1, \ldots, x_p$ be generators of M. Let

$$C = \left\{x \in \mathbb{N}^I \mid \exists \lambda_1, \ldots, \lambda_p \in \mathbb{Q}_+ \cap [0, 1] : x = \sum \lambda_i x_i\right\}\,.$$

Then C contains each x_i and is a set of generators for M'. Indeed, if $nx = \sum \lambda_i x_i \in M$ for some $n \geq 1$ and some $\lambda_i \in \mathbb{N}$, then

$$x = \sum \left\lfloor \frac{\lambda_i}{n} \right\rfloor x_i + \sum \left(\frac{\lambda_i}{n} - \left\lfloor \frac{\lambda_i}{n} \right\rfloor \right) x_i \,,$$

where $\lfloor z \rfloor$ is the integral part of z. Thus, it suffices to show that C is finite.

Let E be the subvector space of $\mathbb{R}^I$ generated by M'. Since E has finite dimension, there exists a finite subset J of I such that the $\mathbb{R}$-linear function

$$p_J : E \to \mathbb{R}^J$$

(p_J is the projection $\mathbb{R}^I \to \mathbb{R}^J$) is injective. The image of C by p_J is contained in $\mathbb{N}^J$, and it is also contained in the set

$$K = \left\{ y \in \mathbb{R}^J \mid \exists \lambda_1, \ldots, \lambda_p \in [0, 1] : y = \sum \lambda_i y_i \right\} ,$$

where $y_i = p_J(x_i)$. Now K is compact and $\mathbb{N}^J$ is discrete and closed. Thus $K \cap \mathbb{N}^J$ is finite. It follows that C is finite. □

Proof of Theorem 2.2. Let S be a $\mathbb{Q}_+$-rational series with coefficients in $\mathbb{N}$. We use systematically Proposition 1.5.1. There exists a finitely generated stable $\mathbb{Q}_+$-submodule in $\mathbb{Q}_+\langle\langle A \rangle\rangle$ that contains S. Denote it by $M_{\mathbb{Q}_+}$. Similarly, the series S is $\mathbb{Q}$-rational with coefficients in $\mathbb{Z}$, and therefore S is $\mathbb{Z}$-rational. Thus, there is a finitely generated $\mathbb{Z}$-submodule in $\mathbb{Z}\langle\langle A \rangle\rangle$ that contains S, say $M_\mathbb{Z}$. Then $M = M_{\mathbb{Q}_+} \cap M_\mathbb{Z}$ is a stable $\mathbb{N}$-submodule of $\mathbb{N}\langle\langle A \rangle\rangle$ containing S, and it suffices to show that M is finitely generated.

Let $T_1, \ldots, T_r$ be series in $M_{\mathbb{Q}_+}$ generating it as a $\mathbb{Q}_+$-module, and let

$$M'_{\mathbb{Q}_+} = \sum \mathbb{N} T_i \,.$$

This is a finitely generated $\mathbb{N}$-module. Since $M_\mathbb{Z}$ is also a finitely generated $\mathbb{N}$-module, the $\mathbb{N}$-module

$$M' = M_\mathbb{Z} \cap M'_{\mathbb{Q}_+} \subset \mathbb{N}\langle\langle A \rangle\rangle$$

is finitely generated (this follows from Lemma 2.3, noting that $\mathbb{N}$-module = commutative monoid). Consequently,

$$\overline{M} = \{T \in \mathbb{N}\langle\langle A \rangle\rangle \mid \exists n \geq 1, nT \in M'\}$$

is, in view of Lemma 2.5, a finitely generated $\mathbb{N}$-module. Finally, the $\mathbb{N}$-module $\overline{M} \cap M_\mathbb{Z}$ is finitely generated by Lemma 2.3. We claim that $M = \overline{M} \cap M_\mathbb{Z}$, which implies the theorem.

In order to prove the claim, let T be in M. Then $T \in M_{\mathbb{Q}_+} \cap M_\mathbb{Z}$. Thus $T = \sum_{i=1}^r \alpha_i T_i$ with $\alpha_i \in \mathbb{Q}_+$. It follows that, for some $n \geq 1$, $nT \in M'_{\mathbb{Q}_+}$ and since T is also in $M_\mathbb{Z}$, $nT \in M'_{\mathbb{Q}_+} \cap M_\mathbb{Z} = M'$. Consequently, $T \in \overline{M}$ and finally $T \in \overline{M} \cap M_\mathbb{Z}$.

Conversely, let $T \in \overline{M} \cap M_\mathbb{Z}$. Since $M' \subset M_{\mathbb{Q}_+}$, we have $\overline{M} \subset M_{\mathbb{Q}_+}$ and we see that $T \in M_{\mathbb{Q}_+} \cap M_\mathbb{Z} = M$. □

We now give two examples of extensions which are not Fatou extensions.

Example 2.1 *The ring* $\mathbb{Z}$ *is not a Fatou extension of the semiring* $\mathbb{N}$. Consider the series

$$S = \sum_{w \in \{a,b\}^*} (|w|_a - |w|_b)^2 w \,.$$

This series is $\mathbb{Z}$-rational (it is the Hadamard square of the series considered in Example 3.4.1) and has coefficients in $\mathbb{N}$. However, it is not $\mathbb{N}$-rational, since otherwise its support would be a rational language (Lemma 3.1.4), and also the complement of its support. In Example 3.4.1, it was shown that this set is not the support of any rational series.

Example 2.2 *The semiring* $\mathbb{R}_+$ *is not a Fatou extension of* $\mathbb{Q}_+$. Let $\alpha = (1 + \sqrt{5})/2$ be the golden ratio and let S be the series

$$S = \sum_{w \in \{a,b\}^*} (\alpha^{2(|w|_a - |w|_b)} + \alpha^{-2(|w|_a - |w|_b)}) w \,.$$

Since $S = (\alpha^2 a + \alpha^{-2} b)^* + (\alpha^{-2} a + \alpha^2 b)^*$, the series S is $\mathbb{R}_+$-rational. Moreover, since α is an algebraic integer over $\mathbb{Z}$ and $-1/\alpha$ is its conjugate, one has for all $n \in \mathbb{N}$

$$\alpha^{2n} + \alpha^{-2n} \in \mathbb{Z} \,.$$

Consequently, S has coefficients in $\mathbb{N}$. Assume that S is $\mathbb{Q}_+$-rational. Then by Theorem 2.2, it is $\mathbb{N}$-rational. However, the language $S^{-1}(2) = \{w \mid (S, w) = 2\}$ is

$$S^{-1}(2) = \{w \in \{a, b\}^* \mid |w|_a = |w|_b\}$$

since $x + 1/x > 2$ for all $x > 0, x \neq 1$. Since the language $S^{-1}(2)$ is not rational, the series S is not $\mathbb{N}$-rational (Corollary 3.2.7). Thus S is not $\mathbb{Q}_+$-rational.

To end this section, we prove the the following result about series with nonnegative coefficients.

Theorem 2.6 (Schützenberger 1970) *If* $S \in \mathbb{N}\langle\langle A \rangle\rangle$ *is an* $\mathbb{N}$*-rational series, then*

$$S - \underline{\mathrm{supp}(S)} \in \mathbb{N}\langle\langle A \rangle\rangle$$

is $\mathbb{N}$*-rational.*

Recall that $\underline{L}$ is the characteristic series of the language L.

Proof. We follow Salomaa and Soittola (1978, page 51). In view of Proposition 1.5.1, there exist rational series $S_1, \ldots, S_n$ such that the $\mathbb{N}$-submodule of $\mathbb{N}\langle\langle A \rangle\rangle$ they generate is stable and contains S. By Lemma 3.1.4, the supports $\mathrm{supp}(S_1), \ldots, \mathrm{supp}(S_n)$ are rational languages. Let $\mathbf{L}$ be the family of languages obtained by taking all intersections of $\mathrm{supp}(S_1), \ldots, \mathrm{supp}(S_n)$. Then $\mathbf{L}$ is a finite set of rational languages. The set $\mathbf{L}' = \{u^{-1}L \mid u \in A^*, L \in \mathbf{L}\}$ is also a finite set of rational languages (Corollary 3.2.8). Let $\mathbf{T}$ be the set of characteristic series of the languages in $\mathbf{L}'$.

Let M be the finitely generated $\mathbb{N}$-submodule of $\mathbb{N}\langle\langle A\rangle\rangle$ generated by $\mathbf{T}$ and by the series

$$S_i' = S_i - \underline{\operatorname{supp}(S_i)}$$

for $i = 1, \ldots, n$. We claim that if $a_j \in \mathbb{N}$ and $T = \sum a_j S_j$, then $T - \underline{\operatorname{supp}(T)}$ is in M.

Indeed, $S_j = S_j' + \underline{\operatorname{supp}(S_j)}$, thus $T = \sum a_j S_j' + U$, where $U = \sum a_j \underline{\operatorname{supp}(S_j)}$. Note that $\operatorname{supp}(S_j') \subset \operatorname{supp}(S_j)$, hence $\operatorname{supp}(T) = \operatorname{supp}(U)$. We may write $U = \sum b_k T_k$ where each integer b_k is ≥ 1 and the $T_k \in \mathbf{T}$ have disjoint supports. This is done by keeping only the j's with $a_j \geq 1$ and by making the necessary intersections of supports. Hence $U - \underline{\operatorname{supp}(U)} = \sum(b_k - 1)T_k \in M$ and $T - \underline{\operatorname{supp}(T)} = \sum a_j S_j' + U - \underline{\operatorname{supp}(U)} \in M$.

Since S is an $\mathbb{N}$-linear combination of the S_j, the series $S - \underline{\operatorname{supp}(S)}$ is in M by the claim. We show that M is stable, which will end the proof by Proposition 1.5.1. Indeed, let $u \in A^*$. Then $u^{-1}T \in \mathbf{T}$ by construction, hence is in M, for any T in $\mathbf{T}$. Consider $u^{-1}S_i' = u^{-1}S_i - \underline{\operatorname{supp}(u^{-1}S_i)}$. Since $u^{-1}S_i$ is an $\mathbb{N}$-linear combination of the S_j, we deduce by the claim that $u^{-1}S_i'$ is in M. □

7.3 Polynomial identities and rationality criteria

Let K be a commutative ring and let $\mathfrak{M}$ be a K-algebra. Recall that $\mathfrak{M}$ satisfies a *polynomial identity* if for some set X of noncommuting variables and some nonzero polynomial $P(x_1, \ldots, x_k) \in K\langle X\rangle$, one has

$$\forall m_1, \ldots, m_k \in \mathfrak{M}, \quad P(m_1, \ldots, m_k) = 0\,.$$

The *degree* of the identity is $\deg(P)$. The identity is called *admissible* if the support of P contains some word of length $\deg(P)$ whose coefficient is invertible in K.

Classical examples of polynomial identities are the following ones. Let

$$S_k(x_1, \ldots, x_k) = \sum_{\sigma \in \mathfrak{S}_k} (-1)^\sigma x_{\sigma 1} x_{\sigma 2} \cdots x_{\sigma k}$$

where $\mathfrak{S}_k$ denotes the set of permutations of $\{1, \ldots, k\}$ and $(-1)^\sigma$ is the signature of the permutation σ. Then, if $\mathfrak{M}$ is a K-module spanned by $k-1$ generators, it satisfies the admissible polynomial identity $S_k = 0$, see Exercise 3.1.

There is another interesting case: suppose that $\mathfrak{M} = K^{n\times n}$. Then, by the previous remark, $\mathfrak{M}$ satisfies the identity $S_{n^2+1} = 0$. Actually, according to the theorem of Amitsur–Levitzki, $K^{n\times n}$ satisfies the identity $S_{2n} = 0$, see Procesi (1973), Rowen (1980) or Drensky (2000).

Theorem 3.1 (Shirshov) *Let $\mathfrak{M}$ be a K-algebra satisfying an admissible polynomial identity of degree n. Suppose that $\mathfrak{M}$ is generated as K-algebra by a finite set E. If each element of $\mathfrak{M}$ which is a product of at most $n-1$ elements taken in E is integral over K, then $\mathfrak{M}$ is a finitely generated K-module.* □

For a proof, see Rowen (1980), Lothaire (1983) or Drensky (2000).

A *ray* is a subset of A^* of the form uw^*v for some words u, v, w; the word w is the *pattern* of the ray. Given a ray $R = uw^*v$ and a series S, we define the one variable series $S(R) = \sum_{n\geq 0}(S, uw^n v)x^n$.

Theorem 3.2 *Let K be a commutative ring and let $S \in K\langle\langle A\rangle\rangle$. Then S is rational if and only if there exists an integer $d \geq 1$ such that the syntactic algebra of S satisfies an admissible polynomial identity of degree d, and moreover, for any word w of length $< d$, the series $S(R)$, for all rays R with pattern w, satisfy a common linear recurrence relation.*

Proof. Suppose that S is rational. Then by Theorem 2.1.2 its syntactic algebra is a finitely generated K-module, hence it satisfies an identity of the form $S_d = 0$, which is clearly admissible. Moreover, let R be a ray with pattern w and let (λ, μ, γ) be a linear representation of S. Then the series $S(R)$ satisfies the linear recurrence associated to the characteristic polynomial $x^\ell + a_1 x^{\ell-1} + \cdots + a_\ell$ of the matrix μw; indeed the Cayley–Hamilton theorem implies that $\mu w^\ell + a_1 \mu w^{\ell-1} + \cdots + a_\ell = 0$, hence multiplying by $\lambda\mu u \mu w^n$ on the left and by $\mu v \gamma$ on the right we obtain $(S, uw^{n+\ell}v) + a_1(S, uw^{n+\ell-1}v) + \cdots + a_\ell(S, uw^n v) = 0$, which shows that $S(R)$ satisfies the indicated recurrence relation.

Conversely, consider the algebra morphism $\mu : K\langle A\rangle \to \mathfrak{M}$ onto the syntactic algebra $\mathfrak{M}$ of the series S. Then $\mathfrak{M}$ is generated as algebra by the set $\mu(A)$. Let w be a word of length $< d$. By hypothesis, each of the series $S(R) = \sum_{n\geq 0}(S, uw^n v)x^n$, for $u, v \in A^*$, satisfies the same linear recurrence of the form

$$(S, uw^{n+\ell}v) + a_1(S, uw^{n+\ell-1}v) + \cdots + a_\ell(S, uw^n v), \quad n \geq 0,$$

where the coefficients $a_1, \ldots, a_\ell$ depend only on w and not on u, v. This implies that

$$(S, u(w^\ell + a_1 w^{\ell-1} + \cdots + a_\ell)v) = 0$$

for any words u, v. Consequently, by Lemma 2.1.1, $w^\ell + a_1 w^{\ell-1} + \cdots + a_\ell$ is in the syntactic ideal of S. Since the latter is the kernel of μ, we obtain

$$\mu(w)^\ell + a_1\mu(w)^{\ell-1} + \cdots + a_\ell = 0.$$

Thus $\mu(w)$ is integral over K, and $\mathfrak{M}$ is a finitely generated K-module by Shirshov's theorem. Hence S is rational by Theorem 2.1.2. □

This result allows us to establish a rationality criterion for languages.

We say that an element m of a monoid M is *torsion* if m generates a finite submonoid of M; equivalently, $m^k = m^\ell$ for some $1 \leq k < \ell$. We say that M is a *torsion monoid* if each element in M is torsion.

Theorem 3.3 *A language is rational if and only if its syntactic algebra satisfies an admissible polynomial identity and its syntactic monoid is torsion.*

Proof. The necessity of the condition follows from Propositions 3.2.1, 3.3.1 and Theorem 3.2. Conversely, by Theorem 3.2.10, it suffices to show that the characteristic series of the language is a rational series. Now, by Proposition 3.3.2, the syntactic monoid of the language is a multiplicative submonoid of its syntactic algebra and generates the latter as algebra. Since each element m of the monoid satisfies an equation of the form $m^k = m^\ell$ with $k \neq \ell$, the element m is integral over K and the theorem of Shirshov applies: the syntactic algebra is a finitely generated K-module and the series is rational by Theorem 2.1.2. □

A variant of the previous criterion is given by the next result. Before stating it, we introduce a notation. If $x, u_1, \ldots, u_n, y$ are words and σ is a permutation in $\mathfrak{S}_n$, we denote by $xu_\sigma y$ the word $xu_{\sigma 1}u_{\sigma 2}\cdots u_{\sigma n}y$.

Corollary 3.4 *A language L is rational if and only if its syntactic monoid is torsion and if for some $n \geq 2$ and any words $x, u_1, \ldots, u_n, y$, the following condition holds: the number of even permutations σ such that $xu_\sigma y \in L$ is equal to the number of odd permutations σ such that $xu_\sigma y \in L$.*

Proof. Let $\mathfrak{M}$ be the syntactic algebra of the characteristic series of L. We show that the last condition in the statement means that $\mathfrak{M}$ satisfies the polynomial identity $S_n = 0$. Indeed, since S_n is multilinear, it is enough to show that this condition is equivalent to

$$S_n(m_1, \ldots, m_n) = 0 \tag{3.1}$$

for any choice of $m_1, \ldots, m_n$ in some set spanning $\mathfrak{M}$ as a K-module. For this set we take $\mu(A^*)$, where $\mu : K\langle A\rangle \to \mathfrak{M}$ is the natural algebra morphism. Then (3.1) is equivalent to the fact that $S_n(u_1, \ldots, u_n) \in I$ for any words $u_1, \ldots, u_n$ in A^*, where I denotes the syntactic ideal of $\underline{L}$, since $I = \mathrm{Ker}\mu$. By Lemma 2.1.1, this is equivalent to $(\underline{L}, xS_n(u_1, \ldots, u_n)y) = 0$ for all $x, y \in A^*$. The latter equality may be written as

$$\sum_{\sigma \text{ even}} (\underline{L}, xu_\sigma y) = \sum_{\sigma \text{ odd}} (\underline{L}, xu_\sigma y),$$

which is exactly the last condition of the statement.

In order to conclude we apply Theorem 3.3, knowing that if L is rational, then $\mathfrak{M}$ satisfies an identity of the form $S_n = 0$. □

7.4 Fatou ring extensions

Let L be a commutative integral domain, let K be a subring of L, and let G, F be their respective field of fractions, so that we have the embeddings

$$\begin{array}{ccc} K & \hookrightarrow & L \\ \downarrow & & \downarrow \\ F & \hookrightarrow & G \end{array}$$

Theorem 4.1 *L is a Fatou extension of K if and only if each element of F which is integral over L and quasi-integral over K, is integral over K.*

A *weak Fatou ring* is a commutative integral domain with field of fractions F such that F is a Fatou extension of K.

Corollary 4.2 *K is a weak Fatou ring if and only if each element of F which is quasi-integral over K is integral over K.*

Proof. Replace L by F in the theorem and observe that an element of F is always integral over F. □

Corollary 4.3 *Each Noetherian commutative integral domain is a weak Fatou ring.*

Proof. See Exercise 4.1. □

Corollary 4.4 *Each completely integrally closed commutative integral domain is a weak Fatou ring.*

Proof of Theorem 4.1. 1. Suppose that L is a Fatou extension of K. Let $m \in F$ be quasi-integral over K and integral over L. By Proposition 1.3, there exists $d \in K \setminus 0$ such that $dm^n \in K$ for any $n \in \mathbb{N}$. Moreover, for some $\ell_1, \ldots, \ell_d \in L$, one has $m^d = \ell_1 m^{d-1} + \cdots + \ell_d$. Let $S = \sum_{n \geq 0} dm^n x^n \in K[[x]]$ and $Q(x) = 1 - \ell_1 x - \cdots - \ell_d x^d \in L[x]$. Then QS is in $L[x]$, hence S is an L-rational series. Since it has coefficients in K, by assumption it is a K-rational series. Consequently, for some matrix M over K and some row and column vectors λ, γ, one has $dm^n = \lambda M^n \gamma$ for all $n \geq 0$. It follows that the sequence dm^n satisfies the linear recurrence relation associated to the characteristic polynomial of M. Hence, dividing by d, we see that m is integral over K.

2. Conversely, suppose that each element of F which is integral over L and quasi-integral over K is integral over K. Let $S \in K\langle\langle A \rangle\rangle$ be a series which is rational over L. We show that S is rational over K. For this, we will show, using Shirshov's theorem, that the syntactic algebra of S over K is a finitely generated K-module. The claim follows in view of Theorem 2.1.2.

Clearly, the series S is G-rational with coefficients in F, hence it is F-rational by Theorem 2.1. Let (λ, μ, γ) be a minimal linear representation of S over F. Then the algebra $\mu(F\langle A \rangle)$ satisfies a polynomial identity of the form $S_k = 0$, with coefficients $1, -1$, hence admissible (see Section 7.3). The same is true for the subring $\mu(K\langle A \rangle)$. We claim that this latter ring is the syntactic algebra $\mathfrak{M}$ over K of S. Indeed, the kernel of μ, viewed as a morphism $F\langle A \rangle \to F^{r \times r}$, is by Corollary 2.2.2 and Lemma 2.1.1, equal to

$$\{P \in F\langle A \rangle \mid \forall u, v \in A^*,\ (S, uPv) = 0\}.$$

Hence the kernel of $\mu | K\langle A \rangle$ is, by the same lemma, equal to the syntactic algebra of S over K, which proves the claim.

Consequently $\mathfrak{M}$ satisfies an admissible polynomial identity. It is generated, as K-algebra, by the finite set $\mu(A)$. In view of Shirshov's theorem, it suffices to show that each $m \in \mathfrak{M}$ is integral over K. For this, let $R(x) \in F[x]$ be the minimal polynomial of m over F. We show below that the coefficients of R are quasi-integral over K and integral over L. This will imply, in view of the hypothesis, that they are integral over K. Hence m is integral over K.

Since $m \in \mathfrak{M} = \mu(K\langle A\rangle)$, we may write $m = \mu(P)$ for some $P \in K\langle A\rangle$.

(i) Note that r is the rank of S over F. By Corollary 2.2.3, there is a common denominator $d \in K \setminus 0$ to all matrices μw, for $w \in A^*$, hence also for all matrices $m^n = \mu(P^n)$, since $P \in K\langle A\rangle$. This shows that $m^n \in d^{-1}K^{r\times r}$ which is a finitely generated K-module; hence m is quasi-integral over K. Thus its minimal polynomial has quasi-integral coefficients by Lemma 1.6.

(ii) Since S has the same rank over F and over G, the linear representation (λ, μ, γ) is minimal also over G (Theorem 2.1.6). By the same technique as above, we see that $\mu(L\langle A\rangle)$ is the syntactic algebra of S over L. Hence it is a finitely generated L-module by Theorem 2.1.2, since S is L-rational. In particular, each element of $\mu(L\langle A\rangle)$ is integral over L. This holds in particular for the element $m \in \mu(K\langle A\rangle) \subset \mu(L\langle A\rangle)$. Therefore, we have $m^s + \ell_1 m^{s-1} + \cdots + \ell_s = 0$ for some $\ell_i \in L$. Since G is the field of fractions of L, the minimal polynomial of m over G divides $x^s + \ell_1 x^{s-1} + \cdots + \ell_s$, thus the roots of this minimal polynomial are integral over L and so are its coefficients. Since m is a matrix over F, the minimal polynomial $R(x)$ of m over F is equal to the minimal polynomial over the field extension G. Hence the coefficients of R are integral over L. □

Exercises for Chapter 7

1.1 Show that each factorial ring is completely integrally closed.

1.2 Let K be an integral domain and F its field of fractions. Show that if an element of F is integral over K, then it is quasi-integral over K.

Deduce that if K is completely integrally closed, then it is integrally closed.

1.3 Let K be a commutative integral domain and let F be its quotient field. Let $\mathfrak{M}$ be an F-algebra, $m \in \mathfrak{M}$ and let M be a finitely generated K-submodule of $\mathfrak{M}$ such that $K[m] \subset M$. Let FM be the finite dimensional F-vector subspace of $\mathfrak{M}$ spanned by M and let $p : FM \to F[m]$ be an F-linear projection. Show that $p(M)$ is a finitely generated K-submodule of $F[\mathfrak{M}]$ which contains $K[m]$.

2.1 Show that for any rational series $S \in K\langle\langle A\rangle\rangle$, where K is a field, the subfield generated by its coefficients is a finitely generated field.

2.2 Show that if K is a subsemiring of L such that each element in L is a right-linear combination of fixed elements $\ell_1, \ldots, \ell_p$ in L, then each L-rational series may be written $\sum_{i=1}^{p} \ell_i S_i$ for some K-rational series S_i (see Lemma 2.1.3 and Exercise 2.1.6).

2.3 Show that each $\mathbb{Z}$-rational series is the difference of two $\mathbb{N}$-rational series (use Exercise 2.2).

2.4 Show that under the hypothesis of Exercise 2.2, if ϕ is a right K-linear mapping $L \to K$, then for each L-rational series S, the series $\phi(S) = \sum_w \phi((S, w))w$ is K-rational.

2.5 Show that for any semiring K, if S is $K^{n \times n}$-rational, then $S_{i,j} = \sum_{i,j} S(w)_{i,j}$ is K-rational for fixed i, j in $\{1, \ldots, n\}$ (use Exercise 2.4).

3.1 (a) Let $P = \sum_{\sigma \in \mathfrak{S}_k} a_\sigma x_{\sigma 1} x_{\sigma 2} \cdots x_{\sigma k} \in K\langle X \rangle$. Show that the K-algebra $\mathfrak{M}$ satisfies the polynomial identity $P = 0$ if and only if $P(m_1, \ldots, m_k) = 0$ for each choice of $m_1, \ldots, m_k$ in some set spanning $\mathfrak{M}$ as a K-module.

(b) Show that $S_k(m_1, \ldots, m_k) = 0$ if two of the m_i's are equal.

(c) Deduce that if $\mathfrak{M}$ is spanned as K-module by $k-1$ elements, then $S_k = 0$ is a polynomial identity of $\mathfrak{M}$.

3.2 Show that a commutative algebra satisfies a polynomial identity. Prove Shirshov's theorem directly in this case.

3.3 If an algebra $\mathfrak{M}$ satisfies an admissible polynomial identity, it satisfies a multilinear one, of the form

$$m_1 m_2 \cdots m_n = \sum_{\substack{\sigma \in \mathfrak{S}_n \\ \sigma \neq \mathrm{id}}} a_\sigma m_{\sigma 1} m_{\sigma 2} \cdots m_{\sigma n}, \quad \forall m_1, \ldots, m_n \in \mathfrak{M}$$

where the a_σ are in K and depend only on $\mathfrak{M}$ (see Procesi (1973), Rowen (1980), Lothaire (1983), Drensky (2000)). Show that if $\mathfrak{M}$ is the syntactic algebra of the series S, then $\mathfrak{M}$ satisfies the previous identity if and only if for any words $x, u_1, \ldots, u_n, y$, one has

$$(S, xu_1 \cdots u_n y) = \sum_{\substack{\sigma \in \mathfrak{S}_n \\ \sigma \neq \mathrm{id}}} a_\sigma (S, xu_{\sigma 1} \cdots u_{\sigma n} y).$$

(*Hint*: Use Lemma 2.1.1.)

4.1 Suppose that K is a Noetherian integral domain with field of fractions F. Using Proposition 1.3, show that for $m \in F$ which is quasi-integral over K, the module $K[m]$ is finitely generated, and deduce that m is integral over K.

4.2 Show that if L is an integral domain with subring K, and if moreover K is a weak Fatou ring, then L is a Fatou extension of K.

4.3 Let k be a field and consider the algebra $k[x, y]$ of commutative polynomials in x, y over k. Let K be its k-subalgebra generated by the monomials $x^{n+1}y^n$ for $n \geq 0$.

Show that K is not a weak Fatou ring. (*Hint*: Consider the element xy of the field of fractions of K.)

Notes to Chapter 7

Fliess, in Fliess (1974a), calls *strong Fatou ring* a ring K satisfying Theorem 1.1. Sontag and Rouchaleau (1977) show that for a principal ring K, the ring $K[t]$ is a strong Fatou ring. In the case of one variable, the class of strong Fatou rings is completely characterized by Theorem 1.5. (The formulation is different, but it is equivalent by the results of Section 6.1.) For several variables, a complete characterization of strong Fatou rings is still lacking. Karhumäki (1977) has characterized those polynomials $P \in \mathbb{Z}[x_1, \ldots, x_n]$ such that the rational series over $A = \{x_1, \ldots, x_n\}$

$$\sum_{w \in A^*} P(|w|_{x_1}, \ldots, |w|_{x_n})w$$

is $\mathbb{N}$-rational.

Sections 7.3 and 7.4 follow Reutenauer (1980a). In the case of one variable, the analogue of Theorem 4.1 is due to Cahen and Chabert (1975). Corollary 4.3 appears in Salomaa and Soittola (1978), Exercise 2 of Section II.6. Exercise 4.3 is from Bourbaki (1964), Chapitre 5, exercice 2.

8

Positive series in one variable

This chapter contains several results on rational series in one variable with nonnegative coefficients.

In the first section, poles of positive rational series are described. In Section 8.2 series with polynomial growth are characterized.

The main result (Corollary 3.2) is a characterization of K_+-rational series in one variable when $K = \mathbb{Z}$ or K is a subfield of $\mathbb{R}$.

The star height of positive series is the concern of the last section. It is shown that each K_+-rational series in one variable has star height at most 2, and that the arguments of the stars are quite simple series.

8.1 Poles of positive rational series

In this section, we start the study of series with nonnegative coefficients. Consider series of the form

$$\sum a_n x^n$$

with all coefficients in $\mathbb{R}_+$. If such a series is the expansion of a rational function, it does not imply in general that it is $\mathbb{R}_+$-rational (see Exercise 1.2). We shall characterize those rational functions over $\mathbb{R}$ whose series expansion is $\mathbb{R}_+$-rational. We call them $\mathbb{R}_+$*-rational functions.*

Theorem 1.1 (Berstel 1971) *Let $f(x)$ be an $\mathbb{R}_+$-rational function which is not a polynomial, and let ρ be the minimum of the moduli of its poles. Then ρ is a pole of f, and any pole of f of modulus ρ has the form $\rho\theta$, where θ is a root of unity.*

Observe that the minimum of the moduli of the poles of a rational function is the radius of convergence of the associated series. We start with a lemma.

Lemma 1.2 *Let $f(x)$ be a rational function which is not a polynomial and with a series expansion $\sum a_n x^n$ having nonnegative coefficients. Let ρ be the minimum of the moduli of the poles of f. Then ρ is a pole of f, and the multiplicity of any pole of f of modulus ρ is at most that of ρ.*

Proof. Let $z \in \mathbb{C}$, $|z| < \rho$. Then

$$|f(z)| = \left|\sum a_n z^n\right| \leq \sum a_n |z|^n = f(|z|)\,. \tag{1.1}$$

Let z_0 be a pole of modulus ρ, and let π be its multiplicity. Assume that the multiplicity of ρ as a pole of f is less than π. Then the function

$$g(z) = (\rho - z)^\pi f(z)$$

is analytic in the neighborhood of ρ, and $g(\rho) = 0$, whence

$$\lim_{r \to 1, r<1} (\rho - \rho r)^\pi f(\rho r) = 0\,. \tag{1.2}$$

The function

$$h(z) = (z_0 - z)^\pi f(z)$$

is analytic at z_0 and $h(z_0) \neq 0$. Thus

$$\lim_{z \to z_0, |z|<z_0} |(z_0 - z)^\pi f(z)| > 0\,.$$

In particular, setting $z = rz_0$, with $0 \leq r < 1$, this implies

$$\lim_{r \to 1, r<1} |z_0^\pi (1 - r)^\pi f(rz_0)| > 0\,.$$

In view of Equation (1.1), this shows that

$$\lim_{r \to 1, r<1} \rho^\pi (1 - r)^\pi f(r\rho) > 0$$

contradicting (1.2). □

Proof of Theorem 1.1. Let **S** be the set of polynomials with nonnegative coefficients and of rational functions with series expansions having nonnegative coefficients and satisfying the conclusions of the statement. It suffices to show that **S** is closed for sum, product and star. Recall that the star operation is

$$f \mapsto f^* = \sum_{n \geq 0} f^n = (1 - f)^{-1}\,.$$

Let $f = \sum a_n x^n$ and g be in **S**. Let ρ_f be the radius of convergence of f. Recall that $\rho_f = \sup\{r \in \mathbb{R}_+ \mid \sum a_n r^n < \infty\}$. Since the associated series has nonnegative coefficients, one has $\rho_{f+g} = \min(\rho_f, \rho_g)$ and, if $f, g \neq 0$, $\rho_{fg} = \min(\rho_f, \rho_g)$ (see Exercise 1.1). Thus, according to Lemma 1.2, $f + g$ and fg are in **S**, since each pole of $f + g$ and of fg is a pole of f or of g.

Now, let $f(x) = \sum_{n \geq 1} a_n x^n \in$ **S**, and assume $f \neq 0$. The poles of $f^* = (1 - f)^{-1}$ are the zeros of $1 - f$. Observe that $\sum a_n \rho_f^n = \infty$ since otherwise $\lim_{r \to \rho_f} f(r)$ would exist and this is impossible because f has a pole at ρ_f. The coefficients a_n being nonnegative, the function $r \mapsto \sum a_n r^n$ is strictly increasing from 0 to ∞ when r ranges from 0 to ρ_f, and consequently there is a unique real number r with $0 < r < \rho_f$ such that $f(r) = 1$. Thus r is a pole of f^*. Let z be a pole of

f^* of modulus $\leq r$. We prove that $z = r\theta$ for some root of unity θ. Indeed, the relations

$$1 = \sum a_n z^n = \mathrm{Re}\left(\sum a_n z^n\right) = \sum a_n \mathrm{Re}(z^n)$$
$$\leq \sum a_n |z|^n \leq \sum a_n r^n = 1$$

show that equality holds everywhere. Consequently, $a_n \mathrm{Re}(z^n) = a_n r^n$ for all $n \geq 0$. Let n be an integer with $a_n \neq 0$ (it exists because $f \neq 0$). Then $\mathrm{Re}(z^n) = r^n$ and $|z| \leq r$ imply $z^n = r^n$ whence $z = r\theta$ for θ some n-th root of unity. Thus f^* is in **S**. □

8.2 Polynomially bounded series over $\mathbb{Z}$ and $\mathbb{N}$

A series $S \in \mathbb{Z}\langle\langle A \rangle\rangle$ has *polynomial growth* or is *polynomially bounded* if there exist an integer $q \geq 0$ and a real number C such that

$$|(S, w)| \leq C|w|^q$$

for all nonempty words w.

Proposition 2.1 *Let $S = \sum_{n \geq 0} a_n x^n$ be a $\mathbb{Z}$-rational series which has polynomial growth. If the coefficients a_n are in $\mathbb{N}$, then S is $\mathbb{N}$-rational.*

Proof. The result is true if S is a polynomial. Assume S is not a polynomial. We may assume that S is strict, by Proposition 6.1.6. The proof is in three steps.

1. We first show that the moduli of the eigenvalues of S are bounded by 1. Let C and p be such that $|a_n| \leq Cn^p$ for all n large enough. The radius of convergence of the series $\sum n^p x^n$ is 1, since indeed $\limsup(n^p)^{1/n} = 1$, so the radius of convergence ρ of S is at least 1. Since S is strict, we have by Theorem 6.2.1

$$a_n = \sum_{i=1}^{r} P_i(n)\lambda_i^n\,. \tag{2.1}$$

Since the radius of convergence ρ of S is $\rho = \max\{1/|\lambda_1|, \ldots, 1/|\lambda_r|\}$, it follows that $|\lambda_i| \leq 1$ for $i = 1, \ldots, r$.

2. Next, we show that all λ_i in (2.1) are roots of unity. Consider indeed the series $S' = \sum b_n x^n$ with

$$b_n = \sum_{i=1}^{r} \lambda_i^n\,. \tag{2.2}$$

Note that b_n is symmetric in the λ_i's. Since S is rational over $\mathbb{Z}$, its minimal polynomial has coefficients in $\mathbb{Z}$. Hence the λ_i's are algebraic integers and consequently b_n is in $\mathbb{Z}$ and S' is rational over $\mathbb{Z}$, by Fatou's lemma.

Let $Q = \prod(x - \lambda_i)$. The sequence (b_n) satisfies the linear recurrence associated to Q.

In view of (2.2), the sequence (b_n) is bounded, and since the b_n are integers, it is periodic. Indeed, the sequence (b_n) satisfies a linear recurrence relation of length r say, and since the number of distinct r-tuples $(b_n, b_{n+1}, \ldots, b_{n+r-1})$ is bounded, there are two indices $m < m'$ such that $(b_m, b_{m+1}, \ldots, b_{m+r-1}) = (b_{m'}, b_{m'+1}, \ldots, b_{m'+r-1})$, and one gets that $b_{m+r} = b_{m'+r}$ and, with $h = m' - m$, $b_n = b_{n+h}$ for all large enough n. It follows that Q divides $x^h - 1$, showing that all roots λ_i are roots of unity.

3. We now show that we may apply the next proposition. In view of the preceding computation, all λ_i in (2.1) are roots of unity. If $\lambda_i^h = 1$ for $i = 1, \ldots, r$, then the sequences $(a_{nh+k})_{n\geq 0}$ for $0 \leq k \leq h-1$ have the form

$$a_{nh+k} = R_k(n) \quad n \geq 0$$

for polynomials R_k defined by

$$R_k(t) = \sum_{i=0}^{r} \lambda_i^k P_i(ht+k)\,.$$

In view of the next proposition, each polynomial $R_k(t+\ell)$, for some $\ell \in \mathbb{N}$, is a linear combination, with coefficients in $\mathbb{N}$, of binomial polynomials $\binom{t}{d} = \frac{t(t-1)\cdots(t-d+1)}{d!}$. Note that (see Exercise 2.7)

$$\frac{x^d}{(1-x)^{d+1}} = \sum_{n\geq 0} \binom{n}{d} x^n\,.$$

Since these series are obviously $\mathbb{N}$-rational, each series $\sum R_k(n)x^n$ is $\mathbb{N}$-rational. This proves the proposition. □

Proposition 2.2 *Let $P(x)$ be a complex polynomial of degree d such that $P(n) \in \mathbb{Z}$ for all $n \in \mathbb{N}$. If $P(n) \in \mathbb{N}$ for all large enough $n \in \mathbb{N}$, then there exists $\ell \geq 0$ and $a_0, \ldots, a_d \in \mathbb{N}$ such that*

$$P(x+\ell) = a_0\binom{x}{d} + a_1\binom{x}{d-1} + \cdots + a_d\,.$$

Proof. We may assume that P is nonzero. It is easily seen (Exercise 2.2) that

$$P(x) = \sum_{i=0}^{d} a_i \binom{x}{d-i}$$

for some nonzero $a_0 \in \mathbb{N} \setminus 0$ and $a_1, \ldots, a_d \in \mathbb{Z}$. If all $a_1, \ldots, a_d$ are in $\mathbb{N}$, we are done. Assume the contrary and let h be the smallest index such that $a_h < 0$. Set $\ell = \max\{1+h, -a_h\}$.

We use Vandermonde's convolution formula that holds for binomial polynomials. For $\ell, m \in \mathbb{N}$:

$$\binom{x+\ell}{m} = \sum_{j+k=m} \binom{\ell}{j}\binom{x}{k}\,.$$

It follows that

$$P(x+\ell) = \sum_{i=0}^{d} a_i \binom{x+\ell}{d-i} = \sum_{i=0}^{d} a_i \sum_{j+k=d-i} \binom{\ell}{j}\binom{x}{k}$$
$$= \sum_{k=0}^{d} \Big(\sum_{i+j=d-k} a_i \binom{\ell}{j} \Big) \binom{x}{k} = \sum_{k=0}^{d} \Big(\sum_{i+j=k} a_i \binom{\ell}{j} \Big) \binom{x}{d-k}$$
$$= \sum_{k=0}^{d} b_k \binom{x}{d-k},$$

where for $k = 0, \ldots, d$

$$b_k = \sum_{i=0}^{k} a_i \binom{\ell}{k-i}.$$

Clearly $b_0, \ldots, b_{h-1} \geq 0$, and

$$b_h = a_0 \binom{\ell}{h} + \cdots + a_h \geq a_0 \ell + a_h \geq 0,$$

since $\ell \geq -a_h$ and $a_0 \geq 1$. Thus $P(x+\ell)$ has nonnegative coefficients $b_0, \ldots, b_h$. Arguing by induction on h, the result follows. □

8.3 Characterization of K_+-rational series

Theorem 1.1 gives a necessary condition for a rational function to be $\mathbb{R}_+$-rational. We now give a sufficient condition in the general case. For this, we go back to the vocabulary of formal series.

A rational series with complex coefficients is said to have a *dominating eigenvalue* if there is among its eigenvalues (in the sense of Section 6.1) a unique eigenvalue having maximal modulus. It is equivalent to say that the associated rational function is either a polynomial or has a unique pole of minimal modulus.

Theorem 3.1 (Soittola 1976) *Let $K = \mathbb{Z}$ or K be a subfield of $\mathbb{R}$. If a K-rational series has a dominating eigenvalue and nonnegative coefficients, then it is K_+-rational.*

Corollary 3.2 *A series over K_+ is K_+-rational if and only if it is the merge of polynomials and of rational series having a dominating eigenvalue.*

Proof. If S is K_+-rational, it is by Proposition 6.2.4 the merge of K_+-rational series which are simple. In particular, no quotient of eigenvalues of such a series is a root of unity $\neq 1$. Hence by Theorem 1.1, these series have a dominating eigenvalue. For the converse, one uses Theorem 3.1 and Exercise 6.2.5. □

Corollary 3.3 *If a $\mathbb{R}_+$-rational series has coefficients in K_+, then it is K_+-rational.*

Proof. Let S be a $\mathbb{R}_+$-rational series which has coefficients in K_+. By the preceding corollary, S is a merge of polynomials and of rational series having a dominating eigenvalue. By Soittola's theorem 3.1, each of the series of the merge is K_+-rational. By Corollary 3.2, S itself is K_+-rational. □

Let $S = \sum_{n\geq 0} a_n x^n$ be a series which is not a polynomial. We know by Section 6.2 that there exists an exponential polynomial for a_n that is

$$a_n = \sum_i P_i(n)\lambda_i^n$$

for n large enough. Suppose that λ_1 is the dominating eigenvalue of S. Then we call *dominating coefficient* of S the highest nonzero coefficient α of P_1. Observe that when $n \to \infty$

$$a_n \sim \alpha n^{\deg(P_1)} \lambda_1^n . \tag{3.1}$$

In particular, $a_n \neq 0$ for large n. Moreover

$$\frac{a_{n+1}}{a_n} \sim \lambda_1 . \tag{3.2}$$

Lemma 3.4 *Let S, S' be real rational series which are not polynomials and which have the same dominating eigenvalue λ_1 with dominating coefficients α, α'.*

(i) *The series SS' has also the dominating eigenvalue λ_1 with dominating coefficient positively proportional to $\alpha\alpha'$.*

(ii) *The coefficients of S are ultimately positive if and only if λ_1 and α are positive real numbers.*

(iii) *If S is the inverse of a polynomial P with $P(0) = 1$, and if λ_1 is a positive real number, then $\alpha > 0$.*

Proof. (i) We write S as a $\mathbb{C}$-linear combination of partial fractions, as in the proof of Theorem 6.2.1. Let β be the coefficient of $1/(1-\lambda_1 x)^{k+1}$ in this combination, where $k = \deg(P_1)$. Since, by Exercise 2.7, $1/(1-\lambda_1 x)^{k+1} = \sum_{n\geq 0} \binom{n+k}{k} \lambda_1^n x^n$ and $\binom{n+k}{k} = \frac{n^k}{k!} + \cdots$, the dominating term of $P_1(n)$ is $\beta \frac{n^k}{k!}$, and $\alpha = \beta/k!$. If we do similarly for S', we obtain a dominating term of the form $\beta' \frac{n^\ell}{\ell!}$ and $\alpha' = \beta'/\ell!$. The product SS' has the eigenvalue λ_1 with multiplicity $k + \ell + 2$, the dominating term is $\beta\beta' \frac{n^{k+\ell+1}}{(k+\ell+1)!}$, so the dominating coefficient is $\alpha\alpha' k!\ell!/(k+\ell+1)!$. This gives the result.

(ii) If the a_n are ultimately positive, then $\lambda_1 \geq 0$ by (3.2), and $\lambda_1 \neq 0$ since S is not a polynomial. Moreover, α is positive by (3.1). Conversely, if $\lambda_1, \alpha > 0$, then $a_n > 0$ for n large enough by (3.1).

(iii) We have $P(x) = \prod_{i=1}^d (1 - \lambda_i x) \in \mathbb{R}[x]$ with $\lambda_i \in \mathbb{C}$, $\lambda_1 = \cdots = \lambda_k > |\lambda_{k+1}|, \ldots, |\lambda_d|$, for some k with $1 \leq k \leq d$. In order to compute the dominating coefficient α of P^{-1}, we write P^{-1} as a $\mathbb{C}$-linear combination of series $1/(1-\lambda_i x)^j$.

Then $\alpha = \beta/(k-1)!$ where β is the coefficient of $1/(1-\lambda_1 x)^k$ in this linear combination. To compute β, multiply the linear combination by $(1-\lambda_1 x)^k$ and put then $x = \lambda_1^{-1}$. Since only fractions $1/(1-\lambda_1 x)^j$ with $j \le k$ occur, this is well defined and gives

$$\beta = \frac{1}{\prod_{i=k+1}^{d} \left(1 - \frac{\lambda_i}{\lambda_1}\right)} .$$

Now, the numbers λ_i^{-1}, for $i = k+1, \ldots, d$ are the roots of the real polynomial $\prod_{i=k+1}^{d}(1-\lambda_i x)$. Hence, either λ_i is real and then $|\lambda_i| < \lambda_1$ and thus $1 - \dfrac{\lambda_i}{\lambda_1} > 0$, or λ_i is not real and then there is some j such that λ_i, λ_j are conjugate. Then so are $1 - \dfrac{\lambda_i}{\lambda_1}$ and $1 - \dfrac{\lambda_j}{\lambda_1}$, so that their product is positive. This shows that α is positive. □

Given an integer $d \ge 1$ and numbers $B, G_1, \ldots, G_d$ in $\mathbb{R}_+$, we set

$$G(x) = \sum_{i=1}^{d-1} G_i x^i$$

and we call *Soittola denominator* a polynomial of the form

$$D(x) = (1 - Bx)(1 - G(x)) - G_d x^d . \tag{3.3}$$

If $d = 1$, we agree that $B = 0$. In this limit case, $D(x) = 1 - G_1 x$. The numbers $B, G_1, \ldots, G_d$ are called the *Soittola coefficients* of $D(x)$ and B is called its *modulus*.

Note that setting

$$D(x) = 1 - g_1 x - \cdots - g_d x^d$$

the expression (3.3) is equivalent to

$$\begin{aligned} g_1 &= B + G_1 \\ g_i &= G_i - BG_{i-1} , \quad i = 2, \ldots, d . \end{aligned} \tag{3.4}$$

Likewise, we call *Soittola polynomial* a polynomial of the form

$$x^d - g_1 x^{d-1} - \cdots - g_d \tag{3.5}$$

with the g_i as above.

Lemma 3.5 *Let*

$$P(x) = \prod_{i=1}^{d} (1 - \lambda_i x)$$

be a polynomial in $\mathbb{R}[x]$ *with* $\lambda_i \in \mathbb{C}$, $\lambda_1 > 1$ *and* $\lambda_1 > |\lambda_2|, \ldots, |\lambda_d|$. *Let*

$$P_n(x) = \prod_{i=1}^{d} (1 - \lambda_i^n x) .$$

For n large enough, $P_n(x)$ is a Soittola denominator with modulus $< \lambda_1^n$ and with Soittola coefficients in the subring generated by the coefficients of P.

Proof. Let $e_{i,n} = \sum_{j_1 < \cdots < j_i} \lambda_{j_1}^n \cdots \lambda_{j_i}^n$ be the i-th elementary symmetric function of $\lambda_1^n, \ldots, \lambda_d^n$. By the fundamental theorem of symmetric functions, $e_{i,n}$ is in the ring generated by the functions $e_{i,1}$, for $1 \le i \le d$, hence in the ring generated by the coefficients of $P = P_1$ (see also Exercise 3.2).

Clearly $e_{1,n} \sim \lambda_1^n$ when $n \to \infty$. Note that for $i \ge 2$, each term in $e_{i,n}$ is a product of i factors taken in the λ_j's, and containing at least one factor with modulus $< \lambda_1$. Therefore $e_{i,n}/\lambda_1^{in} \to 0$ when $n \to \infty$.

We may assume $d \ge 2$. Define $B = \lfloor e_{1,n}/2 \rfloor$ and $G_1, \ldots, G_d$ by the formulas $G_1 = e_{1,n} - B$ and $G_i - BG_{i-1} = (-1)^{i-1} e_{i,n}$ for $i = 2, \ldots, d$ (we do not indicate the dependence on n which is understood). When $\lambda_1^n \to \infty$, we have $B \sim \lambda_1^n/2 \sim G_1$. Arguing by induction on i, suppose that $G_i \sim \lambda_1^{in}/2^i$. We have $G_{i+1} = (-1)^i e_{i+1,n} + BG_i$. Now $BG_i \sim \lambda_1^{(i+1)n}/2^{i+1}$ and we know that $e_{i+1,n}/\lambda_1^{(i+1)n} \to 0$. Thus $G_{i+1} \sim \lambda_1^{(i+1)n}/2^{i+1}$. The lemma follows. □

We call *Perrin companion matrix* of the Soittola polynomial (3.5) the matrix

$$P = \begin{pmatrix} B & 1 & 0 & \cdots & 0 \\ 0 & 0 & \ddots & \ddots & \vdots \\ \vdots & & \ddots & \ddots & 0 \\ 0 & \cdots & \cdots & 0 & 1 \\ G_d & \cdots & \cdots & G_2 & G_1 \end{pmatrix}. \tag{3.6}$$

It differs from a usual companion matrix by the entry 1, 1 which is not 0 but B. In the limit case $d = 1$, one sets $P = (G_1)$.

Lemma 3.6 *Let $D(x)$ be the Soittola denominator* (3.5). *Given $S = \sum_{n \ge 0} a_n x^n$, define $T = \sum_{n \ge 0} t_n x^n$ and $U = \sum_{n \ge 0} u_n x^n$ by*

$$T = DS \quad \text{and} \quad U = (1 - Bx)S.$$

Then for $n \ge 0$,

$$P \begin{pmatrix} a_n \\ u_{n+1} \\ \vdots \\ u_{n+d-1} \end{pmatrix} + \begin{pmatrix} 0 \\ \vdots \\ 0 \\ t_{n+d} \end{pmatrix} = \begin{pmatrix} a_{n+1} \\ u_{n+2} \\ \vdots \\ u_{n+d} \end{pmatrix}. \tag{3.7}$$

Moreover, if T is a polynomial of degree $\le h$, then for any m in $\mathbb{N}$

$$a_{m+h} = (1, 0, \ldots, 0) P^m (a_h, u_{h+1}, \ldots, u_{h+d-1})^T.$$

Proof. Note that in the limit case $d = 1$, the first relation must be read as $G_1 a_n + t_{n+1} = a_{n+1}$, which is easy to verify, since one has by convention $D = 1 - G_1 x$.

We may therefore assume that $d \geq 2$. The first matrix product is equal to

$$\begin{pmatrix} Ba_n + u_{n+1} \\ u_{n+2} \\ \vdots \\ u_{n+d-1} \\ \alpha \end{pmatrix}$$

where

$$\alpha = G_d a_n + \sum_{i=1}^{d-1} G_i u_{n+d-i} \, .$$

Observe next that

$$T = (1 - Bx)(1 - G(x))S - G_d x^d S = (1 - G(x))U - G_d x^d S \, .$$

Thus

$$t_{n+d} = u_{n+d} - \sum_{i=1}^{d-1} G_i u_{n+d-i} - G_d a_n \, ,$$

showing that $\alpha + t_{n+d} = u_{n+d}$. This proves the first identity. Suppose now that T is a polynomial of degree $\leq h$. Then $0 = t_{h+d} = t_{h+d+1} = \cdots$ Using induction on m and (3.7) for $n = h, h+1, \ldots$, we obtain

$$P^m \begin{pmatrix} a_h \\ u_{h+1} \\ \vdots \\ u_{h+d-1} \end{pmatrix} = \begin{pmatrix} a_{m+h} \\ u_{m+h+1} \\ \vdots \\ u_{m+h+d-1} \end{pmatrix}$$

which implies the second identity. □

Proof of Soittola's theorem. 1. We may assume that S is not a polynomial. By Lemma 3.4 (ii), the dominating eigenvalue λ_1 of S is positive. We may assume that $\lambda_1 > 1$. Indeed, if K is a subfield of $\mathbb{R}$, then we replace $S(x)$ by $S(\alpha x)$ for α in $\mathbb{N}$ large enough; then the eigenvalues are multiplied by α and we are done. If $K = \mathbb{Z}$ and $\lambda_1 \leq 1$, then by Proposition 2.1, S is $\mathbb{N}$-rational.

2. Write $S(x) = N(x)/D(x)$ where D is the smallest denominator with $D(0) = 1$. Then $N, D \in K[x]$. Let m be the multiplicity of the eigenvalue λ_1 of S. Since K is a factorial subring of $\mathbb{R}$, we may write $D(x) = D_1(x) \cdots D_m(x)$, where each polynomial $D_i(x)$ has coefficients in K, has the simple factor $1 - \lambda_1 x$ and satisfies $D_i(0) = 1$.

Decompose S as a merge $S = \sum_{0 \leq i < p} x^i S_i(x^p)$. Then the eigenvalues of S_i are the p-th powers of those of S (equivalently the poles of S_i are the p-th powers of those of S). Hence, if p is chosen large enough, Lemma 3.5 shows that we may assume that D_1 is a Soittola denominator of the form

$$D_1(x) = (1 - Bx)\left(1 - \sum_{i=1}^{d-1} G_i x^i\right) - G_d x^d$$

with $d \geq 1$, $B, G_i \in K_+$ and $B < \lambda_1$. Since $a_{n+1}/a_n \sim \lambda_1$ we see that $u_{n+1} = a_{n+1} - Ba_n \geq 0$ for n large enough.

3. Let

$$T = \sum_{n\geq 0} t_n x^n = D_1 S\,.$$

Suppose first that λ_1 is simple, that is $m = 1$. Then T is a polynomial and Lemma 3.6 shows that $\sum_{n\geq 0} a_{n+h} x^n$ is K_+-rational for h large enough. Hence S is K_+-rational. Suppose next that $m \geq 2$ and argue by induction on m. Note that S, D_1^{-1} and T have the dominating eigenvalue λ_1, the latter with multiplicity $m - 1$. Lemma 3.4(iii) and (ii) show that D_1^{-1} and S have positive dominating coefficient. Thus by Lemma 3.4(i), since $D_1^{-1}T = S$, the series T also has positive dominating coefficient. This implies by Lemma 3.4(ii) that T has ultimately positive coefficients and consequently that for h large enough, the series $\sum_{n\geq 0} t_{n+h+d} x^n$ is K_+-rational, by induction on m.

Thus $t_{n+h+d} = \nu N^n \gamma$ for some representation (ν, N, γ) over K_+. Define a representation (ℓ, M, c) over K_+ by

$$\ell = (1, 0, \ldots, 0)\,, \quad M = \begin{pmatrix} P & Q \\ 0 & N \end{pmatrix}, \quad c = \begin{pmatrix} a_h \\ u_{h+1} \\ \vdots \\ u_{h+d-1} \\ \gamma \end{pmatrix}$$

where h is chosen large enough and where all rows of Q are 0 except the last which is ν. We prove that

$$M^n c = \begin{pmatrix} a_{h+n} \\ u_{h+n+1} \\ \vdots \\ u_{h+n+d-1} \\ N^n \gamma \end{pmatrix}.$$

This is true for $n = 0$ by definition. Admitting it holds for n, the equality for $n + 1$ follows from Lemma 3.6 (Equation (3.7) where n is replaced by $n + h$), since $QN^n\gamma$ is a column vector whose components are all 0 except the last one which is $\nu N^n \gamma = t_{n+h+d}$. We deduce that $\ell M^n c = a_{n+h}$ and $S = \sum_{i=0}^{h-1} a_i x^i + x^h \sum_{n\geq 0} a_{n+h} x^n$ is therefore K_+-rational. □

8.4 Star height 2

We consider now the star height of K_+-rational series.

Theorem 4.1 *Let K be a subfield of $\mathbb{R}$ or $K = \mathbb{Z}$. Any K_+-rational series is in the subsemiring of $K_+[[x]]$ generated by $K_+[x]$ and by the series of the form*

$$(Bx^p)^* \quad or \quad \left(\sum_{i=1}^{d-1} G_i x^i + G_d x^d (Bx^p)^*\right)^*$$

with $p, d \geq 1$, $B, G_i \in K_+$. In particular, they have star height at most 2.

Proof. Denote by $\mathcal{L}$ this semiring. It is clearly closed under the substitution $x \mapsto \alpha x^q$ for $q \geq 1$, $\alpha \in K_+$. Thus it is also closed under the merge of series.

So, if we follow the proof of Soittola's theorem, we may continue after steps 1. and 2. We start with a notation. Given a series $V = \sum_{n\geq 0} v_n x^n$ and an integer $h \geq 0$, we write $V^{(h)} = \sum_{n>h} v_n x^n$ and $V_{(h)} = \sum_{n\leq h} v_n x^n$. It follows from $U = (1 - Bx)S$ that

$$U^{(h)} = S^{(h)} - BxS^{(h-1)} = S^{(h)}(1 - Bx) - Ba_h x^{h+1},$$

$$U_{(h)} = S_{(h)} - BxS_{(h-1)} = S_{(h-1)}(1 - Bx) + a_h x^h.$$

We show below the existence of a polynomial P_h with coefficients in K_+, for h large enough, such that

$$U^{(h)} = \left(P_h + T^{(h)} + a_h G_d x^{h+d} (Bx)^*\right) H^*$$

where

$$H = G + G_d x^d (Bx)^*.$$

If $m = 1$, we take h large enough and $T^{(h)} = 0$. If $m \geq 2$, we conclude by induction on m that $T^{(h)}$ is in $\mathcal{L}$. Thus the series $U^{(h)}$ is in $\mathcal{L}$, and since $(1 - Bx)S^{(h)} = Ba_h x^{h+1} + U^{(h)}$ the series

$$S = \sum_{i=0}^{h} a_i x^i + (Bx)^*(Ba_h x^{h+1} + U^{(h)})$$

is in $\mathcal{L}$. Now, from

$$T = D_1 S = (1 - Bx)(1 - H)S = U(1 - H),$$

we get

$$\begin{aligned} T^{(h)} &= \left(U(1 - H)\right)^{(h)} = \left(U^{(h)}(1 - H)\right)^{(h)} + \left(U_{(h)}(1 - H)\right)^{(h)} \\ &= U^{(h)}(1 - H) + U_{(h)}^{(h)} - \left(U_{(h)} H\right)^{(h)} \\ &= U^{(h)}(1 - H) - \left(U_{(h)} H\right)^{(h)}. \end{aligned}$$

Next

$$\left(U_{(h)}H\right)^{(h)} = \left(U_{(h)}G\right)^{(h)} + \left(U_{(h)}G_d x^d (Bx)^*\right)^{(h)}$$

Recall that $G = \sum_{i=1}^{d-1} G_i x^i$. The first term of the right-hand side is

$$\left(U_{(h)}G\right)^{(h)} = \sum_{\substack{0\le j\le h \\ 0<\ell<d \\ j+\ell>h}} u_j G_\ell x^{j+\ell} .$$

Setting $j+\ell = h+i$ with $0 < i < d$, this can be rewritten as $\sum_{i=1}^{d-1} w_i x^{h+i}$ with

$$w_i = \sum_{\substack{0\le j\le h \\ 0<\ell<d \\ j+\ell=h+i}} u_j G_\ell .$$

Now note that in this sum, since $\ell < d$, we have $j > h - d$, hence $u_j \ge 0$ for h large enough. This shows that $\left(U_{(h)}G\right)^{(h)}$ is a polynomial with coefficients in K_+.

To compute the second term, recall that $U_{(h)} = S_{(h-1)}(1 - Bx) + a_h x^h$. Consequently

$$U_{(h)}(Bx)^* = S_{(h-1)} + a_h x^h (Bx)^* .$$

So the term $(U_{(h)}G_d x^d (Bx)^*)^{(h)}$ reduces to the sum of a polynomial with coefficients in K_+ and of the series $G_d a_h x^{h+d}(Bx)^*$. Thus we obtain, for h large enough

$$T^{(h)} = U^{(h)}(1-H) - G_d a_h x^{h+d}(Bx)^* - P_h$$

with $P_h \in K_+[x]$. □

Exercises for Chapter 8

1.1 Let $f = \sum a_n x^n$, $g = \sum b_n x^n$ with $a_n, b_n \in \mathbb{R}_+$. Denote by ρ_f the radius of convergence of f, that is

$$\rho_f = \sup\{r \in \mathbb{R}_+ \mid \sum a_n r^n < \infty\} .$$

(a) Show that if $r \in [0, \min(\rho_f, \rho_g)[$, then $\sum_{n\ge 0}(a_n + b_n) < \infty$. Conclude that $\min(\rho_f, \rho_g) \le \rho_{f+g}$.

(b) Show that if $r \in [0, \rho_{f+g}[$, then $\sum_n a_n r^n < \infty$. Deduce that $\rho_{f+g} \le \rho_f$ and finally that $\rho_{f+g} = \min(\rho_f, \rho_g)$.

(c) Show that if $r \in [0, \min(\rho_f, \rho_g)[$, then $\sum_n \big(\sum_{i+j=n} a_i b_j\big) r^n < \infty$. Conclude that $\min(\rho_f, \rho_g) \le \rho_{fg}$.

(d) Suppose now that $g \ne 0$. Show that if $r \in [0, \rho_{fg}[$, then $\sum_n a_n r^n < \infty$. Deduce that $\rho_{fg} \le \rho_f$. Finally, deduce that if $f, g \ne 0$, then $\rho_{fg} = \min(\rho_f, \rho_g)$.

1.2 (a) Let θ be a real number. Show that the series $S = \sum_{n\ge 0}(\cos^2 n\theta)x^n$ is a $\mathbb{C}$-rational series. (Give an expression for S as a rational function by using the formula $\cos n\theta = 1/2(e^{in\theta} + e^{-in\theta})$.)

(b) Let $0 < a < c$ be integers and let θ be a real number with $0 < \theta < \pi/2$, such that $\cos\theta = a/c$. Show that the numbers $c^n \cos n\theta$ are integers. Show that the series $T = \sum(c^{2n} cos^2 n\theta)x^n$ is $\mathbb{Z}$-rational with coefficients in $\mathbb{N}$.

(c) Show that if $c \neq 2a$, then $z = e^{i\theta}$ is not a root of unity. (*Hint*: Show that z is an algebraic number of degree ≤ 2, and use the fact that if z is a n-th primitive root of 1, then $\phi(n) \leq 2$ where ϕ is Euler's function, so $n = 1, 2, 3, 4$ or 6. Show that this is impossible.) Show that T is not $\mathbb{R}_+$-rational (use Theorem 1.1, see Berstel (1971), and also Eilenberg (1974, Chapter VIII, Example 6.1)).

1.3 Show that the $\mathbb{Z}$-rational series

$$\frac{x + 5x^2}{1 + x - 5x^2 - 125x^3} = \sum_{n \geq 0}(2\cdot 5^n - (3+4i)^n - (3-4i)^n)x^n$$
$$= x + 4x^2 + x^3 + 144x^4 + \cdots$$

has positive coefficients but is not $\mathbb{N}$-rational. (*Hint*: Use the fact that $3 + 4i$ and $3 - 4i$ have norm 5.)

1.4 Let $c > d$ be integers such that $d \pm i\sqrt{c^2 - d^2}$ are not roots of unity, and define a sequence a_n by

$$a_n = b_1 c^n + b_2\left(d + i\sqrt{c^2 - d^2}\right)^n + b_3\left(d - i\sqrt{c^2 - d^2}\right)^n$$

for integers $b_1 \geq b_2 + b_3$. Show that $\sum a_n x^n$ is $\mathbb{Z}$-rational with nonnegative coefficients and is not $\mathbb{N}$-rational. (*Hint*: Use the fact that $d \pm i\sqrt{c^2 - d^2}$ have norm c. This exercise extends Exercise 1.3.) Example: for $c = 3, d = 2, b_1 = 2, b_2 = b_3 = 1$, one gets

$$\sum a_n x^n = \frac{4 - 12x + 24x^2}{1 - 5x + 15x^2 - 27x^3} = 4 + 8x + 4x^2 + 8x^3 + \cdots$$

1.5 Let $S = \sum a_n x^n = P(x)/Q(x)$ be a rational series over $\mathbb{R}$, where $P(x)$ and $Q(x)$ have no common root, and $Q(x)$ is a polynomial of degree 2 with $Q(0) = 1$ and $P(x)$ is of degree ≤ 1. Set $Q(x) = 1 - ax - bx^2$ and $P(x) = c - dx$. Set further $Q(x) = (1 - \alpha x)(1 - \beta x)$.

(a) Show that $a_0 = c$, $a_1 = ac - d$ and for $n \geq 2$

$$a_n = \begin{cases} \dfrac{1}{\alpha - \beta}\left((\alpha c - d)\alpha^n - (\beta c - d)\beta^n\right) & \text{if } \alpha \neq \beta\,, \\ \alpha^{n-1}\left((\alpha c - d)n + \alpha c\right) & \text{if } \alpha = \beta\,. \end{cases}$$

(b) Assuming that $a_n \geq 0$ for $n \geq 0$, show successively that $c \geq 0$, $ac - d \geq 0$, $a \geq 0$, $a^2 + 4b \geq 0$ and $\alpha c - d > 0$.

(c) Show that conversely, if these five conditions are fulfilled, then $a_n \geq 0$ for $n \geq 0$.

1.6 Let a and b be integers, $a \geq 0$, and let $k \geq 2$. Let

$$f(x) = \frac{1}{1 - ax + bx^k}$$

and set $r = a(k-1)/k$ and $\delta = r^k/(k-1) - b$. The aim of the exercise is to show that $f(x)$ is $\mathbb{N}$-rational if and only if $\delta \geq 0$. Note that for $k = 2$, this reduces to the condition that the discriminant $a^2 - 4b$ is nonnegative.

(a) Show that

$$1 - ax + bx^k = (1 - rx)(1 - g(x)) + \frac{\delta}{r}x^{k-1},$$

with

$$g(x) = \sum_{i=1}^{k-2} \frac{r^i x^i}{k-1} + \frac{b}{r}x^{k-1}.$$

Conclude that if $\delta \geq 0$, then

$$f(x) = (rx)^*(g(x) + \frac{\delta}{r}x^{k-1}(rx))^*$$

is $\mathbb{R}_+$-rational, and thus is $\mathbb{N}$-rational.

(b) Set $p(x) = x^k - ax^{k-1} + b$, and suppose $b > 0$ and $k \geq 3$. Show that for even k, the polynomial $p(x)$ has no real root when $\delta < 0$, and has two real roots otherwise. Show that when k is odd, then $p(x)$ has one real root which is negative if $\delta < 0$ and three real roots, with two of them positive otherwise. Conclude that if $f(x)$ is $\mathbb{N}$-rational, then $\delta \geq 0$. (*Hint*: Show that $p'(x)$ has the two roots 0 and r, and that $p''(r) > 0$.)

1.7 The aim of this exercise is to prove that there exist $\mathbb{N}$-rational series of star height exactly 2. Let a, b be positive integers, and set $Q(x) = 1 - ax + bx^2$. Assume the the discriminant $a^2 - 4b$ is nonnegative. Then $Q(x)$ has two real roots, and they are equal if and only if $a^2 - 4b = 0$. It follows from Exercise 1.6 that $f(x) = 1/Q(x)$ is $\mathbb{N}$-rational.

(a) Show that if the polynomial $Q(x)$ is reducible over $\mathbb{Q}$, then $f(x)$ has star height 1.

From now on, we suppose that $Q(x)$ is irreducible over $\mathbb{Q}$.

(b) Show that if $f(x)$ has star height 1, then there exist polynomials $P(x) \in \mathbb{Z}[x]$ and $N(x) \in \mathbb{N}[x]$ with $N(0) = 0$ such that $Q(x)P(x) = 1 - N(x)$.

(c) Show that the roots of $Q(x)$ are both real positive.

(d) Show that a polynomial of the form $1 - N(x)$, with $N(x) \in \mathbb{N}[x]$ and $N(0) = 0$, has exactly one real positive root which is simple. Conclude that $f(x)$ has star height 2.

(e) Show that if $a \geq 2 + b$, then $Q(x)$ is irreducible. Taking $a = 3$, $b = 1$, conclude that the series

$$\frac{1}{1 - 3x + x^2} = \sum F_{2n+1}x^n = \frac{1}{1 - x - \dfrac{1}{1 - x}},$$

where F_n is the n-th Fibonacci number (see also Example 3.2.1), has star height 2.

2.1 Show that for any polynomial P of degree d, the series $\sum_{n\geq 0} P(n)x^n$ is a rational series and that the sequence $(P(n))_{n\geq 0}$ satisfies the linear recurrence relation associated to the polynomial $(x-1)^{d+1}$.

2.2 Let $P(x)$ be a complex polynomial of degree d, and assume that $P(n) \in \mathbb{Z}$ for $n \in \mathbb{N}$. Show that there exist integers $a_0, \ldots a_d$ with $a_d \neq 0$ such that $P(x) = \sum_{i=0}^{d} a_i \binom{x}{i}$.

3.1 Let $S = \sum a_n x^n = P(x)/Q(x)$ be a rational series over $\mathbb{R}$, where $P(x)$ and $Q(x)$ have no common root, and $Q(x)$ is a polynomial of degree 2 with $Q(0) = 1$. Show that an S is $\mathbb{R}_+$-rational if and only if all coefficients a_n are nonnegative. (*Hint*: Set $Q(x) = (1-\alpha x)(1-\beta x)$ and use Exercise 1.5 to show that if all a_n are nonnegative, then α and β are real, and that at least one is positive. Then, use Soittola's theorem.)

3.2 Let K be a subring of some field and $P \in K[x]$ with $P(0) = 1$. Let M be the companion matrix of P. With the notations of Lemma 3.5, show that $P_n = \det(1 - M^n x)$. Deduce that the coefficients of P_n are in the subring generated by the coefficients of P.

3.3 Show that the characteristic polynomial of a Perrin companion matrix is the corresponding Soittola polynomial (see Perrin (1992)).

3.4 Show that the inverse of a Soittola denominator is an $\mathbb{R}_+$-rational series. Show that $\dfrac{1}{D(x)} = (Bx)^*(G(x) + G_d x^d (Bx)^*)^*$.

3.5 Let M be a square matrix over some subsemiring K of a commutative ring. Show that $\det(1 - Mx)^{-1}$ is a K-rational series. (*Hint*: Let M_i be the submatrix corresponding to the first i rows and columns. Show that $\det(1 - M_{i-1})/\det(1 - M_i x)$ is K-rational and then take the product.)

3.6 (a) Let $S = \sum_{n\geq 0} a_n x^n \in \mathbb{C}[[x]]$ be rational with a dominating eigenvalue λ. Let $S = P/Q(1-\lambda x)$, with $P, Q \in \mathbb{C}[x]$ and $Q(0) = 1$, in lowest terms. Show that $(x^{-n}S)Q(1-\lambda x)$ is a polynomial of degree ultimately equal to $\deg(Q)$ and that $\lim_{n\to\infty}(x^{-n}S)Q(1-\lambda x)/a_n = Q$, with coefficientwise limit.

(b) Modify Lemma 3.5 so that the conclusion includes the property that $(Bx)^* \prod_{i=2}^{d}(1-\lambda_i x)$ has positive coefficients.

(c) Let $S(x) = N(x)/D(x)$, with $D(x)$ equal to the Soittola denominator (3.3), with the condition that $(Bx)^* E$ has positive coefficients, where $D(x) = (1-\lambda x)E(x)$ and λ is the dominating root. Define $x^{-n}S = a_n R_n(x)/D(x)$. Show that $(Bx)^* R_n(x)$ has positive coefficients for n large enough. Deduce that S is K_+-rational.

(d) Deduce an alternative proof of Soittola's theorem in the case where the dominant eigenvalue is simple. See Katayama *et al.* (1978).

3.7 By drawing the weighted automaton associated to a Perrin companion matrix, give another proof of Theorem 4.1, see Perrin (1992).

3.8 Show that, for integers $n, N \geq 2$, the series

$$\frac{1}{1-nt+t^N} \quad \text{and} \quad \frac{1}{1-nt+t^N(1-t)/(1-t^N)}$$

are $\mathbb{N}$-rational. (*Hint*: Set $Q(t) = 1-(n-1)t - \cdots - (n-1)t^{N-1}$ and compute $(1-t)(1-Q(t))$.)

4.1 Let $A = \{a, b\}$. A *Dyck word* over A is a word w such that $|w|_a = |w|_b$ and $|u|_a \geq |u|_b$ for each prefix u of w. The *height* of a Dyck word w is $\max\{|u|_a - |u|_b\}$, where u ranges over the prefixes of w. The first Dyck words are

$$1, ab, aabb, abab, aaabbb, aababb, aabbab, abaabb, ababab, \ldots$$

The words $aabb$, $aababb$, $abaabb$ have height 2. Denote by D the set of Dyck words over A.

(a) Show that $\underline{D} = 1 + a\underline{D}b\underline{D}$.

(b) Denote by D_h the set of Dyck words of height at most h. In particular $D_0 = \{1\}$ is just composed of the empty word. Show that for $h \geq 0$, $\underline{D}_{h+1} = 1 + a\underline{D}_h b\underline{D}_{h+1}$.

Set $f(x) = \sum_{n\geq 0} \mathrm{Card}(D \cap A^{2n})x^n$, and $f_h(x) = \sum_{n\geq 0} \mathrm{Card}(D_h \cap A^{2n})x^n$. These are the generating functions of the number of Dyck words (Dyck words of height at most h).

(c) Show that $f = (xf)^*$ and that $f_{h+1} = (xf_h)^*$ for $h \geq 0$.

(d) Show that $f_h = q_{h-1}/q_h$ for $h \geq 0$, where $q_{h+1} = q_h - xq_{h-1}$ for $h \geq 0$, with $q_0 = q_{-1} = 1$.

(e) Give an expression of star height at most 2 for f_3, f_4, f_5.

Notes to Chapter 8

A proof of Theorem 1.1 based on the Perron–Frobenius theorem has been given by Fliess (1975).

The proof of Theorem 3.1 given here is based on Soittola (1976), Perrin (1992). The proof of Theorem 3.1 by Katayama *et al.* (1978) seems to have a serious gap, see the final comments in Berstel and Reutenauer (2008a); however it works in the case of a simple dominant eigenvalue, and this is summarized in Exercise 3.6. Recently, algorithmic aspects of the construction have been considered in Barcucci *et al.* (2001) and in Koutschan (2005, 2008). The example of Exercise 1.3 is from Gessel (2003), Exercise 1.4 is from Koutschan (2008). Exercises 1.5 and 3.1 are from an unpublished paper of the late C. Birger, 1971, see also Salomaa and Soittola (1978). In Halava *et al.* (2006) it is shown that it is decidable, for second order linear recurrences, whether all terms are nonnegative. This has been extended to third order recurrences in Laohakosol and Tangsupphathawat (2009). A related result is given in Bell and Gerhold (2007): the authors consider the density of the nonnegative terms. Exercise 1.6 is from Lavallée (2008). Exercise 1.7 is adapted from a result of Bassino (1997).

An open problem, in relation to Theorem 3.1, is the characterization of those polynomials whose inverses are $\mathbb{N}$-rational or $\mathbb{R}_+$-rational series. Polynomials of the form $\det(1 - Mx)$, where M is a square matrix with coefficients in $\mathbb{N}$, are examples of such polynomials (see Exercise 3.5). There are polynomials whose inverse is $\mathbb{N}$-rational and which are not of this kind. An example is $1 - 3x + 5x^2 - 8x^3$: its inverse is $\mathbb{N}$-rational (see Barcucci *et al.* (2001)) but is not of the form $\det(1 - Mx)$ (see Lavallée (2009)).

Exercise 3.5 was suggested by Aaron Lauve, following ideas of Gelfand *et al.* (2005) and Konvalinka and Pak (2007). The fractions in Exercise 3.8 are from Berger (2008).

There is a general metamathematical principle that goes back to M.-P. Schützenberger and that states the following: whenever a rational series in one variable counts a class of objects, then the series is $\mathbb{N}$-rational. This phenomenon has been observed on a large number of examples: generating series and zeta functions in combinatorics, Hilbert series of graded or filtered algebras, growth series of monoids or of groups. See the introduction of Reutenauer (1997).

Part III

Applications

9

Matrix semigroups and applications

In the first section, we show that the size of a finite semigroup of matrices can be bounded (Theorem 1.1). This implies that the finiteness is decidable for a matrix semigroup. As a consequence, one can decide whether the image of a rational series is finite.

In Section 9.2, series with polynomial growth are studied. We give deep results of Schützenberger characterizing these series (Theorem 2.5 and Corollary 2.6). To give a flavor of these results, we consider the following example:

$$\sum_{n\geq 0} n^2 x^n = \frac{x}{1-x} + 3\Big(\frac{x}{1-x}\Big)^2 + 2\Big(\frac{x}{1-x}\Big)^3.$$

This identity shows that the rational series on the left-hand side, which is polynomially bounded, belongs to the subalgebra of series generated by the rational series with coefficients 0 and 1; moreover, the degree of growth of the series is 2, and at the right one performs only products of $2 + 1 = 3$ series with coefficients 0 and 1. The results of Schützenberger reported in this chapter extend, to the noncommutative case, this kind of identity. To do this, Schützenberger uses tools from noncommutative algebra. A particular case of these tools is the Burnside problem for matrix semigroups.

To complete the chapter, we give Simon's results on finite groups of matrices over the tropical semiring and their consequences on limited languages.

9.1 Finite matrix semigroups and the Burnside problem

We first give a result concerning finite monoids of matrices. Recall that for a given word w, we denote by w^* the submonoid generated by w.

Theorem 1.1 (Jacob 1978, Mandel and Simon 1977) *Let $\mu : A^* \to \mathbb{Q}^{n\times n}$ be a monoid morphism such that, for all $w \in A^*$, the monoid μw^* is finite. Then there exists an effectively computable integer N, depending only on* $\operatorname{Card} A$ *and n, such that* $\operatorname{Card} \mu(A^*) \leq N$.

As we shall see, the function $(\operatorname{Card} A, n) \mapsto N$ grows extremely rapidly. There exists however one case where there is a reasonable bound (which moreover does not depend on Card A), namely the case described in the lemma below.

A set E of matrices in $\mathbb{Q}^{n\times n}$ is called *irreducible* if there is no subspace of $\mathbb{Q}^{1\times n}$ other than 0 and $\mathbb{Q}^{1\times n}$ which is invariant for all matrices in E (the matrices act on the right on $\mathbb{Q}^{1\times n}$).

Lemma 1.2 (Schützenberger 1962b) *Let $M \subset \mathbb{Q}^{n\times n}$ be an irreducible monoid of matrices such that all nonvanishing eigenvalues of matrices in M are roots of unity. Then* $\operatorname{Card} M \le (2n+1)^{n^2}$.

Proof. Let $m \in M$. The eigenvalues $\neq 0$ of m are roots of unity, whence algebraic integers over $\mathbb{Z}$. Hence $\operatorname{tr}(m)$ is an algebraic integer. Since $\operatorname{tr}(m) \in \mathbb{Q}$ and $\mathbb{Z}$ is integrally closed, this implies that $\operatorname{tr}(m) \in \mathbb{Z}$. The norm of each eigenvalue is 0 or 1. Thus $|\operatorname{tr}(m)| \le n$. This shows that $\operatorname{tr}(m)$ takes at most $2n+1$ distinct values for $m \in M$.

Let $m_1, \ldots, m_k \in M$ be a basis of the subspace N of $\mathbb{Q}^{n\times n}$ generated by M. Clearly $k \le n^2$. Define an equivalence relation $\sim$ on M by

$$m \sim m' \iff \operatorname{tr}(mm_i) = \operatorname{tr}(m'm_i) \ \text{ for } i = 1, \ldots, k\,.$$

The number of equivalence classes of this relation is at most $(2n+1)^k$. In order to prove the lemma, it suffices to show that $m \sim m'$ implies $m = m'$.

Let $m, m' \in M$ be such that $m \sim m'$. Set $p = m - m'$, and assume $p \neq 0$. There exists a vector $v \in \mathbb{Q}^{1\times n}$ such that $vp \neq 0$. It follows that the subspace vpN of $\mathbb{Q}^{1\times n}$ is not the null space. Since it is invariant under M and M is irreducible, one has $vpN = \mathbb{Q}^{1\times n}$. Consequently, there exists some $q \in N$ such that $vpq = v$. This shows that pq has the eigenvalue 1. Now, for all integers $j \ge 1$,

$$\operatorname{tr}((pq)^j) = \operatorname{tr}(pq(pq)^{j-1}) = 0$$

because $q(pq)^{j-1}$ is a linear combination of the matrices $m_1, \ldots, m_k$, and by assumption $\operatorname{tr}(pr) = 0$ for $r \in M$. Newton's formulas imply that all eigenvalues of pq vanish. This yields a contradiction. □

For the proof of Theorem 1.1, we need another lemma.

Lemma 1.3 (Schützenberger 1962b)

(i) *Let α be a morphism from A^* into a finite monoid M. Then, for each word w of length* $\ge \operatorname{Card}(M)^2$, *there exists a factorization $w = x'zx''$ with $z \neq 1$, $\alpha x' = \alpha(x'z)$ and $\alpha(zx'') = \alpha x''$.*

(ii) *Let $\mu : A^* \to \mathbb{Q}^{n\times n}$ be a multiplicative morphism of the form $\begin{pmatrix} \mu' & \nu \\ 0 & \mu'' \end{pmatrix}$, and let $w = x'zx'' \in A^*$ be such that $\mu'x' = \mu'(x'z)$ and $\mu''(zx'') = \mu''x''$. Then for any n in $\mathbb{N}$,*

$$\begin{aligned} \mu'x'\nu z^n\mu''x'' &= n\,\mu'x'\nu z\mu''x''\,, \\ \nu(x'z^nx'') &= \nu(x'x'') + n\,\mu'x'\nu z\mu''x''\,. \end{aligned} \tag{1.1}$$

Proof. (i) Indeed, the set $\{(x, y) \in (A^*)^2 \mid w = xy\}$ has at least $1 + \mathrm{Card}(M)^2$ elements, and therefore there exist two distinct factorizations

$$w = x'y' = y''x''$$

such that

$$\alpha x' = \alpha y'' \quad \text{and} \quad \alpha y' = \alpha x''\,.$$

We may assume that $|x'| < |y''|$. Then there is a word $z \neq 1$ such that $y'' = x'z$ and $y' = zx''$. Thus $w = x'zx''$ with the required properties.

(ii) One has the identity

$$\begin{pmatrix} a & b \\ 0 & c \end{pmatrix}^n = \begin{pmatrix} a^n & \sum_{k+\ell=n-1} a^k b c^\ell \\ 0 & c^n \end{pmatrix}.$$

Thus

$$\nu(z^n) = \sum_{k+\ell=n-1} \mu'(z^k)\nu z \mu''(z^\ell)\,.$$

Multiplying on the left by $\mu' x'$ and on the right by $\mu'' x''$, we obtain

$$\begin{aligned} \mu' x' \nu z^n \mu'' x'' &= \sum \mu' x' \mu'(z^k) \nu z \mu''(z^\ell) \mu'' x'' \\ &= \sum \mu'(x' z^k) \nu z \mu''(z^\ell x'') = n\, \mu' x' \nu z \mu'' x''\,. \end{aligned}$$

Finally by considering the product $\mu(x'z^n x'') = \mu x' \mu z^n \mu x''$, we obtain

$$\begin{aligned} \nu(x'z^n x'') &= \nu x' \mu''(z^n x'') + \mu' x' \nu(z^n) \mu'' x'' + \mu'(x' z^n) \nu x'' \\ &= \nu x' \mu'' x'' + n\, \mu' x' \nu z \mu'' x'' + \mu' x' \nu x'' \\ &= \nu(x' x'') + n\, \mu' \nu z \mu'' x''\,. \end{aligned}$$

□

Corollary 1.4 (Schützenberger 1962b) *Let $\mu : A^* \to \mathbb{Q}^{n\times n}$ be a morphism into a monoid of matrices which are triangular by blocks*

$$\mu = \begin{pmatrix} \mu' & \nu \\ 0 & \mu'' \end{pmatrix}.$$

Assume that $\mu' A^*$ and $\mu'' A^*$ are finite, and that μw^* is finite for any word w. Then

$$\mathrm{Card}(\nu A^*) \leq \sum_{0 \leq i < (H'H'')^2} \mathrm{Card}\, \nu A^i\,,$$

where $H' = \mathrm{Card}\, \mu' A^*$ and $H'' = \mathrm{Card}\, \mu'' A^*$.

Proof. In Lemma 1.3(i), take $\alpha = (\mu', \mu'')$. Then each word w of length $\geq (H'H'')^2$ has a factorization $w = x'zx''$ with $z \neq 1$ and the relations (1.1) hold. Thus, since μz^* is finite, $\nu(x'z^*x'')$ is also finite and we must have $\mu' x' \nu z \mu'' x'' = 0$ and $\nu w = \nu(x'x'')$. Since $|x'x''| < |w|$, the corollary follows. □

Proof of Theorem 1.1. Assume first that the monoid μA^* is irreducible, and consider any matrix $\mu w \in \mu A^*$. Since μz^* is finite, there are integers $0 \leq i < j$ with

$\mu w^i = \mu w^j$. This implies that the eigenvalues of w are 0 or roots of unity. The theorem thus follows from Corollary 1.4.

If μA^* is not irreducible, there is some subspace V of $\mathbb{Q}^{1\times n}$ which is invariant under μA^*. Consider a supplementary space W of V. In a basis which is adapted to the decomposition $\mathbb{Q}^{1\times n} = W \oplus V$, the morphism μ admits the form described in Lemma 1.3. Arguing by induction on the dimension of the representation, the result follows from Corollary 1.4. □

Recall that an element s of a semigroup S is a *torsion element* if s generates a finite subsemigroup of S; equivalently, $s^k = s^\ell$ for some $1 \le k < \ell$. We say that S is a *torsion semigroup* if each element in S is a torsion element.

Corollary 1.5 (McNaughton and Zalcstein 1975) *Every finitely generated torsion semigroup of square matrices over $\mathbb{Q}$ is finite.* □

Recall that a *ray* is a subset of A^* of the form uv^*w, with $u, v, w \in A^*$.

Corollary 1.6 *Let $S \in \mathbb{Q}\langle\langle A\rangle\rangle$ be a rational series such that for any ray R, the set $\{(S, w) \mid w \in R\}$ is finite. Then the set of coefficients of S is finite.*

Proof. Let (λ, μ, γ) be a minimal linear representation of S. By Corollary 2.2.3, there exist polynomials $P_1, \ldots, P_n, Q_1, \ldots, Q_n$ such that for all words w,

$$\mu w = ((S, P_i w Q_j))_{1\le i,j\le n} \,.$$

By assumption, the set $\{(S, uw^m v) \mid m \in \mathbb{N}\}$ is finite for all words u, v, w. The same holds for the set $\{(S, Pw^m Q) \mid m \in \mathbb{N}\}$ where P, Q are polynomials. This shows that μw^* is finite for any word w. By Corollary 1.5, the monoid μA^* is finite, and in particular

$$\{(S, w) \mid w \in A^*\}$$

is finite, since $(S, w) = \lambda\mu w\gamma$. □

Corollary 1.7 (Jacob 1978) *It is decidable whether a finite set of matrices over $\mathbb{Q}$ generates a finite monoid.*

Proof. By Theorem 1.1, there is an upper bound on the size of such a monoid if it is finite. Let E be a finite set of matrices, M the monoid generated by E, and let N be the upper bound given in Theorem 1.1. Then M is finite if and only if every product of N matrices in E equals a product of at most $N - 1$ matrices in E. This last condition is clearly decidable. □

Recall that the image of a series is the set of its coefficients.

Corollary 1.8 (Jacob 1978) *It is decidable whether a rational series has a finite image.* □

9.2 Polynomial growth

We now turn our attention to questions concerning growth of rational series over $\mathbb{Z}$. Recall that a series $S \in \mathbb{Z}\langle\langle A\rangle\rangle$ has *polynomial growth* or is *polynomially bounded* if there exist a real number $q \geq 0$ and a real number C such that

$$|(S, w)| \leq C|w|^q$$

for all nonempty words w. The smallest of these q, if it exists, is called the *degree of growth* of S. Observe that series with degree of growth 0 are precisely the series with finite image.

In the sequel, we shall consider morphisms $\mu : A^* \to \mathbb{Q}^{n\times n}$ which have the block-triangular form

$$\mu = \begin{pmatrix} \mu_0 & \nu_1 & \star & \cdots & \star \\ 0 & \ddots & \ddots & \ddots & \vdots \\ \vdots & & \ddots & \ddots & \star \\ \vdots & & & \ddots & \nu_q \\ 0 & \cdots & \cdots & 0 & \mu_q \end{pmatrix} \tag{2.1}$$

where each μ_i is square and is therefore itself a morphism.

Theorem 2.1 *Let $S \in \mathbb{Z}\langle\langle A\rangle\rangle$ be a rational series and let (λ, μ, γ) be a minimal linear representation of S. Then S has polynomial growth if and only if the set $\{\mathrm{tr}(\mu w) \mid w \in A^*\}$ is finite.*

Proof. Suppose first that S has polynomial growth. Then there exist, by Corollary 2.2.3, real numbers C, q such that for all i, j, $|(\mu w)_{i,j}| \leq C|w|^q$ for all nonempty words w. It follows that, for any $r \in \mathbb{N} \setminus 0$, we have $|(\mu w^r)_{i,j}| \leq Cr^q|w|^q$. Consequently, for every eigenvalue ρ of μw one has

$$|\rho|^r \leq C'r^q$$

for some constant C'. Thus $|\rho| \leq 1$. This implies that $-n \leq \mathrm{tr}(\mu w) \leq n$, where n is the dimension of μ. Since S is $\mathbb{Z}$-rational, there exists a minimal linear representation with coefficients in $\mathbb{Z}$ (Theorem 7.1.1). This representation is similar to (λ, μ, γ) by Theorem 2.2.4 and consequently, the trace of any matrix μw is an integer. Hence each $\mathrm{tr}(\mu w)$ is in $\{-n, \ldots, n\}$.

Conversely, suppose that the set $\{\mathrm{tr}(\mu w) \mid w \in A^*\}$ is finite. Let w be a word and let $\lambda_1, \ldots, \lambda_n$ be the eigenvalues of μw with their multiplicities. The sequence

$$a_p = \sum_{1\leq i\leq n} \lambda_i^p = \mathrm{tr}(\mu w^p)$$

takes only a finite number of distinct values. Since it satisfies a linear recurrence relation (by Exercise 6.1.2), it is ultimately periodic, and there is a relation

$$a_{p+h} = a_{p+k} \quad p \geq 0$$

for some $h, k \in \mathbb{N}$, $h > k$. The minimal polynomial (see Section 6.1) of the rational series $\sum_{p \in \mathbb{N}} a_p x^p$ divides the polynomial $x^h - x^k$. Consequently, the eigenvalues of this series (in the sense defined in Section 6.1) are roots of unity or 0. In view of the uniqueness of the exponential polynomial (Section 6.2), the λ_i are therefore roots of unity or 0.

Next, if the monoid μA^* is not irreducible, then μ can be put, by changing the basis, into the form

$$\mu = \begin{pmatrix} \mu' & \nu \\ 0 & \mu'' \end{pmatrix}.$$

Arguing by induction, μ is equivalent to a morphism of the form (2.1) with each $\mu_i A^*$ irreducible. By Lemma 1.2 and by the previous remarks, all monoids $\mu_i A^*$ are finite. To complete the proof, it suffices to apply the following two lemmas. □

Lemma 2.2 *Let K be a commutative semiring.*

(i) *Let*

$$\mu = \begin{pmatrix} \mu' & \nu \\ 0 & \mu'' \end{pmatrix}$$

be a morphism $A^* \to K^{n \times n}$, where μ', μ'' are themselves morphisms. Every series recognized by μ is a linear combination of series recognized by μ' or by μ'' and of series of the form $S'aS''$, where S' is recognized by μ', $a \in A$ and S'' is recognized by μ''.

(ii) *If $\mu : A^* \to K^{n \times n}$ has the form* (2.1) *with each μ_i of finite image, then each series recognized by μ is a linear combination of products of at most $k+1$ characteristic series of rational languages.*

Proof. (i) A series recognized by μ is a linear combinations of series of the form

$$\sum_w (\mu w)_{i,j} w \tag{2.2}$$

with $0 \leq i, j \leq n$. It suffices to show that when i, j are coordinates corresponding to ν, the series (2.2) is a linear combination of series of the form $S'aS''$. This is a consequence of the formula

$$\nu w = \sum_{w = xay} \mu' x \nu a \mu'' y\,.$$

(ii) Using (i) iteratively, we see that a series recognized by μ is a K-linear combination of series of the form $S_0 a_1 S_1 a_2 \cdots a_\ell S_\ell$, with $\ell \leq k$, where $a_i \in A$ and each S_i is recognized by some μ_j. Since $\mu_j(A^*)$ is a finite monoid, each language $\mu_j^{-1}(m)$ is rational by Theorem 3.1.1 (Kleene's theorem). Hence a series recognized by μ_j is a linear combination of characteristic series of rational languages and this concludes the proof. □

Lemma 2.3 (i) *Let S, T be two series over $\mathbb{R}$ and $p, q \in \mathbb{N}$. If S has degree of growth q and T has degree of growth p, then ST has degree of growth at most $p + q + 1$.*

(ii) *The product of $q + 1$ characteristic series of rational languages has degree of growth at most q.*

Proof. (i) We have $|(S, w)| \leq C\binom{|w|+q}{q}$ and $(T, w)| \leq D\binom{|w|+p}{p}$ for suitable constants C, D. Since $(ST, w) = \sum_{w=uv}(S, u)(T, v)$, it follows that

$$|(ST, w)| \leq CD \sum_{w=uv} \binom{|u|+q}{q}\binom{|v|+p}{p}.$$

The summation is equal to the coefficient of $x^{|w|}$ in the product

$$\sum_i \binom{i+q}{q} x^i \sum_j \binom{j+p}{p} x^j.$$

Since $\sum_i \binom{i+q}{q} x^i = 1/(1-x)^{q+1}$, we obtain that this coefficient is $\binom{|w|+p+q+1}{p+q+1}$. Since this is a polynomial in $|w|$ of degree $p + q + 1$, the assertion follows.

(ii) follows from (i) by induction. □

Corollary 2.4 *It is decidable whether a rational series $S \in \mathbb{Z}\langle\langle A\rangle\rangle$ has polynomial growth.*

Proof. A minimal linear representation (λ, μ, γ) of S can effectively be computed. Then according to Theorem 2.1, the series S has polynomial growth if and only if the series

$$\sum_w \mathrm{tr}(\mu w)w$$

has a finite image. This series is rational (Lemma 2.1.3) and it is decidable, by Corollary 1.8 whether a rational series has a finite image. □

The main result of this section is the following theorem.

Theorem 2.5 (Schützenberger 1962b) *Let S be a $\mathbb{Z}$-rational series which has polynomial growth. Then S has a minimal linear representation (λ, μ, γ) whose coefficients are in $\mathbb{Z}$, and such that μ has the block-triangular form* (2.1) *where each $\mu_i A^*$ is a finite monoid. Moreover, let q be the smallest integer for which this holds. Then the degree of growth of S exists and is equal to q and there exist words $x_0, \ldots, x_q$, $y_1, \ldots, y_q$ such that $(S, x_0 y_1^n x_1 \cdots y_q^n x_q)$ is a polynomial in n of degree q.*

Corollary 2.6 (Schützenberger 1962b) *The degree of growth of a polynomially bounded Z-rational series S is equal to the smallest integer q such that S belongs to the submodule of $\mathbb{Z}\langle\langle A\rangle\rangle$ spanned by the products of at most $q + 1$ characteristic series of rational languages.*

Proof. Suppose that the degree of growth of S is q. Then, by the theorem, there exists a linear representation (λ, μ, γ) of S with μ of the form (2.1) with each $\mu_i A^*$ finite. By Lemma 2.2(ii), we get that the series S is a $\mathbb{Z}$-linear combination of products of no more than $q+1$ characteristic series of rational languages.

Conversely, suppose that S is of this form. Then by Lemma 2.3, the series S has degree of growth $\leq q$, and this proves the corollary. □

Recall that, given a ring K, two representations $\mu, \mu' : A^* \to K^{n\times n}$ are called *similar* if, for some invertible matrix P over K, one has

$$\mu' w = P^{-1} \mu w P$$

for any word w. In other words, μ' is obtained from μ after a change of basis over K. When several rings occur, we will emphasize this by saying *similar over K*.

In the proof of Lemma 2.8, we use the following result.

Theorem 2.7 (see (Lang 1984), Proposition III.7.8) *For any submodule V of $\mathbb{Z}^n$, there is a basis $e_1, \ldots, e_n$ of the $\mathbb{Z}$-module $\mathbb{Z}^n$ and integers $d_1, \ldots, d_k \geq 1$ such that $d_1e_1, \ldots, d_ke_k$ is a basis of the $\mathbb{Z}$-module V.*

Lemma 2.8 *Let $\mu : A^* \to \mathbb{Z}^{n\times n}$ be a representation. Suppose that μ is similar over $\mathbb{Q}$ to a representation $\mu' : A^* \to \mathbb{Q}^{n\times n}$ which has the block-triangular form*

$$\mu' = \begin{pmatrix} \mu_0 & \star & \cdots & \star \\ 0 & \mu_1 & \ddots & \vdots \\ \vdots & \ddots & \ddots & \star \\ 0 & \cdots & 0 & \mu_q \end{pmatrix}$$

with square diagonal blocks. Then μ is similar over $\mathbb{Z}$ to a representation $\mu'' : A^* \to \mathbb{Z}^{n\times n}$ having the same form and such that the corresponding diagonal blocks of μ' and μ'' are similar over $\mathbb{Q}$.

Proof. The hypothesis means that there is a basis of the $\mathbb{Q}$-vector space $\mathbb{Q}^{n\times 1}$ of column vectors of the form $B_0 \cup \cdots \cup B_q$ such that for any word w, the matrix μw sends the subspace E_i spanned by $B_0 \cup \cdots \cup B_i$ into itself, and that $\mu_i w$ represents the action of μw on B_i modulo E_{i-1}. We put $E_{-1} = 0$.

It suffices therefore to show the existence of a $\mathbb{Z}$-basis of $\mathbb{Z}^{n\times 1}$ of the form $C_0 \cup \cdots \cup C_q$ such that E_i is also spanned over $\mathbb{Q}$ by $C_0 \cup \cdots \cup C_i$. Then C_i, as is B_i, will be a $\mathbb{Q}$-basis of E_i modulo E_{i-1} and therefore the diagonal blocks will be similar over $\mathbb{Q}$, as in the statement.

If V is a submodule of $\mathbb{Z}^n$, then by Theorem 2.7 it has a basis $d_1e_1, \ldots, d_ke_k$ for some basis $e_1, \ldots, e_n$ of $\mathbb{Z}^n$ and some nonzero integers $d_1, \ldots, d_k$. If V is *divisible* (that is, $dv \in V$ and $d \in \mathbb{Z}, d \neq 0$ imply $v \in V$), then one may choose $d_1 = \cdots = d_k = 1$. In other words, given a divisible submodule V of a finitely generated free $\mathbb{Z}$-module F, there exists a free submodule W such that $F = V \oplus W$.

Let $V_i = E_i \cap \mathbb{Z}^{n\times 1}$. These submodules of $\mathbb{Z}^{n\times 1}$ are all divisible and $0 = V_{-1} \subseteq V_0 \subseteq \cdots \subseteq V_q = \mathbb{Z}^{n\times 1}$. Thus we may find free submodules W_i of $\mathbb{Z}^{n\times 1}$ such that

$V_i = V_{i-1} \oplus W_i$ for $i = 0, \dots, q$. Let C_i be a $\mathbb{Z}$-basis of W_i. Then $C_0 \cup \dots \cup C_i$ is a $\mathbb{Z}$-basis of V_i and therefore E_i is spanned over $\mathbb{Q}$ by $C_0 \cup \dots \cup C_i$. □

Proof of Theorem 2.5, first part. Let $S \in \mathbb{Z}\langle\langle A \rangle\rangle$ be a rational series having polynomial growth, and let (λ, μ, γ) be a minimal linear representation of S. We may assume, by Theorem 7.1.1, that (λ, μ, γ) has integral coefficients. The proof of Theorem 2.1 shows that, after a change of the basis of $\mathbb{Q}^{1\times n}$, μ has a decomposition of the form (2.1) where each $\mu_i A^*$ is finite. By Lemma 2.8, the change of basis can be done in $\mathbb{Z}^{1\times n}$. □

Lemma 2.9 (Schützenberger 1962b) *Let $\mu : A^* \to \mathbb{Z}^{n\times n}$ be a representation of the form*

$$\mu = \begin{pmatrix} \mu' & \nu \\ 0 & \mu'' \end{pmatrix},$$

where μ', μ'' are representations of finite image. If $(\nu A^*)v$ is finite for some nonnull vector v of the appropriate size, then μ is similar over $\mathbb{Z}$ to a representation

$$\overline{\mu} = \begin{pmatrix} \mu_1 & \overline{\nu} \\ 0 & \mu_2 \end{pmatrix},$$

where μ_1 and μ_2 are representations of finite image and with $\dim(\mu_1) > \dim(\mu')$.

Proof. By Lemma 2.8, we may work over $\mathbb{Q}$. Let $F = \{u \in \mathbb{Q}^{n\times 1} \mid (\mu A^*)u \text{ is finite}\}$. Then F is invariant under each μw. Let also E', E'' be the subspaces of $\mathbb{Q}^{n\times 1}$ corresponding to μ' and μ''. Then $E' \subseteq F$. Moreover, E'' is a direct sum $E'' = (E'' \cap F) \oplus E''_1$. Taking a basis of E'' corresponding to this direct sum, we see that μ'' is similar to a representation of the form $\begin{pmatrix} \mu''_1 & \nu' \\ 0 & \mu''_2 \end{pmatrix}$. Thus μ is similar to a representation of the form

$$\begin{pmatrix} \mu' & \nu_1 & \nu_2 \\ 0 & \mu''_1 & \nu' \\ 0 & 0 & \mu''_2 \end{pmatrix}.$$

We have

$$F = E' \oplus (E'' \cap F), \tag{2.3}$$

since $E' \subseteq F$ and $\mathbb{Q}^{n\times 1} = E' \oplus E''$. Consequently, for any vector u in F, the set of vectors $\begin{pmatrix} \mu' A^* & \nu_1 A^* \\ 0 & \mu''_1 A^* \end{pmatrix} u$ is finite. Hence $\begin{pmatrix} \mu' & \nu_1 \\ 0 & \mu''_1 \end{pmatrix}$ has finite image. Moreover, μ''_2 also has finite image, since it is a subrepresentation of μ''. Taking

$$\mu_1 = \begin{pmatrix} \mu' & \nu_1 \\ 0 & \mu''_1 \end{pmatrix}, \quad \overline{\nu} = \begin{pmatrix} \nu_2 \\ \nu' \end{pmatrix}, \quad \mu_2 = \mu''_2,$$

we see that μ is similar to $\begin{pmatrix} \mu_1 & \overline{\nu} \\ 0 & \mu_2 \end{pmatrix}$.

Now, if $(\nu A^*)v$ is finite for some nonnull vector v, then F is strictly larger than E' and consequently $\dim(\mu_1) = \dim(\mu') + \dim(\mu_1'') > \dim(\mu')$, since $\dim(\mu_1'') = \dim(E'' \cap F) > 0$ by (2.3), because $\dim F > \dim E'$. □

Lemma 2.10 (Schützenberger 1962b) *Let $\mu : A^* \to \mathbb{Q}^{n\times n}$ be a representation of the form*

$$\mu = \begin{pmatrix} \mu' & \nu \\ 0 & \mu'' \end{pmatrix},$$

where μ', μ'' are representations of finite image, and let $\alpha : A^* \to M$ be a morphism of A^* into a finite monoid M. Let v be a vector such that $(\nu A^*)v$ is infinite. Then there exist words x', z, x'' in A^* such that $\mu' x' \nu z \mu'' x'' v \neq 0$, $\alpha(x'z) = \alpha x'$, $\alpha(zx'') = \alpha x''$ and $\alpha(z^2) = \alpha z$.

Proof. We claim that for each vector v with $(\nu A^*)v$ infinite, there exist words x', z, x'' in A^* such that $\alpha(x'z) = \alpha x'$, $\alpha(zx'') = \alpha x''$ and $\mu' x' \nu z \mu'' x'' v \neq 0$. Indeed, arguing by contradiction, let w be a word of length greater than or equal to $(\mathrm{Card}(M)\,\mathrm{Card}(\mu' A^*)\,\mathrm{Card}(\mu'' A^*))^2$. Then by Lemma 1.3(i), there exists a factorization $w = x'zx''$ with z nonempty and $\varphi(x'z) = \varphi(x')$, $\varphi(zx'') = \varphi(x'')$, where $\varphi = (\alpha, \mu', \mu'')$. Then, by assumption, we have $\mu' x' \nu z \mu'' x'' v = 0$. By Lemma 1.3(ii), $\nu(w)v = \nu(x'zx'')v = \nu(x'x'')v$, and since $x'x''$ is shorter than w, we contradict the hypothesis that $(\nu A^*)v$ is infinite, and the claim is proved.

Now $\alpha(z^n)$ is idempotent for some $n \geq 1$. Since $\mu' x' \nu z^n \mu'' x'' = n\, \mu' x' \nu z \mu'' x''$ by Lemma 1.3(ii), the lemma is proved by replacing z by z^n. □

In the sequel, we will consider matrices having an upper triangular form

$$m = \begin{pmatrix} m_{0,0} & m_{0,1} & \cdots & m_{0,q} \\ 0 & m_{1,1} & & \vdots \\ \vdots & \ddots & \ddots & \vdots \\ 0 & \cdots & 0 & m_{q,q} \end{pmatrix} \tag{2.4}$$

where each $m_{i,j}$ is a matrix of fixed size depending on i and j, with $m_{i,i}$ square. We denote by $\mathcal{M}$ this set of matrices. Also, we call *matrix polynomial in n* a matrix of the form

$$m_0 + nm_1 + \cdots + n^d m_d\,,$$

where the m_i are matrices of the same size. If $m_d \neq 0$, then d is the *degree* of this matrix polynomial. If $d = 0$ we say that the polynomial is *constant*.

More generally, we consider also matrix polynomials in several commuting variables $n, n_1, n_2, \ldots$. We denote by $\mathcal{P}$ the set of matrices $m \in \mathcal{M}$ such that each $m_{i,j}$ is a matrix polynomial in n over $\mathbb{Q}$ of degree at most $j - i$.

Lemma 2.11 (i) *$\mathcal{P}$ is a ring.*

(ii) *Let $M_1(n), \ldots, M_q(n) \in \mathcal{P}$. Write $M_k = (m_{i,j}^{(k)})$ in accordance with (2.4). Then the product $M_1(nn_1) \cdots M_q(nn_q)$ is a matrix polynomial in $n, n_1, \ldots, n_q$ and the coefficient of $n^q n_1 \cdots n_q$ in this polynomial is a matrix of the form (2.4) with all $m_{i,j} = 0$ except $m_{0,q} = \bar{m}_{0,1}^{(1)} \bar{m}_{1,2}^{(2)} \cdots \bar{m}_{q-1,q}^{(q-1)}$ where $m_{i-1,i}^{(i)} = \bar{m}_{i-1,i}^{(i)} n +$ constant, for $i = 1, \ldots, q$.*

Proof. (i) Let m, n be of the form (2.4). Then $p = mn$ has the same form, and for $i \leq k$,

$$p_{i,k} = \sum_{i \leq j \leq k} m_{i,j} n_{j,k} \,.$$

If $m_{i,j}$ (resp. $n_{j,k}$) is a matrix polynomial of degree $\leq j - i$ (resp. $k - j$), then $p_{i,k}$ is a matrix polynomial of degree $\leq j - i + k - j = k - i$. Thus $\mathcal{P}$ is a ring.

(ii) Let $p = (p_{i,j}) = M_1(nn_1) \cdots M_q(nn_q)$. We have

$$p_{i,j} = \sum_{i_0 \leq i_1 \leq \cdots \leq i_q} m_{i_0,i_1}^{(1)} m_{i_1,i_2}^{(2)} \cdots m_{i_{q-1},i_q}^{(q)} \,,$$

with $i_0 = i$, $i_q = j$. By hypothesis, $m_{i,j}^{(k)}$ is a matrix polynomial in nn_k of degree $\leq j - i$. Thus a $p_{i,j}$ involving the monomial $n^q n_1 \cdots n_q$ can only be the one with $i = 0$, $j = q$, and the coefficient of this monomial is as indicated. □

Lemma 2.12 (Schützenberger 1962b) *Let a, b, c in $\mathcal{M}$ be such that $a_{i,i} b_{i,i} = a_{i,i}$, $b_{i,i}^2 = b_{i,i}$, $b_{i,i} c_{i,i} = c_{i,i}$ for any $i = 0, \ldots, q$. Set $m^{(n)} = ab^n c$. Then $m^{(n)} \in \mathcal{P}$ and its $i, i+1$ block is $m_{i,i+1}^{(n)} = n a_{i,i} b_{i,i+1} c_{i+1,i+1} + C$, where C is some constant.*

Proof. (i) We compute the n-th power of the matrix b. We first compute its block of coordinates $0, q$. The latter is the sum of all labels of paths of length n from 0 to q in the directed graph with vertices $0, 1, \ldots, q$ and edges $i \to j$, for $i \leq j$, labeled $b_{i,j}$. Such a path has a unique decomposition (abusing slightly the notation)

$$b_{i_0,i_0}^{n_0} b_{i_0,i_1} b_{i_1,i_1}^{n_1} b_{i_1,i_2} \cdots b_{i_{k-1},i_k} b_{i_k,i_k}^{n_k} \,, \tag{2.5}$$

for some vertices $0 = i_0 < i_1 < i_2 < \cdots < i_{k-1} < i_k = q$, $0 \leq k \leq q$, and some exponents $n_0, n_1, \ldots, n_k$ with $n_0 + n_1 + \cdots + n_k + k = n$. Note that $b_{i,i}^h = b_{i,i}$ for $h \geq 1$. Hence, for a fixed k, the sum of the labels of the paths (2.5) is a matrix polynomial of degree $\leq k$ (see Exercise 2.1). Hence the sum of all labels is a polynomial of degree at most q.

Assume now that $q = 1$. Then the paths of (2.5) are of the form $b_{0,0}^{n_0} b_{0,1} b_{1,1}^{n_1}$ with $n_0 + 1 + n_1 = n$. Hence this block of b^n is equal to $nb_{0,0} b_{0,1} b_{1,1} +$ a constant.

Finally, it is easy to generalize this: the i, j-block of b^n is a matrix polynomial of degree $\leq j - i$, and if $j = i + 1$, it is equal to $nb_{i,i} b_{i,i+1} b_{i+1,i+1} +$ some constant.

(ii) We now compute the product $m^{(n)} = ab^n c$. Set $b^n = (d_{i,j})$. Then the u, v-block of the product is

$$m_{u,v}^{(n)} = \sum_{u \leq i \leq j \leq v} a_{u,i} d_{i,j} c_{j,v} \,,$$

which is a sum of matrix polynomials of degree $\leq j-i \leq v-u$, and we are done. In the special case $v = u+1$, the sum is

$$a_{u,u}d_{u,u}c_{u,u+1} + a_{u,u}d_{u,u+1}c_{u+1,u+1} + a_{u,u+1}d_{u+1,u+1}c_{u+1,u+1}\,.$$

The two extreme terms are constants and the middle term is

$$a_{u,u}(nb_{u,u}b_{u,u+1}b_{u+1,u+1} + C)c_{u+1,u+1} = na_{u,u}b_{u,u+1}c_{u+1,u+1} + C'$$

for some constants C and C', since $a_{i,i}b_{i,i} = a_{i,i}$ and $b_{i,i}c_{i,i} = c_{i,i}$. □

Proof of Theorem 2.5, second part.

We may choose, among the linear minimal representations of S of the form (2.1) and with coefficients in $\mathbb{Z}$, a representation having, in lexicographic order from left to right, the largest possible vector $(\dim \mu_0, \dim \mu_1, \ldots, \dim \mu_q)$. This shows, in view of Lemma 2.9, that for $i = 1, \ldots, q$, all the morphisms $\begin{pmatrix} \mu_i & \nu_{i+1} \\ 0 & \mu_{i+1} \end{pmatrix}$ have the property that, for any nonnull vector v_{i+1}, the set $(\nu_{i+1}A^*)v_{i+1}$ is infinite.

Hence, for any such v_{i+1}, there exist by Lemma 2.10, some words x_i', z_{i+1}, x_{i+1}'' such that $\mu_i x_i' \nu_{i+1} z_{i+1} \mu_{i+1} x_{i+1}'' v_{i+1} \neq 0$, and $\overline{\mu}(x_i' z_{i+1}) = \overline{\mu} x_i'$, $\overline{\mu}(z_{i+1} x_{i+1}'') = \overline{\mu} x_{i+1}''$, $\overline{\mu}(z_{i+1}^2) = \overline{\mu} z_{i+1}$, where $\overline{\mu} = (\mu_0, \ldots, \mu_q)$.

Let v_q be some nonzero vector having the size of the last diagonal block. Then we know from the preceding argument the existence of words x_{q-1}', z_q, x_q'' such that $v_{q-1} = \mu_{q-1} x_{q-1}' \nu_q z_q \mu_q x_q'' v_q \neq 0$. Suppose we have defined $v_{i+1}, x_i', z_{i+1}, x_{i+1}''$ such that $v_i = \mu_i x_i' \nu_{i+1} z_{i+1} \mu_{i+1} x_{i+1}'' v_{i+1} \neq 0$. We thus find x_{i-1}', z_i, x_i'' with the above properties such that $v_{i-1} = \mu_{i-1} x_{i-1}' \nu_i z_i \mu_i x_i'' v_i \neq 0$. Finally, we obtain the existence of words $x_0', \ldots, x_{q-1}', z_1, \ldots, z_q, x_1'', \ldots, x_q''$ such that

$$\mu_0 x_0' \nu_1 z_1 \mu_1 x_1'' \mu_1 x_1' \nu_2 z_2 \cdots \mu_{q-1} x_{q-1}' \nu_q z_q \mu_q x_q'' \neq 0\,. \tag{2.6}$$

By Lemma 2.12, the matrix $\mu x_i' \mu z_{i+1}^n \mu x_{i+1}''$ is in $\mathcal{P}$, and its $i, i+1$-block is equal to $n\mu_i x_i' \nu_{i+1} z_{i+1} \mu_{i+1} x_{i+1}'' +$ some constant. This is still true if we replace n by nn_i, with $n_i \geq 1$.

Choose some q-tuple $(n_1, \ldots, n_q)$ of positive integers and form the product

$$\mu x_0' \mu z_1^{nn_1} \mu x_1'' \mu x_1' \mu z_2^{nn_2} \mu x_2'' \mu x_2' \cdots \mu x_{q-1}' \mu z_q^{nn_q} \mu x_q''\,.$$

Since $\mathcal{P}$ is closed under product, this matrix is in $\mathcal{P}$. Consider its $0, q$-block, which is the only one that can have degree q exactly. Viewing it as a matrix polynomial in $n, n_1, \ldots, n_q$, we see by Lemma 2.11(ii) (where we choose $M_i(n) = \mu x_{i-1}' \mu z_i^n \mu x_i''$) that the coefficient of $n^q n_1 n_2 \cdots n_q$ is the left-hand side of (2.6). Thus, we may choose $n_1, \ldots, n_q$ (in the infinite set of positive integers) in such a way that this block has degree q exactly in n.

Now, let $y_i = z_i^{n_i}$ for $i = 1, \ldots, q$ and $x_i = x_i'' x_i'$ for $i = 1, \ldots, q-1$. Then $\mu(x_0' y_1^n x_1 \cdots y_q^n x_q'')$ is a matrix polynomial of degree q exactly, and it follows that $(S, x_0' y_1^n x_1 \cdots y_q^n x_q'')$ is a polynomial in n of degree $\leq q$. Moreover, for any words u, v, $\mu(u x_0' y_1^n x_1 \cdots y_q^n x_q'' v)$ is a matrix polynomial of degree $\leq q$ and therefore $(S, u x_0' y_1^n x_1 \cdots y_q^n x_q'' v)$ is a polynomial of degree $\leq q$. By Corollary 2.2.3, $\mu(x_0' y_1^n x_1 \cdots y_q^n x_q'')$ is a linear combination of polynomials $(S, u x_0' y_1^n x_1 \cdots y_q^n x_q'' v)$

for some words u, v. Hence one of these polynomials in n must have degree exactly q, and we put $x_0 = ux'_0$, $x_q = x''_q v$.

This shows that S has degree of growth at least q, and to conclude the proof, we use Lemma 2.2(ii) and Lemma 2.3(ii). □

The constructions of Sections 9.1 and 9.2 suggest the following problem: given several matrices in $\mathbb{Q}^{n\times n}$, is it decidable whether they have a common stable subspace? can it be effectively computed? These problems may be undecidable, although the first one is decidable if one seeks a subspace in $\mathbb{C}^n$.

9.3 Limited languages and the tropical semiring

Let $L \subset A^*$ be a language. Recall that L^* denotes the submonoid generated by L. Equivalently, $L = \bigcup_{n\geq 0} L^n$. The language L is called *limited* if there exists an integer $m \geq 0$ such that

$$L^* = 1 \cup L \cup \cdots \cup L^m .$$

Suppose that L is a recognizable language, recognized by the automaton $\mathcal{A} = (Q, I, E, T)$, where I, T (the sets of initial and terminal states) are subsets of Q and E is a subset of $Q \times A \times Q$. Let q_0 be a new state, set $Q_0 = q_0 \cup Q$ and let $\mathcal{A}^* = (Q_0, q_0, E_0, q_0)$ be the automaton defined by:

(i) E_0 contains E;
(ii) for each edge $p \xrightarrow{a} q$ in $\mathcal{A}$ with $q \in T$, $p \xrightarrow{a} q_0$ is an edge in $\mathcal{A}^*$;
(iii) for each edge $p \xrightarrow{a} q$ in $\mathcal{A}$ with $p \in I$, $q_0 \xrightarrow{a} q$ is an edge in $\mathcal{A}^*$;
(iv) for each edge $p \xrightarrow{a} q$ in $\mathcal{A}$ with $p \in I, q \in T$, $q_0 \xrightarrow{a} q_0$ is an edge in $\mathcal{A}^*$.

It is easily verified that $\mathcal{A}^*$ recognizes the language L^*.

We show now how to encode the limitedness problem for L into a finiteness problem for a certain semigroup of matrices over the *tropical semiring*. First, we define the latter. It is the semiring, denoted $\mathbb{T}$, whose underlying set is $\mathbb{N} \cup \infty$, with addition $(a, b) \mapsto \min(a, b)$ and product $(a, b) \mapsto a + b$ with $a + \infty = \infty$ and $\min(a, \infty) = a$. Addition and multiplication in $\mathbb{T}$ are commutative and have respective neutral elements ∞ and 0. In particular, the identity matrix over $\mathbb{T}$ has 0 on the diagonal and ∞ elsewhere.

Coming back to the previous automaton, we associate to it a monoid morphism α from A^* into the multiplicative monoid $\mathbb{T}^{Q_0\times Q_0}$ of square matrices over $\mathbb{T}$ indexed by Q_0, defined as follows. For a letter a,

$$(\alpha a)_{p,q} = \begin{cases} \infty & \text{if } p \xrightarrow{a} q \text{ is not an edge of } \mathcal{A}^*; \\ 0 & \text{if } p \xrightarrow{a} q \text{ is an edge of } \mathcal{A}^* \text{ and } q \neq q_0; \\ 1 & \text{if } p \xrightarrow{a} q \text{ is an edge of } \mathcal{A}^* \text{ and } q = q_0. \end{cases}$$

With these notations and definitions, one has the following result.

Proposition 3.1 *A rational language is limited if and only if the associated representation α has finite image.*

Proof. 1. We define the *weight* $\omega(c)$ of a path c in $\mathcal{A}^*$ as the number of edges in c that end at q_0. In particular, the weight of any empty path is 0. We claim that for any word w in A^*, and any $p, q \in Q_0$,

$$(\alpha w)_{p,q} = \min\{\omega(c) \mid c : p \xrightarrow{w} q\}, \tag{3.1}$$

that is, the minimum of the weights of the paths labeled w from p to q (we use here the convention that $\min(\emptyset) = \infty$).

Indeed, if w is the empty word, then the right-hand side of (3.1) is ∞ if $p \neq q$, and is 0 if $p = q$, and this proves (3.1) in this case. If $w = a \in A$, then the right-hand side of (3.1) is ∞ if $p \xrightarrow{a} q$ is not an edge in $\mathcal{A}^*$, it is 0 if $p \xrightarrow{a} q$ is an edge and $q \neq q_0$, and is 1 if it is an edge and $q = q_0$; this is exactly the definition of $(\alpha a)_{p,q}$. Now, let $w = uv$, where u, v are shorter than w, so by induction Equation (3.1) holds for u and v. Then, translating into $\mathbb{N} \cup \infty$ the operations in $\mathbb{T}$, we have

$$(\alpha w)_{p,q} = \min_{r \in Q_0}\big((\alpha u)_{p,r} + (\alpha v)_{r,q}\big).$$

By induction, this is equal to

$$\min_{r \in Q_0}\big(\min\{\omega(d) \mid d : p \xrightarrow{u} r\} + \min\{\omega(e) \mid e : r \xrightarrow{v} q\}\big).$$

Since the minimum is distributive with respect to addition, and since the weight of a path de is the sum of the weights of the paths d and e, we obtain that

$$(\alpha w)_{p,q} = \min_{r \in Q_0}\{\omega(de) \mid d : p \xrightarrow{u} r\,, e : r \xrightarrow{v} q\},$$

and this is equal to the right-hand side of (3.1), as was to be shown.

2. From Equation (3.1), it follows that $(\alpha w)_{q_0,q_0}$ is equal to the least m such that $w \in L^m$, and is ∞ if $w \notin L^*$. Thus L is limited if and only if the set

$$\{(\alpha w)_{q_0,q_0} \mid w \in A^*\} \tag{3.2}$$

is finite.

Now, let $p, q \in Q_0$ and suppose that $(\alpha w)_{p,q} = m \neq \infty$. By (3.1), this means that there is a path $p \xrightarrow{w} q$ in A^* having m edges ending in q_0, and that no other path $p \xrightarrow{w} q$ has fewer such edges. Hence, we find a subpath $q_0 \xrightarrow{u} q_0$, for some factor u of w, having $m - 1$ such edges, and such that no other path $q_0 \xrightarrow{u} q_0$ has fewer such edges. This implies by (3.1) that $(\alpha u)_{q_0,q_0} = m - 1$. We conclude that if the set (3.2) is finite, then so is the set $\{(\alpha w)_{p,q} \mid w \in A^*\}$. Thus L is limited if and only if $\alpha(A^*)$ is finite. □

We need to consider another semiring, denoted $\mathbb{T}_0$, whose underlying set is $\{0, 1, \infty\}$, with the same operations as $\mathbb{T}$, that is: addition in $\mathbb{T}_0$ is the $\min(a, b)$ operation, and multiplication is the usual addition.

Let $\psi : \mathbb{T} \to \mathbb{T}_0$ be the mapping which sends 0 to 0, ∞ to ∞ and any $a \in \mathbb{T} \setminus \{0, \infty\}$ to 1. It is easily verified that ψ is a semiring morphism. Moreover, let ι

be the injective mapping that sends 0, 1 and ∞ in $\mathbb{T}_0$ to themselves in $\mathbb{T}$. Note that ι is not a semiring morphism. However

$$\psi\iota = \mathrm{id}_{\mathbb{T}_0} .$$

The mappings ψ and ι are naturally extended to matrices over $\mathbb{T}$ and $\mathbb{T}_0$.

Theorem 3.2 (Simon 1978) *The following conditions are equivalent for a finitely generated subsemigroup S of $\mathbb{T}^{n\times n}$:*

(i) *S is finite;*
(ii) *S is a torsion semigroup;*
(iii) *for any idempotent e in ψS, one has $(\iota e)^2 = (\iota e)^3$.*

Corollary 3.3 *It is decidable whether a finite subset of $\mathbb{T}^{n\times n}$ generates a finite subsemigroup, and whether a rational language is limited.*

Proof. Since ψ is a monoid morphism and since $\mathbb{T}_0^{n\times n}$ is finite, condition (iii) of the theorem is decidable.

For a rational language L, the limitedness problem is reduced by Proposition 3.1 to the finiteness of a certain finitely generated submonoid of $\mathbb{T}^{n\times n}$, hence to the preceding question. □

We use the natural ordering $\leq$ on $\mathbb{T}$ that extends the natural ordering of $\mathbb{N}$, together with the natural condition that $t \leq \infty$ for all $t \in \mathbb{T}$. This ordering is compatible with the semiring structure since if $a \leq b$, then $\min(a, x) \leq \min(b, x)$ and $a + x \leq b + x$. We extend this ordering to matrices over $\mathbb{T}$, by setting $(a_{i,j}) \leq (b_{i,j})$ if and only if $a_{i,j} \leq b_{i,j}$ for all i, j. Then again, this ordering is compatible with sum and product of matrices over $\mathbb{T}$.

For any subset X of a semigroup S, we denote by X^+ the subsemigroup of S generated by X.

Lemma 3.4 *Let X be a finite subset of the multiplicative semigroup $\mathbb{T}^{n\times n}$ and let $Y = \iota\psi X$. Then X^+ is finite if and only Y^+ is finite.*

Note that $y = \iota\psi x$ is obtained from x by replacing each nonzero finite entry in x by 1, 0 and ∞ being unchanged. Hence, the entries equal to 0 or ∞ in x and y are the same.

Proof. We may assume that some entry of some matrix in X is finite. Let M be the maximum of these finite entries. Let $x_1, \ldots, x_p \in X$, set $y_k = \iota\psi x_k$. We show below that for $i, j \in \{1, \ldots, n\}$, the following hold:

(i) $(x_1 \cdots x_p)_{i,j} = \infty \iff (y_1 \cdots y_p)_{i,j} = \infty$;
(ii) if the entries $(x_1 \cdots x_p)_{i,j}$ and $(y_1 \cdots y_p)_{i,j}$ are finite, then

$$(y_1 \cdots y_p)_{i,j} \leq (x_1 \cdots x_p)_{i,j} \leq M(y_1 \cdots y_p)_{i,j} ,$$

where the right-hand side product is taken in $\mathbb{N}$.

These two properties imply the lemma. For the proof of (i), observe that, by definition of $\mathbb{T}$

$$(x_1 \cdots x_p)_{i,j} = \min\big((x_1)_{i,k_1} + (x_2)_{k_1,k_2} + \cdots + (x_p)_{k_{p-1},j}\big)\,, \tag{3.3}$$

where the minimum is taken over all $k_1, \ldots, k_{p-1}$ in $\{1, \ldots, n\}$ and the sum is taken in $\mathbb{N} \cup \infty$. A similar formula holds for the y_k's.

Now, if $(x_1 \cdots x_p)_{i,j} = \infty$, then for each $k_1, \ldots, k_{p-1}$, the sum in the right-hand side of (3.3) must be ∞ and therefore at least one term $(x_j)_{k_{j-1},k_j}$ is equal to ∞; by the definition of ψ and ι, we obtain that $(y_1 \cdots y_p)_{i,j} = \infty$. The converse is similar, implying (i).

For (ii), the first inequality follows from the properties of the order $\leq$ on $\mathbb{T}^{n\times n}$ and the fact that $\iota\psi x \leq x$. For the second, knowing that $(x_1 \cdots x_p)_{i,j}$ is finite, we may restrict the minimum in (3.3) to those $k_1, \ldots, k_{p-1}$ such that the sum in the right-hand side is finite. Then each term $(x_\ell)_{k_{j-1},k_j}$ is finite and therefore is less than or equal to $M(y_\ell)_{k_{j-1},k_j}$ by the definition of ψ and ι. This implies the second equality in (ii). □

Lemma 3.5 *Let e be idempotent in the multiplicative monoid $\mathbb{T}_0^{n\times n}$ and set $f = \iota e$. For any i, j in $\{1, \ldots, n\}$, one of the following statements holds:*

(i) $(f^m)_{i,j} = f_{i,j}$ *for any* $m \geq 1$;
(ii) $f_{i,j} = 1$ *and* $(f^m)_{i,j} = 2$ *for any* $m \geq 2$;
(iii) $(f^m)_{i,j} = m$ *for any* $m \geq 1$.

Proof. 1. Note that $f_{i,j} \in \{0, 1, \infty\}$. We have $e = \psi\iota e = \psi f$, hence for any $m \geq 1$, $\psi(f^m) = \psi(f)^m = e^m = e$, and therefore

$$e_{i,j} = 0 \iff (f^m)_{i,j} = 0\,;$$
$$e_{i,j} = 1 \iff (f^m)_{i,j} = 1, 2, 3\ldots\,;$$
$$e_{i,j} = \infty \iff (f^m)_{i,j} = \infty\,,$$

by definition of ψ.

2. Suppose that $(f^p)_{i,j} = 0$ for some $p \geq 1$. Then by step 1 one has $e_{i,j} = 0$ and therefore $(f^m)_{i,j} = 0$ for all $m \geq 1$ by step 1 again.

3. Suppose next that $(f^p)_{i,j} = 1$ for some $p \geq 2$. We show that $(f^m)_{i,j} = 1$ for any $m \geq 1$. Indeed by step 1, $e_{i,j} = 1$, hence $f_{i,j} = 1$ again by step 1. Moreover, we have $(f^m)_{i,j} \neq 0$ for any $m \geq 1$ by step 2. Since $f^p = f^{p-1}f$, there exists an index k such that either $(f^{p-1})_{i,k} = 0$ and $f_{k,j} = 1$ or $(f^{p-1})_{i,k} = 1$ and $f_{k,j} = 0$.

In the first case, $(f^m)_{i,k} = 0$ for any $m \geq 1$ by step 2. Hence $(f^m)_{i,j} \leq (f^{m-1})_{i,k} + f_{k,j} \leq 1$ for all $m \geq 2$.

In the second case, we have $(f^m)_{k,j} = 0$ for any $m \geq 1$ by step 2, and by step 1 we get $f_{i,k} = 1$. Hence $(f^m)_{i,j} \leq f_{i,k} + (f^{m-1})_{k,j} \leq 1$ for all $m \geq 2$.

4. We now show that if $2 \leq (f^p)_{i,j} < p$ for some $p \geq 3$, then $(f^m)_{i,j} = 2$ for any $m \geq 2$ and moreover $f_{i,j} = 1$. This latter equality follows from step 1, since we must have $e_{i,j} = 1$, thus $f_{i,j} = 1$.

Let $q = (f^p)_{i,j}$. By the definition of the operations in $\mathbb{T}$ and $\mathbb{T}^{n\times n}$ we have (with addition in $\mathbb{N} \cup \infty$)

$$q = f_{k_0,k_1} + f_{k_1,k_2} + \cdots + f_{k_{p-1},k_p} \tag{3.4}$$

for some $i = k_0, k_1, \ldots, k_{p-1}, k_p = j$. Since $q < \infty$, each term in (3.4) is 0 or 1. Let $0 < h < p$. Then we deduce that $(f^h)_{k_0k_h} < \infty$, hence $f_{k_0,k_h} < \infty$ by step 1, and it follows that $f_{k_0,k_h} \leq 1$; similarly $f_{k_h,k_p} \leq 1$.

Moreover, $q < p$ and therefore (3.4) implies that $f_{k_\ell,k_{\ell+1}} = 0$ for some $0 \leq \ell < p$. Then $(f^m)_{k_\ell,k_{\ell+1}} = 0$ for any $m \geq 1$ by step 2. Suppose that $\ell = 0$. Then $(f^{p-1})_{k_0,k_1} = 0$ and $f_{k_1,k_p} \leq 1$ imply that $(f^p)_{i,j} = (f^p)_{k_0,k_p} \leq 1$, a contradiction; likewise $\ell = p-1$ implies this contradiction. Hence $0 < \ell < p-1$.

We deduce that for any $m \geq 3$, $(f^m)_{i,j} = (f^m)_{k_0,k_p} \leq f_{k_0,k_\ell} + (f^{m-2})_{k_\ell,k_{\ell+1}} + f_{k_{\ell+1},k_p} \leq 1 + 0 + 1 = 2$. Also $(f^2)_{i,j} = (f^2)_{k_0,k_p} \leq f_{k_0,k_1} + f_{k_1,k_p} \leq 1 + 1 = 2$.

Now, we cannot have $(f^m)_{i,j} \leq 1$ for some $m \geq 2$ since this would imply, by steps 2 and 3, that $(f^p)_{i,j} \leq 1$. Thus $(f^m)_{i,j} = 2$ for any $m \geq 2$ and $f_{i,j} = 1$.

5. Suppose now that neither (i) nor (ii) holds. This implies, by steps 2–4 that $(f^p)_{i,j} \geq p$ for all $p \geq 1$. Indeed, if $(f^p)_{i,j} < p$ for some $p \geq 1$, then either $(f^p)_{i,j} = 0$ and (i) holds by step 1, or $(f^p)_{i,j} \geq 1$, which implies $p \geq 2$ (since otherwise $p = 1$ and $f_{i,j} < 1$, contradiction); then either $(f^p)_{i,j} = 1$ and (i) holds by step 3, or $(f^p)_{i,j} \geq 2$ (since otherwise $p = 2$ and $(f^2)_{i,j} < 2$, contradiction), hence $p \geq 3$; then (ii) holds by step 4.

Since the finite entries of f are equal to 0 or 1, the finite entries of f^p are $\leq p$. Thus $(f^p)_{i,j} = p$ or ∞. To conclude, assume that $(f^p)_{i,j} = \infty$ for some $p \geq 1$. Then, by step 1, $e_{i,j} = \infty$. If $(f^m)_{i,j} \neq \infty$ for some $m \geq 1$, then again by step 1, $e_{i,j} \neq \infty$. Consequently $(f^m)_{i,j} = \infty$ for all $m \geq 1$, contradicting that (i) does not hold, and (iii) follows. □

We use the following result. A semigroup S is called *locally finite* if each finite subset of S generates a finite subsemigroup.

Theorem 3.6 (Brown 1971) *Let $\varphi : S \to T$ be a morphism of a semigroups S onto a locally finite semigroup T. If the semigroup $\varphi^{-1}(e)$ is locally finite for each idempotent e in T, then S is locally finite.*

Proof of Theorem 3.2. The implication (i) $\Longrightarrow$ (ii) is clear.

(ii) $\Longrightarrow$ (iii). We have $e = \psi s$ for some $s \in S$. Then $\iota e = \iota\psi s$. Since s is torsion, so is ιe by Lemma 3.4. Let $i, j \in \{1, \ldots, n\}$. Then by Lemma 3.5, condition (iii) of this lemma cannot hold. Hence (i) or (ii) holds and consequently $(\iota e)^2 = (\iota e)^3$.

(iii) $\Longrightarrow$ (i). In view of Brown's theorem, it is enough to show that for any idempotent e in $\mathbb{T}_0^{n\times n}$, the semigroup $\psi^{-1}(e) \cap S$ is locally finite. So, consider a finite subset X of $\psi^{-1}(e) \cap S$. We may suppose that e is in $\psi(S)$. Then by hypothesis

$(\iota e)^2 = (\iota e)^3$. Let $Y = \iota\psi X$. Since $\psi X = \{e\}$, we have $Y = \{\iota e\}$ and consequently Y^+ is finite. Hence X^+ is finite by Lemma 3.4, and we can conclude that $\psi^{-1}(e) \cap S$ is locally finite. □

Exercises for Chapter 9

1.1 Let $S \in \mathbb{Q}\langle\langle A \rangle\rangle$ be a rational series such that, for every ray R, almost all coefficients (S, w), $w \in R$, vanish. Show that S is a polynomial.

1.2 Let $S \in \mathbb{N}\langle\langle A \rangle\rangle$ be an $\mathbb{N}$-rational series having polynomial growth. Show that S is in the $\mathbb{N}$-subalgebra of $\mathbb{N}\langle\langle A \rangle\rangle$ generated by the characteristic series of rational languages (use a rational expression for S and the fact that if $T \in \mathbb{N}\langle\langle A \rangle\rangle$ is not the characteristic series of a code, then the growth of T^* is not polynomial).

1.3 Show that Corollary 2.6 holds when $\mathbb{Z}$ is replaced by $\mathbb{N}$.

2.1 A *composition* of m of length k is a k-tuple of positive integers $(m_1, \ldots, m_k)$ such that $m_1 + \cdots + m_k = m$. Show that the number of such compositions is $\binom{m-1}{k-1}$. (*Hint*: Associate to the composition the subset $\{m_1, m_1 + m_2, \ldots, m_1 + \cdots + m_{k-1}\}$ of $\{1, \ldots, m-1\}$.)

3.1 Show that $\mathbb{T}$ is indeed a semiring by verifying all the axioms given in Section 1.1.

3.2 Show that $L = a \cup (a^2)^* \cup (a^*b)^*$ is limited and find the smallest m such that $L^* = 1 \cup L \cup \cdots \cup L^m$.

3.3 Show that $\mathbb{T}_0$ is indeed a semiring and that $\psi : \mathbb{T} \to \mathbb{T}_0$ is a semiring morphism.

3.4 Show that ι is not a semiring morphism and that $\psi\iota = \mathrm{id}_{\mathbb{T}_0}$.

3.5 Show that the ordering of matrices over $\mathbb{T}$ is compatible with sum and product.

3.6 Show that $\sum_{n \geq 0} na^n \in \mathbb{T}\langle\langle a \rangle\rangle$ is equal to $(1a)^*$.

Notes to Chapter 9

Most of the results of Section 9.1 hold in arbitrary fields. Theorem 1.1 can be extended, but the bound N then also depends on the field considered. Corollaries 1.5, 1.6 hold in arbitrary fields, and Lemma 1.2 holds in fields of characteristic 0, provided M is finitely generated and the bound $(2n+1)^{n^2}$ is replaced by r^{n^2}, where r is the size of the set $\{\mathrm{tr}(m) \mid m \in M\}$. This set is always finite (under the assumptions of the lemma) for a finite monoid M. Corollaries 1.7, 1.8 extend to "computable" fields.

The results and proofs of Section 9.3 are all due to Simon (1978); he shows also that a rational language L is not limited if and only if there exists a word w in L^* such that for any $m \geq 1$, $w^m \notin 1 \cup L \cup \cdots \cup L^m$. Krob has shown that it is undecidable whether two rational series over $\mathbb{T}$ are equal, see Krob (1994). It is also decidable whether a rational series over the tropical semiring has finite image, see Hashiguchi (1982), Leung (1988), Simon (1988, 1994). For results on matrix semigroups related to the present chapter, see Okniński (1998).

10

Noncommutative polynomials

This chapter deals with algebraic properties of noncommutative polynomials. They are of independent interest, but most of them will be of use in the next chapter.

In contrast to commutative polynomials, the algebra of noncommutative polynomials is not Euclidean, and not even factorial. However, there are many interesting results concerning factorization of noncommutative polynomials: this is one of the major topics of the present chapter.

The basic tool is Cohn's weak algorithm (Theorem 1.1) which is the subject of Section 10.1. This operation constitutes a natural generalization of the classical Euclidean algorithm.

Section 10.2 deals with continuant polynomials which describe the multiplicative relations between noncommutative polynomials (Theorem 2.2).

We introduce in Section 10.3 cancellative modules over the ring of polynomials. We characterize these modules (Theorem 3.1) and obtain, as consequences, results on full matrices, factorization of polynomials and inertia.

The main result of Section 10.4 is the extension of Gauss's Lemma to noncommutative polynomials.

10.1 The weak algorithm

Let K be a field and let A be an alphabet. Recall that the *degree* of a polynomial P in $K\langle A\rangle$ was defined in Section 1.2: we will denote it by $\deg(P)$. We recall the usual facts about the degree, that is

$$\deg(0) = -\infty\,,$$

$$\deg(P+Q) \leq \max(\deg(P), \deg(Q))\,, \tag{1.1}$$

$$\deg(P+Q) = \deg(P), \quad \text{if } \deg(Q) < \deg(P)\,,$$

$$\deg(PQ) = \deg(P) + \deg(Q)\,. \tag{1.2}$$

Note that the last equality shows that $K\langle A\rangle$ is an *integral domain*, that is

$$PQ = 0 \quad \text{implies} \quad P = 0 \text{ or } Q = 0\,.$$

Definition A finite family $P_1, \ldots, P_n$ of polynomials in $K\langle A\rangle$ is (right) *dependent* if either some $P_i = 0$ or if there exist polynomials $Q_1, \ldots, Q_n$ such that

$$\deg\Bigl(\sum_i P_i Q_i\Bigr) < \max_i(\deg(P_i Q_i)).$$

Definition A polynomial P is (right) *dependent* on the family $P_1, \ldots, P_n$ if either $P = 0$ or if there exist polynomials $Q_1, \ldots, Q_n$ such that

$$\deg\Bigl(P - \sum_i P_i Q_i\Bigr) < \deg(P)$$

and if furthermore for any $i = 1, \ldots, n$

$$\deg(P_i Q_i) \leq \deg(P).$$

Note that if P is dependent on $P_1, \ldots, P_n$ then the family $P, P_1, \ldots, P_n$ is dependent. The converse is given by the following theorem.

Theorem 1.1 (Cohn 1961) *Let $P_1, \ldots, P_n$ be a dependent family of polynomials with* $\deg(P_1) \leq \cdots \leq \deg(P_n)$. *Then some P_i is dependent on $P_1, \ldots, P_{i-1}$.*

Let P be a polynomial and let u be a word in A^*. We define the polynomial Pu^{-1} as

$$Pu^{-1} = \sum_{w \in A^*} (P, wu)w. \tag{1.3}$$

The operator $P \mapsto Pu^{-1}$ is symmetric to the operator $P \mapsto u^{-1}P$ which was introduced in Section 1.5. It is easy to verify that this operator is linear, and that the following relations hold:

$$\deg(Pu^{-1}) \leq \deg(P) - |u|, \tag{1.4}$$

$$P(uv)^{-1} = (Pv^{-1})u^{-1}. \tag{1.5}$$

Moreover, for any letter a,

$$(PQ)a^{-1} = P(Qa^{-1}) + (Q, 1)Pa^{-1} \tag{1.6}$$

where $(Q, 1)$ denotes as usual the constant term of Q. The last equality is simply the symmetric equivalent of Lemma 1.7.2.

Lemma 1.2 *If P, Q are polynomials and w is a word, then there exists a polynomial P' such that*

$$(PQ)w^{-1} = P(Qw^{-1}) + P'$$

with either $P = P' = 0$ or $\deg(P') < \deg(P)$.

Proof. We may assume $P \neq 0$. If w is the empty word, then $(PQ)w^{-1} = PQ$ and $Qw^{-1} = Q$, so that $(PQ)w^{-1} = P(Qw^{-1})$ and the proof is complete.

Let $w = au$ with a a letter. Then by induction one has

$$(PQ)u^{-1} = P(Qu^{-1}) + P', \quad \deg(P') < \deg(P).$$

Now, by Equation (1.5), one has

$$(PQ)w^{-1} = ((PQ)u^{-1})a^{-1} = (P(Qu^{-1}))a^{-1} + P'a^{-1}.$$

Thus, by Eqs. (1.6) and (1.5), we have

$$\begin{aligned}(PQ)w^{-1} &= P((Qu^{-1})a^{-1}) + (Qu^{-1}, 1)Pa^{-1} + P'a^{-1} \\ &= P(Qw^{-1}) + P''\end{aligned}$$

with $P'' = (Qu^{-1}, 1)Pa^{-1} + P'a^{-1}$. Next, by Equation (1.4), $\deg(Pa^{-1}) < \deg(P)$ and $\deg(P'a^{-1}) \leq \deg(P') - |a| < \deg(P)$. Hence $\deg(P'') < \deg(P)$, as desired. □

Proof of Theorem 1.1. We may suppose that no P_i is equal to 0. Hence $\deg(\sum P_i Q_i) < \max_i(\deg(P_i Q_i))$. Let $r = \max_i(\deg(P_i Q_i))$ and let $I = \{i \mid \deg(P_i Q_i) = r\}$. The polynomial $R = \sum_{i \in I} P_i Q_i$ has degree $\deg(R) < r$. Let $k = \sup(I)$; then $i \in I \implies \deg(P_i) \leq \deg(P_k)$. Let w be a word such that $|w| = \deg(Q_k)$ and $0 \neq (Q_k, w) = \alpha^{-1} \in K$: such a word exists because $Q_k \neq 0$ (otherwise $\deg(R) < r = \deg(P_k Q_k) = -\infty$).

By Lemma 1.2, we have

$$Rw^{-1} = \sum_{i \in I} P_i(Q_i w^{-1}) + \sum_{i \in I} P_i'$$

for some polynomials P_i' with $\deg(P_i') < \deg(P_i)$. Since $Q_k w^{-1} = \alpha^{-1}$,

$$P_k + \alpha \sum_{i \in I \setminus k} P_i(Q_i w^{-1}) = \alpha R w^{-1} - \alpha \sum_{i \in I} P_i'. \tag{1.7}$$

Now, by Equation (1.4)

$$\begin{aligned}\deg(Rw^{-1}) &\leq \deg(R) - |w| < r - |w| \\ &= \deg(P_k Q_k) - \deg(Q_k) = \deg(P_k).\end{aligned}$$

Furthermore, $\deg(P_i') < \deg(P_i) \leq \deg(P_k)$. Consequently, by Equation (1.1), the degree of the right-hand side of Equation (1.7) is $< \deg(P_k)$. Moreover,

$$\begin{aligned}\deg(P_i(Q_i w^{-1})) &= \deg(P_i) + \deg(Q_i w^{-1}) \\ &\leq \deg(P_i) + \deg(Q_i) - \deg(Q_k)\end{aligned}$$

by Equation (1.4). So we have $\deg(P_i(Q_i w^{-1})) \leq r - \deg(Q_k) = \deg(P_k)$. This shows that P_k is dependent on P_i, $i \in I \setminus k$; hence P_k also is dependent on $P_1, \ldots, P_{k-1}$. □

Given two polynomials X, Y, we say that Y is a *weak left divisor* of X if there exist polynomials Q, R such that

$$X = YQ + R \quad \text{with} \quad \deg(R) < \deg(Y).$$

Note that in this case, $Y \neq 0$. Weak left division is not always possible if A has more than one letter (for instance take $X = a$ and $Y = b$ for distinct letters a, b).

Corollary 1.3 *Let X, Y, P, Q_1 be polynomials such that Y is a weak left divisor of $XP + Q_1$ with* $\deg(Q_1) \leq \deg(P)$ *and $P \neq 0$. Then Y is a weak left divisor of X.*

Proof. Note that $Y \neq 0$. If $Y \in K$, the corollary is immediate. Otherwise, we prove it by induction on $\deg(X)$. If $\deg(X) < \deg(Y)$, the proof is immediate. Suppose that $\deg(X) \geq \deg(Y)$. By the assumption, there exist polynomials Q_2 and R_1 such that

$$XP + Q_1 = YQ_2 + R_1\,, \tag{1.8}$$

with $\deg(R_1) < \deg(Y)$. Then

$$\deg(Q_1) \leq \deg(P) < \deg(XP)$$

because $1 \leq \deg(Y) \leq \deg(X)$ and

$$\deg(R_1) < \deg(Y) \leq \deg(X) \leq \deg(XP)$$

because $0 \leq \deg(P)$. Thus, $\deg(Q_1)$ and $\deg(R_1)$ are both $< \max(\deg(XP), \deg(YQ_2))$ and by Equation (1.1),

$$\deg(XP - YQ_2) = \deg(R_1 - Q_1) < \max(\deg(XP), \deg(YQ_2))\,.$$

Hence, X, Y are dependent. In view of Theorem 1.1, X is dependent on Y, hence there exist two polynomials Q_3 and X_1 such that $X = YQ_3 + X_1$ with $\deg(X_1) < \deg(X)$.

Put this expression for X into the Equation (1.8). This gives

$$X_1P + Q_1 = Y(Q_2 - Q_3P) + R_1\,.$$

Thus Y is a weak left divisor of $X_1P + Q_1$. Since $\deg(X_1) < \deg(X)$, we conclude by induction. □

The next result is a particular case of the previous one.

Corollary 1.4 *If X, Y, X', Y' are nonzero polynomials such that $XY' = YX'$, then there exist polynomials Q, R such that $X = YQ + R$ and* $\deg(R) < \deg(Y)$. □

10.2 Continuant polynomials

Definition Let $a_1, \ldots, a_n$ be a finite sequence of elements of a ring. We define the sequences $p_0, \ldots, p_n$ of *continuant polynomials* (with respect to $a_1, \ldots, a_n$) in the following way:

$$p_0 = 1,\ p_1 = a_1\,,$$

and for $2 \leq i \leq n$,

$$p_i = p_{i-1}a_i + p_{i-2}\,.$$

Example 2.1 The first continuant polynomials are

$$p_2 = a_1a_2 + 1\,,$$

$$p_3 = a_1a_2a_3 + a_1 + a_3\,,$$

$$p_4 = a_1a_2a_3a_4 + a_1a_2 + a_1a_4 + a_3a_4 + 1\,.$$

Notation We shall write $p(a_1, \ldots, a_i)$ for p_i.

It is easy to see that the continuant polynomials may be obtained by the "leap-frog construction": consider the "word" $a_1 \cdots a_n$ and all words obtained by repetitively suppressing some factors of the form a_ia_{i+1} in it. Then $p(a_1, \ldots, a_n)$ is the sum, without multiplicity, of all these "words".

Now, we have by definition, for $n \geq 2$,

$$p(a_1, \ldots, a_n) = p(a_1, \ldots, a_{n-1})a_n + p(a_1, \ldots, a_{n-2})\,. \tag{2.1}$$

The combinatorial construction sketched above shows that symmetrically

$$p(a_1, \ldots, a_n) = a_1 p(a_2, \ldots, a_n) + p(a_3, \ldots, a_n)\,. \tag{2.2}$$

An equivalent but useful relation is

$$p(a_n, \ldots, a_1) = a_n p(a_{n-1}, \ldots, a_1) + p(a_{n-2}, \ldots, a_1)\,. \tag{2.3}$$

Proposition 2.1 (Wedderburn 1932) *The continuant polynomials satisfy, for $n \geq 1$, the relation*

$$p(a_1, \ldots, a_n)p(a_{n-1}, \ldots, a_1) = p(a_1, \ldots, a_{n-1})p(a_n, \ldots, a_1)\,. \tag{2.4}$$

Proof. This is surely true for $n = 1$. Suppose $n \geq 2$. Then by Equation (2.1),

$$p(a_1, \ldots, a_n)p(a_{n-1}, \ldots, a_1)$$
$$= p(a_1, \ldots, a_{n-1})\, a_n\, p(a_{n-1}, \ldots, a_1) + p(a_1, \ldots, a_{n-2})p(a_{n-1}, \ldots, a_1)$$

which is equal by induction to

$$p(a_1, \ldots, a_{n-1})\, a_n\, p(a_{n-1}, \ldots, a_1) + p(a_1, \ldots, a_{n-1})p(a_{n-2}, \ldots, a_1)\,.$$

This is equal, by Equation (2.3), to

$$p(a_1, \ldots, a_{n-1})p(a_n, \ldots, a_1)$$

as desired. □

Theorem 2.2 (Cohn 1969) *Let X, Y, X', Y' be nonzero polynomials in $K\langle A\rangle$ such that $XY' = YX'$. Then there exist polynomials $U, V, a_1, \ldots, a_n$ with $n \geq 1$ such that*

$$X = Up(a_1, \ldots, a_n)\,, \qquad Y' = p(a_{n-1}, \ldots, a_1)V\,,$$

$$Y = Up(a_1, \ldots, a_{n-1})\,, \qquad X' = p(a_n, \ldots, a_1)V\,.$$

Moreover, one has $\deg(a_1), \ldots, \deg(a_{n-1}) \geq 1$, *and if* $\deg(X) > \deg(Y)$, *then* $\deg(a_n) \geq 1$.

Proof. (i) Suppose first that X is a right multiple of Y, that is $X = YQ$. Then the theorem is obvious for $U = Y$, $V = Y'$, $n = 1$, $a_1 = Q$; then indeed

$$X = YQ = Up(a_1),\ Y' = 1 \cdot V, Y = U \cdot 1$$

and $YX' = XY' = YQY'$, whence $X' = QY' = p(a_1)V$. Furthermore, if $\deg(X) > \deg(Y)$, then $\deg(Q) \geq 1$.

(ii) Next, we prove the theorem in the case where $\deg(X) > \deg(Y)$, by induction on $\deg(Y)$. If $\deg(Y) = 0$, then X is a right multiple of Y and we may apply (i). Suppose $\deg(Y) \geq 1$. By Corollary 1.4, $X = YQ + R$ for some polynomials Q and R such that $\deg(R) < \deg(Y)$. If $R = 0$, apply (i). Otherwise, we have $YX' = XY' = YQY' + RY'$, hence $Y(X' - QY') = RY'$; note that $Y, R, Y' \neq 0$, and therefore $X' - QY' \neq 0$. Furthermore, $\deg(R) < \deg(Y)$, and we may apply the induction hypothesis: there exist polynomials $U, V, a_1, \ldots, a_n$ such that

$$\begin{aligned} &Y = Up(a_1, \ldots, a_n)\,, && X' - QY' = p(a_{n-1}, \ldots, a_1)V\,, \\ &R = Up(a_1, \ldots, a_{n-1})\,, && Y' = p(a_n, \ldots, a_1)V\,, \\ &\deg(a_1), \ldots, \deg(a_n) \geq 1\,. \end{aligned} \tag{2.5}$$

This implies

$$\begin{aligned} X &= YQ + R = U\big(p(a_1, \ldots, a_n)Q + p(a_1, \ldots, a_{n-1})\big) \\ &= Up(a_1, \ldots, a_n, Q) \end{aligned}$$

by Equation (2.1). Similarly, $X' = p(Q, a_n, \ldots, a_1)V$. Thus X, Y, X', Y' admit the announced expression. Furthermore, $\deg(Q) \geq 1$; indeed, by Equation (1.2), $\deg(X) = \deg(YQ) = \deg(Y) + \deg(Q)$, and hence $\deg(Q) = \deg(X) - \deg(Y) \geq 1$.

This proves the theorem in the case where $\deg(X) > \deg(Y)$.

(iii) In the general case, one has again $X = YQ + R$ with $\deg(R) < \deg(Y)$ (Corollary 1.4). If $R = 0$, the proof is completed by (i). Otherwise, as above, $Y(X' - QY') = RY'$ with $\deg(Y) > \deg(R)$. Thus we may apply (ii): there exist $U, V, a_1, \ldots, a_n$ such that Equation (2.5) holds. Then we obtain, as in (ii):

$$\begin{aligned} X &= Up(a_1, \ldots, a_n, Q)\,, && Y' = p(a_n, \ldots, a_1)V\,, \\ Y &= Up(a_1, \ldots, a_n)\,, && X' = p(Q, a_n, \ldots, a_1)V\,. \end{aligned}$$ □

Proposition 2.3 *Let* $a_1, \ldots, a_n$ *be polynomials such that* $a_1, \ldots, a_{n-1}$ *have positive degree, and let* Y *be a polynomial of degree* 1 *such that* $p(a_{n-1}, \ldots, a_1)$ *and* $p(a_n, \ldots, a_1)$ *are both congruent to a scalar modulo the right ideal* $YK\langle A\rangle$. *Then for* $i = 1, \ldots, n$

$$p(a_i, \ldots, a_1) \equiv p(a_1, \ldots, a_i) \mod YK\langle A\rangle\,.$$

We prove first a lemma.

Lemma 2.4 *Let $a_1, \dots, a_n$ be polynomials such that $a_1, \dots, a_{n-1}$ have positive degree. Then the degrees of $1, p(a_1), \dots, p(a_{n-1}, \dots, a_1)$ are strictly increasing.*

Proof. Obviously $\deg(1) < \deg(a_1)$. Suppose

$$\deg(p(a_{i-2}, \dots, a_1)) < \deg(p(a_{i-1}, \dots, a_1))$$

for $2 \leq i < n - 1$. From the relation

$$p(a_i, \dots, a_1) = a_i p(a_{i-1}, \dots, a_1) + p(a_{i-2}, \dots, a_1),$$

it follows that the degree of $p(a_i, \dots, a_1)$ is equal to $\deg(a_i p(a_{i-1}, \dots, a_1))$, and

$$\begin{aligned}\deg(a_i p(a_{i-1} \dots, a_1)) &= \deg(a_i) + \deg(p(a_{i-1}, \dots, a_1)) \\ &> \deg(p(a_{i-1}, \dots, a_1))\end{aligned}$$

because $\deg(a_i) \geq 1$. This proves the lemma. □

Lemma 2.5 *Let $a_1, \dots, a_n$ be polynomials. Then*

$$p(a_1, \dots, a_n) = 0 \iff p(a_n, \dots, a_1) = 0.$$

Proof. It is enough to show that $p(a_1, \dots, a_n) = 0$ implies $p(a_n, \dots, a_1) = 0$. We use induction. It is clear for $n = 0, 1$. Let $n \geq 2$. By Equation (2.4),

$$p(a_1, \dots, a_n)p(a_{n-1}, \dots, a_1) = p(a_1, \dots, a_{n-1})p(a_n, \dots, a_1).$$

Suppose $p(a_1, \dots, a_n) = 0$. If $p(a_1, \dots, a_{n-1}) \neq 0$, then $p(a_n, \dots, a_1) = 0$ because $K\langle A\rangle$ is an integral domain. If $p(a_1, \dots, a_{n-1}) = 0$, then $p(a_{n-1}, \dots, a_1) = 0$ by induction. Hence, by Eqs. (2.1) and (2.3) $p(a_1, \dots, a_n) = p(a_1, \dots, a_{n-2})$ and $p(a_n, \dots, a_1) = p(a_{n-2}, \dots, a_1)$. By induction, $p(a_{n-2}, \dots, a_1) = 0$, which proves the lemma. □

Proof of Proposition 2.3 (Induction on n). When $n = 1$, the result is evident. Suppose $n \geq 2$. Note that if the condition on the degrees is fulfilled for $a_1, \dots, a_n$, then a fortiori also $a_1, \dots, a_{n-2}$ have positive degree. By assumption, $p(a_n, \dots, a_1)$ is congruent to some scalar α and $p(a_{n-1}, \dots, a_1)$ is congruent to some scalar β modulo $YK\langle A\rangle$. Suppose $p(a_{n-1}, \dots, a_1) = 0$. Then by Equation (2.3), we have $p(a_n, \dots, a_1) = p(a_{n-2}, \dots, a_1)$. Moreover, by Lemma 2.5, $p(a_1, \dots, a_{n-1}) = 0$, so that by Equation (2.1), $p(a_1, \dots, a_n) = p(a_1, \dots, a_{n-2})$. Thus we conclude by induction in this case.

Suppose $p(a_{n-1}, \dots, a_1) \neq 0$. Then by Equation (2.3),

$$a_n p(a_{n-1}, \dots, a_1) + p(a_{n-2}, \dots, a_1) = YQ + \alpha$$

for some polynomial Q. Since $\deg(p(a_{n-2}, \dots, a_1)) < \deg(p(a_{n-1}, \dots, a_1))$ by Lemma 2.4, we obtain by Corollary 1.3 that $a_n \equiv \gamma \mod YK\langle A\rangle$ for some scalar γ. Using Equation (2.3) again, and the fact that $P \equiv \gamma,\ Q \equiv \beta \implies PQ \equiv \gamma\beta$, we obtain $p(a_{n-2}, \dots, a_1) \equiv \alpha - \gamma\beta$. Then, the induction hypothesis gives

$p(a_1, \ldots, a_{n-2}) \equiv \alpha - \gamma\beta$ and $p(a_1, \ldots, a_{n-1}) \equiv \beta$. Hence, by Equation (2.1), $p(a_1, \ldots, a_n) \equiv \beta\gamma + \alpha - \gamma\beta \equiv p(a_n \ldots, a_1)$, as desired. □

10.3 Inertia

Recall that $K\langle A\rangle^{p\times q}$ denotes the set of p by q matrices over $K\langle A\rangle$. In particular, $K\langle A\rangle^{n\times 1}$ is the set of column vectors of order n over $K\langle A\rangle$. This set has a natural structure of right $K\langle A\rangle$-module. If V is in $K\langle A\rangle^{n\times 1}$, we denote by $(V, 1)$ its *constant term*, that is, setting

$$V = \begin{pmatrix} P_1 \\ \vdots \\ P_n \end{pmatrix}$$

one has

$$(V, 1) = \begin{pmatrix} (P_1, 1) \\ \vdots \\ (P_n, 1) \end{pmatrix} \in K^{n\times 1}.$$

Furthermore, if w is a word in A^*, we denote by Vw^{-1} the vector

$$Vw^{-1} = \begin{pmatrix} P_1 w^{-1} \\ \vdots \\ P_n w^{-1} \end{pmatrix}.$$

We have the following relation which is the analogue of Equation (1.5.1)

$$V = (V, 1) + \sum_{a\in A} (Va^{-1})a. \tag{3.1}$$

The vectors Va^{-1}, for $a \in A$, are uniquely defined by this equality.

Definition A (right) submodule E of $K\langle A\rangle^{n\times 1}$ is *cancellative* if, whenever $V \in E$ and $(V, 1) = 0$, then $Va^{-1} \in E$ for any letter $a \in A$.

The next result characterizes cancellative submodules and will be the key to all the results of this section.

Theorem 3.1 *A submodule E of $K\langle A\rangle^{n\times 1}$ is cancellative if and only if it may be generated, as a right $K\langle A\rangle$-module, by p vectors $V_1, \ldots, V_p$ such that the matrix $((V_1, 1), \ldots, (V_p, 1)) \in K^{n\times p}$ is of rank p. In this case, $p \le n$ and $V_1, \ldots, V_p$ are linearly $K\langle A\rangle$-independent.*

Proof. 1. We begin with the easy part: suppose that E is generated by $V_1, \ldots, V_p$ as indicated. Let $V \in E$ with $(V, 1) = 0$. Then

$$V = \sum_{1 \leq i \leq p} V_i P_i \quad (P_i \in K\langle A \rangle).$$

Taking constant terms, we obtain

$$0 = (V, 1) = \sum (V_i, 1)(P_i, 1).$$

Because of the rank condition, we have $(P_i, 1) = 0$ for any i. It follows that $P_i = \sum_{a \in A} (P_i a^{-1})a$, which shows that

$$V = \sum_{i, a} V_i (P_i a^{-1})a.$$

By Equation (3.1) we obtain

$$V a^{-1} = \sum_i V_i (P_i a^{-1}),$$

hence $V a^{-1} \in E$, as desired.

2. Let E be a cancellative submodule of $K\langle A \rangle$. If $V \in K\langle A \rangle^{n \times 1}$, V may be written $V = \sum_{w \in A^*} (V, w)w$ where the $(V, w) \in K^{n \times 1}$ are almost all zero. Let $\deg(V)$ be the maximal length of a word w such that $(V, w) \neq 0$.

Claim. There are vectors $V_1, \ldots, V_p$ in E such that

(i) $\deg(V_1) \leq \deg(V_2) \leq \cdots \leq \deg(V_p)$.
(ii) The vectors $(V_i, 1)$ form a K-basis of the K-space $(E, 1) = \{(V, 1) \mid V \in E\}$.
(iii) If $V \in E$ and $\deg(V) < \deg(V_i)$ then $(V, 1)$ is a K-linear combination of $(V_1, 1), \ldots, (V_{i-1}, 1)$.

Suppose the claim is true. Then the matrix $((V_1, 1), \ldots, (V_p, 1))$ has rank p. We show by induction on $\deg(V)$ that each $V \in E$ is in $E' = \sum_{1 \leq i \leq p} V_i K\langle A \rangle$.

If $\deg(V) = -\infty$, that is $V = 0$, it is obvious. Let $\deg(V) \geq 0$ and let i be the smallest integer such that $\deg(V) < \deg(V_i)$ (with $i = p + 1$ if such an integer does not exist). Then $\deg(V) \geq \deg(V_1), \ldots, \deg(V_{i-1})$. Moreover, if $i \leq p$ then by (iii), $(V, 1)$ is a linear combination of $(V_1, 1), \ldots, (V_{i-1}, 1)$, and if $i = p + 1$ then by (ii), $(V, 1)$ is also a linear combination of $(V_1, 1), \ldots, (V_{i-1}, 1)$. Let $V' = V - \sum_{1 \leq j \leq i-1} \alpha_j V_j$ $(\alpha_j \in K)$ be such that $(V', 1) = 0$. By the cancellative property of E, $V' a^{-1}$ is in E for any letter a. Now,

$$\deg(V') \leq \max(\deg(V), \deg(\alpha_1 V_1), \ldots, \deg(\alpha_{i-1} V_{i-1})) = \deg(V)$$

hence $\deg(V' a^{-1}) < \deg(V)$. It follows by induction that $V' a^{-1} \in E'$. Now, by Equation (3.1), $V' = \sum_a (V' a^{-1})a$, and V' is in E'. Thus $V = V' + \sum_j \alpha_j V_j$ is in E' as well.

3. *Proof of the claim.* For $d = -1, 0, 1, 2, \ldots$, let $F(d)$ be the subspace of $K^{n \times 1}$ defined by

$$F(d) = \{(V, 1) \mid V \in E, \deg(V) \leq d\}.$$

Then

$$0 = F(-1) \subset F(0) \subset F(1) \subset \cdots \subset F(d) \subset \cdots$$

Let $0 \leq d_1 < \cdots < d_q$ be such that for any i, $F(d_i - 1) \subsetneq F(d_i)$ and such that each $F(d)$ is equal to some $F(d_i)$; in other words, one has

$$0 = F(-1) = \cdots = F(d_1 - 1) \subsetneq F(d_1) = \cdots = F(d_2 - 1)$$
$$\subsetneq F(d_2) \subsetneq \cdots \subsetneq F(d_q) = F(d_q + 1) = \cdots$$

In particular, $F(d_q) = (E, 1)$. Now, let B_1 be a basis of $F(d_1)$, B_2 be a basis of $F(d_2)$ mod $F(d_1), \ldots, B_q$ be a basis of $F(d_q)$ mod $F(d_{q-1})$. By the definition of the F's we may find for each i in $\{1, \ldots, q\}$ vectors $W_{i,1}, \ldots, W_{i,k_i}$ in E of degree $\leq d_i$ such that $\{(W_{i,1}, 1), \ldots, (W_{i,k_i}, 1)\} = B_i$; in fact, the degree of each $W_{i,j}$ is exactly d_i, otherwise $(W_{i,j}, 1) \in F(d_i - 1) = F(d_{i-1})$, which contradicts the fact that B_i is a basis mod $F(d_{i-1})$.

Define $V_1, \ldots, V_p$ by

$$(V_1, \ldots, V_p) = (W_{1,1}, \ldots, W_{1,k_1}, W_{2,1}, \ldots, W_{2,k_2}, \ldots, W_{q,k_q}).$$

Then the condition (i) of the claim is clearly satisfied. Moreover, since $F(d_q) = (E, 1)$, condition (ii) is also satisfied. Let $V \in E$ with $\deg(V) < \deg(V_k)$. Then $V_k = W_{i,j}$ for some i, j, and consequently $\deg(V) < d_i = \deg(W_{i,j})$, which implies that $(V, 1) \in F(d_i - 1) = F(d_{i-1})$ and $(V, 1)$ is a linear combination of $W_{1,1}, \ldots, W_{i-1,k_{i-1}}$, hence of $V_1, \ldots, V_{k-1}$. This proves the claim.

4. We show the last assertion of the theorem. Clearly, $p \leq n$. Suppose $\sum V_i P_i = 0$ where $P_i \in K\langle A\rangle$ are not all zero; choose such a relation with $\sup(\deg(P_i))$ minimum. Then $\sum (V_i, 1)(P_i, 1) = 0$ which shows as in (1) that $(P_i, 1) = 0$ for each i. Now some P_j is $\neq 0$, and therefore $P_j a^{-1} \neq 0$ for some letter a. By Equation (3.1) we obtain $\sum V_i (P_i a^{-1}) = 0$, which is a new relation contradicting the above minimality. Thus the V's are $K\langle A\rangle$-independent. □

Definition An n by n matrix M over $K\langle A\rangle$ is *full* if, whenever $M = M_1 M_2$ for some matrices $M_1 \in K\langle A\rangle^{n\times p}$ and $M_2 \in K\langle A\rangle^{p\times n}$, then $p \geq n$.

Corollary 3.2 (Cohn 1961) *Let M be an n by n matrix over $K\langle A\rangle$. If $S_1, \ldots, S_n$ in $K\langle\langle A\rangle\rangle$ are formal series, not all zero, such that $(S_1, \ldots, S_n)M = (0, \ldots, 0)$, then M is not full.*

Proof. Let E be the set of vectors $V \in K\langle A\rangle^{n\times 1}$ such that $(S_1, \ldots, S_n)V = 0$. Then E is a right submodule of $K\langle A\rangle^{n\times 1}$. Let $V = {}^t(P_1, \ldots, P_n) \in E$ be such that $(V, 1) = 0$. Then $(P_i, 1) = 0$ for any i. Moreover $\sum_i S_i P_i = 0$, so that if a is a letter, one has $\sum_i S_i (P_i a^{-1}) = 0$. This means that $Va^{-1} \in E$; thus E is cancellative. By Theorem 3.1, the right $K\langle A\rangle$-module E admits a basis consisting of p vectors $V_1, \ldots, V_p$ such that $\operatorname{rank}((V_1, 1), \ldots, (V_p, 1)) = p$ and $p \leq n$.

Now suppose that $p = n$. Then the matrix $N = ((V_1, 1), \ldots, (V_n, 1)) \in K^{n\times n}$ is invertible. Next N is the constant matrix of $H = (V_1, \ldots, V_n) \in K\langle A\rangle^{n\times n}$, that is $N = (H, 1)$; this implies that H is invertible in $K\langle\langle A\rangle\rangle^{n\times n}$. Now we have $(S_1, \ldots, S_n)H = 0$ (because $(S_1, \ldots, S_n)V_i = 0$ for all i), hence $(S_1, \ldots, S_n) = 0$ (multiply by H^{-1}), a contradiction.

Thus $p < n$. Let $M = (C_1, \ldots, C_n)$, where C_k is the k-th column of M. Then, by hypothesis, C_k belongs to E, hence $C_k = \sum_{j=1}^{p} V_j P_{j,k}$ for some polynomials $P_{j,k}$. Thus

$$M = (V_1, \ldots, V_p)(P_{j,k})_{1\le j\le p,\, 1\le k\le n}$$

and M is not full. □

Corollary 3.3 (Cohn 1982) *Let P_1, P_2, P_3, P_4 be polynomials such that P_2 is invertible as a formal series, that is $(P_2, 1) \neq 0$, and such that $P_1 P_2^{-1} P_3 = P_4$ holds in $K\langle\langle A\rangle\rangle$. Then there exist polynomials Q_1, Q_2, Q_3, Q_4 such that $P_1 = Q_1 Q_2$, $P_2 = Q_3 Q_2$, $P_3 = Q_3 Q_4$, $P_4 = Q_1 Q_4$.*

Proof. Consider the 2×2 matrix over $K\langle A\rangle$:

$$M = \begin{pmatrix} P_1 & P_4 \\ P_2 & P_3 \end{pmatrix}.$$

By assumption, we have

$$(1, -P_1 P_2^{-1})M = 0.$$

Hence M is not full by Corollary 3.2, and M may be written as

$$M = \begin{pmatrix} Q_1 \\ Q_3 \end{pmatrix} (Q_2, Q_4)$$

for some polynomials Q_i. This proves the corollary. □

Let $S_1, \ldots, S_n, T_1, \ldots, T_n$ be formal series. We say that

$$\sum_j S_j T_j$$

is *trivially a polynomial* if, for each j, either $S_j = 0$, or $T_j = 0$, or both S_j and T_j are polynomials. Note that one has

$$\sum_j S_j T_j = (S_1, \ldots, S_n) \begin{pmatrix} T_1 \\ \vdots \\ T_n \end{pmatrix}.$$

Corollary 3.4 (Inertia Theorem, (Bergman 1968, Cohn 1961))
Let $(S_{i,h})_{i\in I,\, 1\le h\le n}$ and $(T_{h,j})_{1\le h\le n,\, j\in J}$ be two families of formal series such that for each $i \in I$ and $j \in J$, $\sum_h S_{i,h} T_{h,j}$ is a polynomial. Then there exists an invertible

matrix M over $K\langle\langle A\rangle\rangle$ such that for any i and j

$$\left[(S_{i,1},\ldots,S_{i,n})M\right]\left[M^{-1}\begin{pmatrix}T_{1,j}\\ \vdots\\ T_{n,j}\end{pmatrix}\right]$$

is trivially a polynomial.

Proof. 1. We prove the theorem first in the case where each $T_{h,j}$ is a polynomial. Let $E=\{V\in K\langle A\rangle^{n\times 1}\mid \forall i\in I,(S_{i,1},\ldots,S_{i,n})V\in K\langle A\rangle\}$. Then E is a cancellative right submodule of $K\langle A\rangle^{n\times 1}$ as may be easily verified (cf. the proof of Corollary 3.2). By Theorem 3.1 there exist p vectors $V_1,\ldots,V_p$ in E which form a basis of E (as a right $K\langle A\rangle$-module) and such that the constant matrix of $(V_1,\ldots,V_p)$ is of rank $p\le n$. By performing a permutation of coordinates, we may assume that

$$(V_1,\ldots,V_p)=\begin{pmatrix}X\\ Y\end{pmatrix},$$

where $(X,1)\in K^{p\times p}$ is invertible. Let

$$M=\begin{pmatrix}X & 0\\ Y & I_{n-p}\end{pmatrix},$$

where I_{n-p} is the identity matrix of order $n-p$. Then $(M,1)\in K^{n\times n}$ is invertible, hence M is invertible in $K\langle\langle A\rangle\rangle^{n\times n}$.

Note that the first p columns of M (that is the V_i's) are in E: this implies, by definition of E, that for any $i\in I$ the first p components of $(S_{i,1},\ldots,S_{i,n})M$ are polynomials.

Moreover, let $1\le h\le p$: then $M^{-1}V_h$ is equal to the h-th column of $M^{-1}M$, that is to the h-th canonical vector $E_h\in K^{n\times 1}$. Now let $j\in J$. Then by assumption $V={}^t(T_{1,j},\ldots,T_{n,j})$ is in E. Hence $V=\sum_{1\le h\le p}V_hP_h$ for some polynomials P_h. Thus $M^{-1}V=\sum_h M^{-1}V_hP_h$ is equal, by the previous remark, to $\sum_h E_hP_h={}^t(P_1,\ldots,P_p,0,\ldots,0)$. This shows that the product

$$\left[(S_{i,1},\ldots,S_{i,n})M\right]\left[M^{-1}\begin{pmatrix}T_{1,j}\\ \vdots\\ T_{n,j}\end{pmatrix}\right]$$

is trivially a polynomial.

2. We come to the general case. Let

$$H=\{h\in\{1,\ldots,n\}\mid \forall j\in J,\ T_{h,j}\in K\langle A\rangle\}.$$

If $H=\{1,\ldots,n\}$, then we are in case 1. Suppose $|H|<n$: we may suppose that $H=\{1,\ldots,p\}$ with $0\le p<n$ (including the case $H=\emptyset$). Suppose that $\forall i\in I,\forall h\notin H,\ S_{i,h}=0$. Then

$$\sum_{h=1}^{n}S_{i,h}T_{h,j}=\sum_{h=1}^{p}S_{i,h}T_{h,j}$$

is a polynomial, so we are reduced to case 1 (with p instead of n).

Otherwise, there is some $i_0 \in I$ such that for some $h_0 \notin H$, $S_{i_0,h_0} \neq 0$. Choose $h_0 \notin H$ such that $\omega(S_{i_0,h_0}) \leq \omega(S_{i_0,h})$ for any $h \notin H$ (for the definition of ω, see Section 1.3). Choose polynomials $R_1, \ldots, R_p$ such that for $1 \leq h \leq p$, $\omega(S_{i_0,h} + R_h) \geq \omega(S_{i_0,h_0})$. Define S'_h by $S'_h = S_{i_0,h} + R_h$ if $1 \leq h \leq p$ and $S'_h = S_{i_0,h}$ if $p < h \leq n$. Then $\omega(S'_{h_0}) \leq \omega(S'_h)$, $S'_{h_0} = S_{i_0,h_0} \neq 0$ and

$$\sum_{1\leq h\leq n} S'_h T_{h,j} = \sum_{h\leq p}(S_{i_0,h} + R_h)T_{h,j} + \sum_{h>p} S_{i_0,h} T_{h,j}$$

$$= \sum_{1\leq h\leq n} S_{i_0,h} T_{h,j} + \sum_{h\leq p} R_h T_{h,j}$$

is a polynomial, by definition of $H = \{1, \ldots, p\}$. Let w be a word of minimal length in the support of S'_{h_0}; then $w^{-1}S'_{h_0}$ is an invertible formal series, and for any h, since $\omega(S'_h) \geq |w|$, one has $w^{-1}(S'_h T_{h,j}) = (w^{-1}S'_h)T_{h,j}$. Hence $\sum_h (w^{-1}S'_h)T_{h,j}$ is a polynomial. Define the matrix $N \in K\langle\langle A\rangle\rangle^{n\times n}$ which coincides with the $n \times n$ identity matrix except in the h_0-th row, where it is equal to $(w^{-1}S'_1, \ldots, w^{-1}S'_n)$; in particular the entry of the coordinate (h_0, h_0) of N is the invertible series $w^{-1}S'_{h_0}$, so N is invertible in $K\langle\langle A\rangle\rangle^{n\times n}$. Let $M = N^{-1}$. Then for any j, $M^{-1}\,{}^t(T_{1,j}, \ldots, T_{n,j}) = N\,{}^t(T_{1,j}, \ldots, T_{n,j})$ is equal to ${}^t(T_{1,j}, \ldots, T_{n,j})$ except in the h_0-th component, where it is equal to $\sum(w^{-1}S'_h)T_{h,j}$: thus the first p and the h_0-th components of $M^{-1}\,{}^t(T_{1,j}, \ldots, T_{n,j})$ are polynomials and we may conclude the proof by induction on $n - p$ because we have increased $|H|$. □

10.4 Gauss's Lemma

We consider in this section polynomials with integer or rational coefficients. Everything would work (with slight changes) with any factorial ring instead of $\mathbb{Z}$.

Definition A polynomial $P \in \mathbb{Q}\langle A\rangle$ is *primitive* if $P \neq 0$, $P \in \mathbb{Z}\langle A\rangle$ and if its coefficients have no nontrivial common divisors in $\mathbb{Z}$.

Definition The *content* of a nonzero polynomial $P \in \mathbb{Q}\langle A\rangle$ is the unique positive rational number $c(P)$ such that $P/c(P)$ is primitive.

Notation $P/c(P)$ will be denoted by $\overline{P}$.

Example 4.1 $c(4/3 + 6a - 2ab) = 2/3$ because $3/2(4/3 + 6a - 2ab) = 2 + 9a - 3ab$ is primitive.

Note that for $P \neq 0$

$$P \text{ primitive} \iff c(P) = 1\,, \tag{4.1}$$

$$P \in \mathbb{Z}\langle A\rangle \iff c(P) \in \mathbb{N}\,. \tag{4.2}$$

Theorem 4.1 (Gauss's Lemma)

(i) *If P, Q are primitive, then so is PQ.*
(ii) *If P, Q are nonzero polynomials, then $c(PQ) = c(P)c(Q)$ and $\overline{PQ} = \overline{P}\,\overline{Q}$.*

Proof. (i) Suppose PQ is not primitive. Then there is some prime number n which divides each coefficient of PQ. This means that the canonical image $\phi(PQ)$ of PQ in $(\mathbb{Z}/n\mathbb{Z})\langle A\rangle$ vanishes. But $\mathbb{Z}/n\mathbb{Z}$ is a field, so $(\mathbb{Z}/n\mathbb{Z})\langle A\rangle$ is an integral domain (Section 1.1); moreover $0 = \phi(PQ) = \phi(P)\phi(Q)$, so $\phi(P) = 0$ or $\phi(Q) = 0$. This means that n divides all coefficients of P or of Q, and contradicts the fact that P and Q are primitive.

(ii) By (i), $PQ/c(P)c(Q) = (P/c(P))(Q/c(Q))$ is primitive. So, by definition of the content of PQ, $c(PQ) = c(P)c(Q)$. Now, $\overline{PQ} = PQ/c(PQ)$ so that $\overline{PQ} = PQ/c(P)c(Q) = \overline{P}\,\overline{Q}$. □

Corollary 4.2 *Let $a_1, \ldots, a_n$ be polynomials. Then the continuant polynomials $p(a_1, \ldots, a_n)$ and $p(a_n, \ldots, a_1)$ are both zero or have the same content.*

Proof. (Induction on n). The result is obvious for $n = 0, 1$. Let $n \geq 2$. By Lemma 2.5, we may suppose that both polynomials are $\neq 0$. We have, by Proposition 2.1,

$$p(a_1, \ldots, a_n)p(a_{n-1}, \ldots, a_1) = p(a_1, \ldots, a_{n-1})p(a_n, \ldots, a_1) . \tag{4.3}$$

By induction, either $p(a_1, \ldots, a_{n-1}) = p(a_{n-1}, \ldots, a_1) = 0$, in which case $p(a_1, \ldots, a_n) = p(a_1, \ldots, a_{n-2})$ by (2.1) and $p(a_n, \ldots, a_1) = p(a_{n-2}, \ldots, a_1)$ by (2.3), and we conclude by induction; or $c(p(a_{n-1}, \ldots, a_1)) = c(p(a_1, \ldots, a_{n-1}))$, which implies by (4.3) and Theorem 4.1 that $c(p(a_1, \ldots, a_n)) = c(p(a_n, \ldots, a_1))$. □

Corollary 4.3 *Let P_1, P_2, P_3, P_4 be nonzero polynomials in $\mathbb{Z}\langle A\rangle$ such that P_2 is invertible in $\mathbb{Q}\langle\langle A\rangle\rangle$ and such that $P_1P_2^{-1}P_3 = P_4$. Then there exist polynomials $R_1, R_2, R_3, R_4 \in \mathbb{Z}\langle A\rangle$ such that*

$$P_1 = R_1R_2,\ P_2 = R_3R_2,\ P_3 = R_3R_4,\ P_4 = R_1R_4 .$$

Proof. By Corollary 3.3 we have

$$P_1 = Q_1Q_2,\ P_2 = Q_3Q_2,\ P_3 = Q_3Q_4,\ P_4 = Q_1Q_4$$

for some polynomials $Q_1, Q_2, Q_3, Q_4 \in \mathbb{Q}\langle A\rangle$.

Let $c_i = c(Q_i)$, $i = 1, 2, 3, 4$. By Theorem 4.1 we have

$$c(P_1) = c_1c_2,\ c(P_2) = c_3c_2,\ c(P_3) = c_3c_4,\ c(P_4) = c_1c_4 .$$

Thus $c(P_4) = c(P_1)c(P_3)/c(P_2)$.

As by hypothesis and Equation (4.2) $c(P_i) \in \mathbb{N}$, there exist positive integers d_1, d_2, d_3, d_4 such that

$$c(P_1) = d_1d_2,\ c(P_2) = d_3d_2,\ c(P_3) = d_3d_4,\ c(P_4) = d_1d_4 .$$

Moreover, by Theorem 4.1,

$$\overline{P}_1 = \overline{Q}_1\overline{Q}_2,\ \overline{P}_2 = \overline{Q}_3\overline{Q}_2,\ \overline{P}_3 = \overline{Q}_3\overline{Q}_4,\ \overline{P}_4 = \overline{Q}_1\overline{Q}_4\,.$$

Put $R_i = d_i\overline{Q}_i$, $i = 1, 2, 3, 4$. Then $R_i \in \mathbb{Z}\langle A\rangle$ and

$$P_1 = c(P_1)\overline{P}_1 = d_1 d_2 \overline{Q}_1\overline{Q}_2 = R_1 R_2\,.$$

Similarly $P_2 = R_3 R_2$, $P_3 = R_3 R_4$ and $P_4 = R_1 R_4$. □

Proposition 4.4 *Let Y be a primitive polynomial of degree 1 which vanishes for some integer values of the variables. Let $P, Q \in \mathbb{Z}\langle A\rangle$ and let $\alpha \in \mathbb{Z}$, $\alpha \neq 0$ be such that $PQ \equiv \alpha \mod Y\mathbb{Z}\langle A\rangle$. Then $P \equiv \beta$, $Q \equiv \gamma \mod Y\mathbb{Z}\langle A\rangle$ for some $\beta, \gamma \in \mathbb{Z}$ such that $\alpha = \beta\gamma$.*

Proof. We have $PQ = YQ_2 + \alpha$ for some polynomial Q_2. As $\alpha \neq 0$, we have $Q \neq 0$ and we may apply Corollary 1.3. This shows that $P = \beta + YT$ for some $\beta \in \mathbb{Q}$ and $T \in \mathbb{Q}\langle A\rangle$. Hence $YQ_2 + \alpha = \beta Q + YTQ$. Since $\alpha \neq 0$ and $\deg(Y) > 0$, we obtain $\beta \neq 0$: indeed, otherwise $YTQ = YQ_2 + \alpha$, implying that Y divides α. This shows that $Q = \gamma + YS$ for some $\gamma \in \mathbb{Q}$ such that $\alpha = \beta\gamma$ and $S \in \mathbb{Q}\langle A\rangle$. Now the assumption on Y and the fact that P, Q have integer coefficients imply that $\beta, \gamma \in \mathbb{Z}$. Since $YT = P - \beta \in \mathbb{Z}\langle A\rangle$, we obtain that $c(Y)c(T) \in \mathbb{N}$ by Equation (4.2) and Theorem 4.1 (ii). But Y is primitive, so $c(Y) = 1$, which shows that $c(T) \in \mathbb{N}$ and $T \in \mathbb{Z}\langle A\rangle$ by (4.2). Similarly, $S \in \mathbb{Z}\langle A\rangle$. □

Exercises for Chapter 10

1.1 Let $P_1, \ldots, P_n, Q_1, \ldots, Q_n$ be polynomials. A relation $\sum_{i=1}^{n} P_i Q_i = 0$ is called *trivial* if for each i, either $P_i = 0$ or $Q_i = 0$. Note that $\sum P_i Q_i$ may be written

$$(P_1, \ldots, P_n)\begin{pmatrix} Q_1 \\ \vdots \\ Q_n \end{pmatrix}.$$

Show that if $\sum_{i=1}^{n} P_i Q_i = 0$, then there exists an invertible n by n matrix M with coefficients in $K\langle A\rangle$ such that the relation

$$\big[(P_1, \ldots, P_n)M\big]\left[M^{-1}\begin{pmatrix} Q_1 \\ \vdots \\ Q_n \end{pmatrix}\right] = 0$$

is trivial (cf. (Cohn 1961)).

1.2 (a) Let X, Y, X', Y' be nonzero formal series such that $XY' = YX'$, with $\omega(X) \geq \omega(Y)$ (cf Chapter 1). Show that there exists a formal series U such that $X = YU$, $X' = UY'$.

(b) Let S be a formal series and let C be its centralizer, that is $C = \{T \in K\langle\langle A\rangle\rangle \mid ST = TS\}$. Show that if $T_1, T_2 \in C$ and $\omega(T_2) \geq \omega(T_1)$, then there exists $T \in C$ such that $T_2 = T_1 T$. (*Hint*: One may suppose $\omega(S) \geq 1$; let n be such that $\omega(S^n) \geq \omega(T_1), \omega(T_2)$: use (a) three times.) Let $T \in C$ such that $\omega(T) \geq 1$ is minimum. Show that $C = K[[T]]$, that is

$$C = \left\{ \sum_{n \in \mathbb{N}} a_n T^n \mid a_n \in K \right\}$$

(see Cohn (1961) and also Lothaire (2002, Theorem 9.1.1)).

2.1 Show that for $n \geq k \geq 1$ the continuant polynomials satisfy the identities

$$p(a_1, \ldots, a_n)p(a_{n-1}, \ldots, a_k) - p(a_1, \ldots, a_{n-1})p(a_n, \ldots, a_k)$$
$$= (-1)^{n+k} p(a_1, \ldots, a_{k-2})$$

with the conventions: $p(a_1, \ldots, a_{k-2}) = 0$ if $k = 1$, $= 1$ if $k = 2$, and $p(a_{n-1}, \ldots, a_k) = 1$ if $k = n$. Show that the number of words in the support of $p(a_1, \ldots, a_n)$ is the n-th Fibonacci number F_n (see Example 3.2.1).

2.2 Show that if $a_1, \ldots, a_n$ are commutative variables, then

$$a_1 + \cfrac{1}{a_2 + \cfrac{1}{a_3 + \cfrac{1}{\cdots + \cfrac{1}{a_n}}}} = \frac{p(a_1, \ldots, a_n)}{p(a_2, \ldots, a_n)}.$$

2.3 Show that for $n \geq 1$

$$\begin{pmatrix} a_1 & 1 \\ 1 & 0 \end{pmatrix} \begin{pmatrix} a_2 & 1 \\ 1 & 0 \end{pmatrix} \cdots \begin{pmatrix} a_n & 1 \\ 1 & 0 \end{pmatrix} = \begin{pmatrix} p(a_1, \ldots, a_n) & p(a_1, \ldots, a_{n-1}) \\ p(a_2, \ldots, a_n) & p(a_2, \ldots, a_{n-1}) \end{pmatrix}.$$

3.1 Let M be an n by n polynomial matrix such that $M = M_1 M_2$ with $M_1 \in K\langle\langle A\rangle\rangle^{n \times p}$ and $M_2 \in K\langle\langle A\rangle\rangle^{p \times n}$. Show that then one may choose M_1, M_2 to be polynomial matrices (use the inertia theorem; see Cohn (1985)).

Notes to Chapter 10

Most of the results of this chapter are due to P. M. Cohn. We have already seen a result concerning noncommutative polynomials in Chapter 2 (Corollary 2.3.3): in P. M. Cohn's terminology, it means that $K\langle A\rangle$ is a *fir* ("free ideal ring"). The terminology "continuant" stems from its relation to continuous fractions (see Exercises 2.2 and 2.3). Corollary 3.2 is a special case of a more general result, stating that every polynomial matrix which is singular over the free field is not full (see Cohn (1961, 2006)).

11

Codes and formal series

The aim of this chapter is to present an application of formal series to the theory of variable-length codes. The main result (Theorem 4.1) states that every finite complete code admits a factorization into three polynomials which reflect its combinatorial structure.

The first section contains some basic facts on codes and prefix codes. These are easily expressed by means of series.

Section 11.2 is devoted to complete codes and their relations to Bernoulli morphisms (Theorem 2.4). Concerning the degree of a code, we give in Section 11.3 only the very basic results needed in Section 11.4.

This last section is devoted to the proof of the main result. It uses the material of the previous section and of Chapter 10.

11.1 Codes

Definition A *code* is a subset C of A^* such that whenever $u_1, \ldots, u_n, v_1, \ldots, v_p$ in C satisfy

$$u_1 \cdots u_n = v_1 \cdots v_p \,, \tag{1.1}$$

then $n = p$ and $u_i = v_i$ for $i = 1, \ldots, n$.

Note that if C is a code, then $C \subset X^+$ $(= X^* \setminus 1)$.

Example 1.1 The set $\{a, ab, ba\}$ is not a code, because the word aba has two factorizations in it:

$$aba = a(ba) = (ab)a \,.$$

Example 1.2 The set $\{a, ab, bb\}$ is a code; indeed, no word in it is a prefix of another, so in each relation of the form (1.1), either u_1 is a prefix of v_1 or vice versa, so one has $u_1 = v_1$ and one concludes by induction on n.

Example 1.3 The set $\{b, ab, a^2b, a^3b, \ldots, a^nb, \ldots\} = a^*b$ is a code, for the same reason as in Example 1.2.

Example 1.4 The set $\{a^3, a^2ba, a^2b^2, ab, ba^2, baba, bab^2, b^2a, b^3\}$ is a code, for the same reason; note that in this case, moreover no word is a suffix of another.

Example 1.5 The set $C = \{a^2, ab, a^2b, ab^2, b^2\}$ is a code. Indeed, let $\underline{C}$ denote its characteristic polynomial; then we have

$$\begin{aligned}1 - \underline{C} &= 1 - a^2 - ab - a^2b - ab^2 - b^2\\ &= (1 - b - a^2 - ab) + (b - b^2 - a^2b - ab^2)\\ &= (1 - b - a^2 - ab)(1 + b)\\ &= ((1 - a - b) + (a - a^2 - ab))(1 + b)\\ &= (1 + a)(1 - a - b)(1 + b).\end{aligned}$$

Thus, in $\mathbb{Z}\langle\langle A\rangle\rangle$, we have

$$(1 - \underline{C})^{-1} = (1 + b)^{-1}(1 - a - b)^{-1}(1 + a)^{-1}.$$

By the results of Section 1.4, for any proper formal series S, $(1 - S)^{-1} = \sum_{n\geq 0} S^n = S^*$ and $(1 - a - b)^{-1} = \underline{A}^* = \underline{A^*}$ is the sum of all words on A (and hence, its nonzero coefficients are all equal to 1). Thus

$$\underline{A^*} = (1 + b)\Big(\sum_{n\geq 0} \underline{C}^n\Big)(1 + a).$$

This shows that the series $\sum_{n\geq 0} \underline{C}^n$ has no coefficient ≥ 2, since otherwise $\underline{A^*}$ would have such a coefficient. From

$$\sum_{n\geq 0} \underline{C}^n = \sum_{n\geq 0} \sum_{u_1,\ldots,u_n\in C} u_1 \cdots u_n$$

we obtain that no word has two distinct factorizations of the form $u_1 \cdots u_n$ ($u_i \in C$), so C is a code.

Recall that for any language X, $\underline{X}$ denotes its characteristic series (considered as an element of $\mathbb{Q}\langle\langle A\rangle\rangle$ in the present chapter). One of the arguments of the last example may be generalized as follows.

Proposition 1.1 *Let C be a subset of A^+ and let $\underline{C}$ be its characteristic series. Then C is a code if and only if one has in $\mathbb{Z}\langle\langle A\rangle\rangle$*

$$(1 - \underline{C})^{-1} = \underline{C}^* = \underline{C^*}. \tag{1.2}$$

Proof. The first equality is always true, as shown in Section 1.4. We have

$$\sum_{n\geq 0} \sum_{u_1,\ldots,u_n\in C} u_1 \cdots u_n = \sum_{n\geq 0} \underline{C}^n = \underline{C}^*.$$

If C is a code, then the words

$$u_1 \cdots u_n \quad (n \geq 0, u_i \in C)$$

are all distinct, so the left-hand side is equal to $\underline{C^*}$. If C is not a code, then two of these words are equal, so the left-hand side is a series with at least one coefficient ≥ 2: it cannot be equal to $\underline{C^*}$, because the latter has only 0, 1 as coefficients. □

The previous result provides an effective algorithm for testing whether a given rational subset of C of A^+ is a code. Indeed, one has merely to check if the rational power series $(\underline{C})^* - \underline{C^*}$ is equal to 0; for this, apply Corollary 2.3.6, or use the effective construction of the minimal representation in Section 2.3.

However, there is a more direct algorithm. We give below, without proof, the algorithm of Sardinas and Patterson (see Lallement 1979, Berstel and Perrin 1985, Berstel *et al.* 2009). Recall that for any language X and any word w, we denote by $w^{-1}X$ the language

$$w^{-1}X = \{u \in A^* \mid wu \in X\}.$$

More generally, if Y is a language, we denote by $Y^{-1}X$ the language

$$Y^{-1}X = \bigcup_{w \in Y} w^{-1}X.$$

Now let C be a subset of A^+. Define a sequence of languages C_n by

$$C_0 = C^{-1}C \setminus 1,$$
$$C_{n+1} = C_n^{-1}C \cup C^{-1}C_n \quad (n \geq 0).$$

Then C *is a code if and only if no* C_n *contains the empty word.* If C is finite, the sequence (C_n) is periodic (because each word in C_n is a factor of some word in C). The same is true if C is rational (see Berstel *et al.* (2009), Prop. I.3.3). Hence in these cases, we obtain an effective algorithm.

Another way to express the fact that a set of words is a code is by means of the so-called unambiguous operations. Let X, Y be languages. Recall that their *union* is *unambiguous* if they are disjoint languages, that their *product* is *unambiguous* if $x, x' \in X$, $y, y' \in Y$, and $xy = x'y'$ implies $x = x'$, $y = y'$ and that the *star* X^* is *unambiguous* if X is a code.

Proposition 1.2 *Let* X, Y *be languages.*

(i) *The union of* X *and* Y *is unambiguous if and only if* $\underline{X \cup Y} = \underline{X} + \underline{Y}$.
(ii) *The product* XY *is unambiguous if and only if* $\underline{XY} = \underline{X}\,\underline{Y}$.
(iii) *If* $1 \notin X$, *then the star* X^* *is unambiguous if and only if* $\underline{X}^* = \underline{X^*}$.

Proof. The first two assertions are a direct consequence of the definitions. The last one is merely a reformulation of Proposition 1.1. □

We have already met a family of codes in Section 2.3: the *prefix codes*. A set is prefix if no word in it is a prefix of another word in it. A prefix set which is not reduced to the empty word is easily seen to be a code, called a prefix code. Symmetrically, one defines *suffix codes*. A code is called *bifix* if it is both prefix and suffix.

Proposition 1.3 *Let C be a code such that for any word v in C^*, one has $v^{-1}C^* \subset C^*$. Then C is a prefix code.*

Note the converse: for any set C and for any word v in C^*, one has $C^* \subset v^{-1}C^*$.

Proof. Suppose $u = vw$, with u, v in C and $w \in A^*$. We have to show that $w = 1$. Now $w = v^{-1}u \in v^{-1}C^* \subset C^*$, hence $w \in C^*$. Therefore $w = c_1 \cdots c_n$ ($c_i \in C$) and $u = vc_1 \cdots c_n \in C$. The only possibility for C to be a code is $n = 0$, that is $w = 1$, and C is a prefix code. □

Proposition 1.4 *Let C be a prefix code such that $CA^* \cap wA^*$ is nonempty for any word w. Let P be the set of proper prefixes of the words in C. Then one has in $\mathbb{Z}\langle\langle A\rangle\rangle$, the equality $\underline{C} - 1 = \underline{P}(\underline{A} - 1)$.*

Proof. Let $P' = A^* \setminus CA^*$. Then, by Proposition 2.3.1, $C = P'A \setminus P'$ because $C \neq \{1\}$. It follows easily that $\underline{C} - 1 = \underline{P}'(\underline{A} - 1)$.

It remains to show that $P = P'$. Let w be in P; then w is a proper prefix of some word in C and so has no prefix in C, C being a prefix code; hence $w \notin CA^* \implies w \in P'$.

Let w be in P'. By assumption, there are words $c \in C$, $u, v \in A^*$ such that $cu = wv$; since $w \notin CA^*$, w must be a proper prefix of c, so $w \in P$. □

Let C be a code. Define, for any word u, the series S_u inductively by

$$S_1 = 1\,,$$
$$S_u = a^{-1}S_v + (S_v, 1)a^{-1}\underline{C}\,, \quad \text{for } u = va \ (a \in A)\,.$$

Note that S_u has nonnegative coefficients. The reader may verify that the support of S_u consists of proper suffixes of C (see Exercise 1.3).

Lemma 1.5 *Let C be a code. Then for any word u, $u^{-1}(\underline{C}^*) = S_u\underline{C}^*$. In particular, S_u is a characteristic series. If C is finite, then S_u is a polynomial.*

Proof. We shall use the formulas of Lemma 1.7.2.

We prove $u^{-1}(\underline{C}^*) = S_u\underline{C}^*$ by induction on $|u|$. If $u = 1$, it is clearly true. Let $u = va$, with $a \in A$. Then by induction $v^{-1}(\underline{C}^*) = S_v\underline{C}^*$. Thus, by Lemma 1.7.2,

$$\begin{aligned} u^{-1}(\underline{C}^*) &= a^{-1}v^{-1}(\underline{C}^*) = a^{-1}(S_v\underline{C}^*) = (a^{-1}S_v)\underline{C}^* + (S_v, 1)(a^{-1}\underline{C}^*) \\ &= (a^{-1}S_v)\underline{C}^* + (S_v, 1)(a^{-1}\underline{C})\underline{C}^* = S_u\underline{C}^*\,. \end{aligned}$$

Now, since $u^{-1}(\underline{C}^*)$ is obviously a characteristic series, the same holds for S_u. It is easily verified by induction that S_u is a polynomial if C is finite. □

One defines symmetrically the series $P_u \in \mathbb{Z}\langle\langle A\rangle\rangle$ by

$$P_1 = 1\,,$$
$$P_{av} = P_v a^{-1} + (P_v, 1)\underline{C}a^{-1}\,, \quad \text{for } a \in A \text{ and } v \in A^*\,.$$

See Equation (10.1.3) for the notation Pa^{-1}. If C is finite, P_v is a polynomial. Now we define for a couple (u, v) of words another series in the following way:

$$F_{u,1} = 0\,,$$
$$F_{u,av} = (P_v, 1)S_u a^{-1} + F_{u,v}a^{-1}\,.$$

As above, the series $F_{u,v}$ clearly has nonnegative coefficients.

We denote below by $u^{-1}Sv^{-1}$ the series $(u^{-1}S)v^{-1} = u^{-1}(Sv^{-1})$, see Exercise 1.5.

Proposition 1.6 *Let C be a code. Then for any words u and v, $u^{-1}(\underline{C}^*)v^{-1} = S_u\underline{C}^*P_v + F_{u,v}$. In particular, $F_{u,v}$ is a characteristic series. If C is finite, then $F_{u,v}$ is a polynomial.*

Proof. (Induction on $|v|$). The result is obvious if $v = 1$ by Lemma 1.5. Let $a \in A$. Then $u^{-1}(\underline{C}^*)(av)^{-1} = [u^{-1}(\underline{C}^*)v^{-1}]a^{-1}$ is equal, by induction and Lemma 1.7.2, to

$$\begin{aligned}
&(S_u\underline{C}^*P_v)a^{-1} + F_{u,v}a^{-1} \\
&\quad = S_u\underline{C}^*(P_v a^{-1}) + (P_v, 1)S_u(\underline{C}^*a^{-1}) + (P_v, 1)S_u a^{-1} + F_{u,v}a^{-1} \\
&\quad = S_u\underline{C}^*(P_v a^{-1}) + (P_v, 1)S_u\underline{C}^*(\underline{C}a^{-1}) + F_{u,av} \\
&\quad = S_u\underline{C}^*P_{av} + F_{u,av}\,.
\end{aligned}$$

This proves the formula.

Now, since $S_u\underline{C}^*P_v$ has nonnegative coefficients and since $u^{-1}(\underline{C}^*)v^{-1}$ is a characteristic series, the same holds for $F_{u,v}$. If C is finite, it is easily seen by induction on the definition that $F_{u,v}$ is a polynomial. □

11.2 Completeness

Definition A language $C \subset A^*$ is *complete* if, for any word w, the set $C^* \cap A^*wA^*$ is nonempty.

Lemma 2.1 *If C is complete, then any word w is either a factor of a word in C or may be written as*

$$w = smp\,,$$

with $m \in C^*$ and where s (p) is a suffix (prefix) of a word of C.

Proof. We have $xwy \in C^*$ for some words x, y. Let us represent a word in C^* schematically by

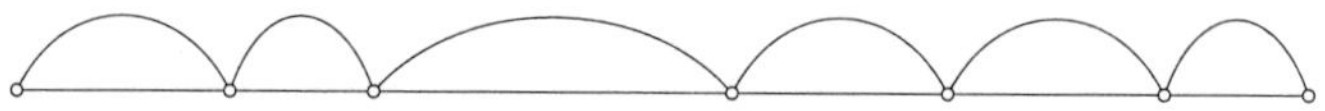

Here an arc represents an element of C. Then we have two cases:

(1)

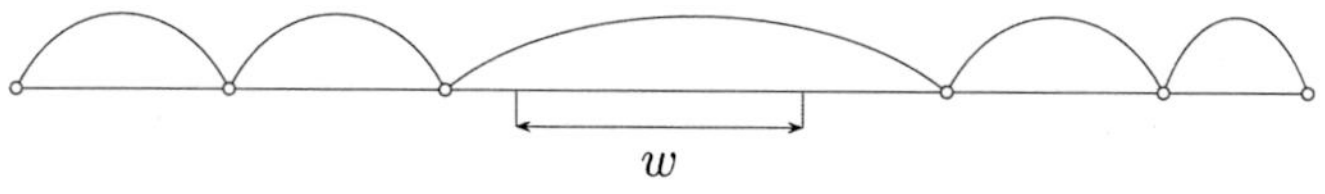

(2)

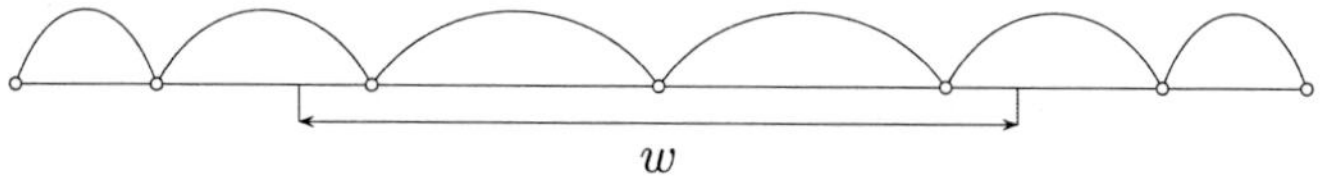

In the first case, w is a factor of a word in C. In the second case, $w = smp$ as in the lemma. □

Definition A *Bernoulli morphism* is a mapping $\pi : A^* \to \mathbb{R}$ such that

(i) $\pi(w) > 0$ for any word w,
(ii) $\pi(1) = 1$,
(iii) $\pi(uv) = \pi(u)\pi(v)$ for any words u, v,
(iv) $\sum_{a \in A} \pi(a) = 1$.

It is called *uniform* if $\pi(a) = 1/|A|$ for any letter a. We define for any language X the *measure* of X by

$$\pi(X) = \sum_{w \in X} \pi(w)$$

(it may be infinite). We shall frequently use the following inequalities:

$$\pi(\bigcup X_i) \leq \sum \pi(X_i), \quad \pi(XY) \leq \pi(X)\pi(Y).$$

Note that, for any n, one has $\pi(A^n) = \sum_{w \in A^n} \pi(w) = \left(\sum_{a \in A} \pi(a)\right)^n = 1$.

Lemma 2.2 *Let C be a code. Then $\pi(C) \leq 1$.*

Proof. Since C is the union of its finite subsets, it is enough to show the lemma in the case where C is finite. Let p be the maximal length of words in C. Then

$$C^n \subset A \cup A^2 \cup \cdots \cup A^{pn}.$$

Thus $\pi(C^n) \leq pn$. Now, since C is a code, each word in C^n has only one factorization of the form $u_1 \cdots u_n$ ($u_i \in C$). Since π is multiplicative, we obtain $\pi(C^n) = \pi(C)^n$. Hence

$$\pi(C)^n \leq pn.$$

This shows that $\pi(C) \leq 1$. □

Lemma 2.3 *Let C be a finite complete language. Then $\pi(C) \geq 1$.*

Proof. By Lemma 2.1, we may write

$$A^* = SC^*P \cup F ,$$

where S, P, F are finite languages. Thus

$$\infty = \pi(A^*) \leq \pi(S)\pi(C^*)\pi(P) + \pi(F) .$$

This shows that $\pi(C^*) = \infty$. Now

$$C^* = \bigcup_{n \geq 0} C^n$$

so that $\pi(C^*) \leq \sum_{n \geq 0} \pi(C^n)$. Moreover, $\pi(C^n) \leq \pi(C)^n$, π being multiplicative. Thus $\infty \leq \sum_{n \geq 0} \pi(C)^n$, which shows that $\pi(C) \geq 1$. □

Theorem 2.4 (Schützenberger and Marcus 1959, Boë *et al.* 1980) *Let C be a finite subset of A^* and let π be a Bernoulli morphism. Then any two of the following assertions imply the third one:*

(i) *C is a code,*
(ii) *C is complete,*
(iii) $\pi(C) = 1$.

Note that this gives an algorithm for testing whether a given finite code is complete. We need another lemma.

Lemma 2.5 *Let X be a language and let w be a word such that $X \cap A^*wA^*$ is empty. Then $\pi(X) < \infty$.*

Proof. Let $\ell = |w|$ and for $i = 0, \ldots, \ell - 1$

$$X_i = \{v \in X \mid |v| \equiv i \bmod \ell\} .$$

Then $X_i \subset A^i(A^\ell \setminus w)^*$. Indeed $v \in X_i$ implies $v = uv_1 \cdots v_n$ with $|u| = i$ and for any j, $|v_j| = \ell$; by assumption, w is not factor of v, hence w is none of the v_j's: thus $v_j \in A^\ell \setminus w$, which proves the claim.

Now

$$\pi(A^\ell \setminus w) = \pi(A^\ell) - \pi(w) = 1 - \pi(w) < 1$$

and

$$\pi[(A^\ell \setminus w)^*] = \pi\left[\bigcup_{n \geq 0}(A^\ell \setminus w)^n\right] \leq \sum_{n \geq 0} \pi[(A^\ell \setminus w)^n]$$

$$\leq \sum_{n \geq 0} [\pi(A^\ell \setminus w)]^n < \infty .$$

Thus $\pi(X_i) = \pi[A^i(A^\ell \setminus w)^*] \leq \pi(A^i)\pi[(A^\ell \setminus w)^*] < \infty$ and since $X = \bigcup_{0 \leq i < \ell} X_i$, we obtain $\pi(X) < \infty$. □

Proof of Theorem 2.4. Lemmas 2.2 and 2.3 show that (i) and (ii) imply (iii).

Let C be a code with $\pi(C) = 1$. Suppose C is not complete. Then for some word w, $C^* \cap A^*wA^*$ is empty. Hence, by Lemma 2.5, $\pi(C^*) < \infty$. Since C is a code, $\pi(C^*)$ is equal to the sum $\sum_{n \geq 0} \pi(C)^n$. The latter being finite, we deduce that $\pi(C) < 1$, a contradiction.

Let C be complete and $\pi(C) = 1$. Then C^n is complete for any n; indeed, for any word w, there are words $u, v, c_1, \ldots, c_p$ ($c_i \in C$) such that $uwv = c_1 \cdots c_p$ (C being complete). Let r be such that $p + r$ is a multiple of n; then $uwvc_1^r = c_1 \cdots c_p c_1^r \in (C^n)^*$, which shows that $(C^n)^* \cap A^*wA^*$ is not empty. This implies that C^n is complete. Thus, by Lemma 2.3, $\pi(C^n) \geq 1$ for any n. But as usual $\pi(C^n) \leq \pi(C)^n = 1$, and therefore $\pi(C^n) = \pi(C)^n$ for any n.

Suppose C is not a code. Then for some words $u_1, \ldots, u_n, v_1 \ldots, v_p$ in C we have $u_1 \cdots u_n = v_1 \cdots v_p$ and $u_1 \neq v_1$. Hence $u_1 \cdots u_n v_1 \cdots v_p = v_1 \cdots v_p u_1 \cdots u_n$, and we have obtained a word in C^{n+p} which has two distinct factorizations. Consequently

$$\begin{aligned}\pi(C^{n+p}) &= \pi\big(\{w_1 \cdots w_{n+p} \mid w_i \in C\}\big) \\ &< \sum_{w_1, \ldots, w_{n+p} \in C} \pi(w_1 \cdots w_{n+p}) = \pi(C^{n+p})\end{aligned}$$

which is a contradiction. □

Let π be a Bernoulli morphism. Since π is multiplicative, it may be extended to an algebra morphism, still denoted by π,

$$\pi : \mathbb{Z}\langle A\rangle \to \mathbb{R}$$

by the formula

$$\pi\Big(\sum_w (P, w)w\Big) = \sum_w (P, w)\pi(w)\,.$$

Note that, because the measure of A is 1, one has

$$\pi(\underline{A} - 1) = 0\,.$$

Theorem 2.6 (Schützenberger 1965) *Let C be a finite code such that for any word w, the set $C^* \cap wA^*$ is nonempty. Then C is a prefix code.*

Proof. Let C' be the set of words in C having no proper prefix in C, that is $C' = C \setminus CA^+$. Clearly C' is a prefix code. Moreover, if w is a word, then for some words $c_1, \ldots, c_n \in C$, $u \in A^*$, one has by assumption

$$c_1 \cdots c_n = wu\,.$$

Then either $c_1 \in C'$, or c_1 has a prefix in C'. Therefore $C'A^* \cap wA^*$ is nonempty.

Let P be the set of proper prefixes of the words in C'. Then by Proposition 1.4, $\underline{C'} - 1 = \underline{P}(\underline{A} - 1)$. Apply the morphism $\pi : \mathbb{Z}\langle A\rangle \to \mathbb{R}$, obtaining $\pi(\underline{C'} - 1) = 0$ because $\pi(\underline{A} - 1) = 0$. Thus $\pi(C') = 1$. Since C is a code, we have by Lemma 2.2, $\pi(C) \leq 1$. But $C' \subset C$ and π is positive. Hence $C = C'$ is prefix. □

Theorem 2.7 (Reutenauer 1985) *Let P in $\mathbb{N}\langle A\rangle$ be without constant term such that $P - 1 = X(\underline{A} - 1)Y$ for some polynomials X, Y in $\mathbb{R}\langle\langle A\rangle\rangle$. Then $P = \underline{C}$ for some finite complete code C. Furthermore, if $Y \in \mathbb{R}$ ($X \in \mathbb{R}$), then C is a prefix (suffix) code.*

Proof. 1. Note that if S, T are formal series, then

$$\operatorname{supp}(ST) \subset \operatorname{supp}(S)\operatorname{supp}(T)\,.$$

Moreover, if S is proper, then

$$\operatorname{supp}(S^*) \subset \operatorname{supp}(S)^*\,.$$

2. We have $1 - P = X(1 - \underline{A})Y$. By assumption, $1 - P$ is invertible in $\mathbb{R}\langle\langle A\rangle\rangle$. The same holds for $1 - \underline{A}$ since its inverse is $\underline{A}^* = \underline{A^*}$. This shows that X and Y are also invertible. So we obtain

$$(1 - P)^{-1} = Y^{-1}(1 - \underline{A})^{-1}X^{-1}$$

which implies

$$(1 - \underline{A})^{-1} = Y(1 - P)^{-1}X\,.$$

Thus

$$\underline{A}^* = YP^*X\,. \tag{2.1}$$

By 1, this implies that each word w may be written as $w = ymx$, with $y \in \operatorname{supp}(Y)$, $m \in \operatorname{supp}(P)^*$ and $x \in \operatorname{supp}(X)$. Let $C = \operatorname{supp}(P)$ and let u be a word such that $|u| > \deg(X), \deg(Y)$. Let v be any word. Then $w = uvu$ may be written $uvu = ymx$ as above, which shows, by the choice of u, that $m = v_1 v v_2$. Hence $C^* \cap A^*vA^*$ is nonempty: we have shown that C is complete. Thus, by Lemma 2.3, $\pi(C) \geq 1$ (where π is some Bernoulli morphism). Now, as $P - 1 = X(\underline{A} - 1)Y$, we obtain $\pi(P) = 1$. Therefore

$$1 \leq \pi(C) \leq \pi(P) = 1$$

because P has nonnegative integer coefficients. This shows, π being positive, that $P = \underline{C}$ and that $\pi(C) = 1$. It follows by Theorem 2.4 that C is a code, and thus a finite complete code.

Suppose now that $Y \in \mathbb{R}$. Then, as above, Equation (2.1) shows that for any word v, one has $vu = mx$ for some words $m \in C^*$, $x \in \operatorname{supp}(X)$ (u being chosen as before). Then, since $|u| > |x|$, we obtain $m = vv_1$ which shows that $C^* \cap vA^*$ is nonempty. We conclude by Theorem 2.6. □

11.3 The degree of a code

Given a monoid M, recall that an *ideal* in M is a nonempty subset J which is closed for left and right multiplication by elements of M. Moreover, an *idempotent* is an element e which is equal to its square, that is $e^2 = e$. Recall also that if M is finite, then M has a minimal ideal, see Appendix 2 of Chapter 12.

Theorem 3.1 *Let C be a finite complete code. There exist a finite monoid M and a surjective morphism $\phi : A^* \to M$ such that $C^* = \phi^{-1}\phi(C^*)$. Let D be the minimal ideal of M. There exists an idempotent e in $D \cap \phi(C^*)$; further $\phi(C^*) \cap eMe$ is a subgroup of the group eMe.*

It will not be shown here that the index of $\phi(C^*) \cap eMe$ in eMe depends only on C; for this, we refer the reader to the book by Berstel *et al.* (2009). This being admitted, we introduce the following definition.

Definition With the notation of Theorem 3.1, the index of $eMe \cap \phi(C^*)$ in eMe is called the *degree* of C.

Proof of Theorem 3.1. Clearly, C^* is a rational subset of A^* (cf. Section 3.1). Hence, by Kleene's theorem (Theorem 3.1.1), it is recognizable. This shows that there exist a finite monoid M, a monoid morphism $\phi : A^* \to M$, and a subset N of M such that $C^* = \phi^{-1}(N)$. Clearly, we may assume that ϕ is surjective; then $N = \phi(C^*)$ and $C^* = \phi^{-1}\phi(C^*)$.

Let D be the minimal ideal of M and w a word in $\phi^{-1}(D)$. Then $C^* \cap A^*wA^*$ is nonempty (because C is complete), hence there exist words u, v such that uwv is in C^*. Now $m = \phi(uwv)$ is in $\phi(C^*)$ and also in D (because $m = \phi(u)\phi(w)\phi(v)$, $\phi(w) \in D$, and D is an ideal). Some power $e = m^n$ with $n \geq 1$ of m is idempotent and still lies in $\phi(C^*) \cap D$. Then eMe is a finite group with neutral element e, by A2.5(ii), Appendix 2 of Chapter 12.

Now, $\phi(C^*)$ is clearly a submonoid of M. Hence, the product of any two elements of $eMe \cap \phi(C^*)$ lies in $eMe \cap \phi(C^*)$. Take $a \in eMe \cap \phi(C^*)$. Then for some $n \geq 2$, $a^n = e$ (eMe being a finite group). Then a^{n-1} is the inverse of a in eMe, and belongs to $eMe \cap \phi(C^*)$. Thus, the latter is a subgroup of eMe. □

11.4 Factorization

Theorem 4.1 (Reutenauer 1985) *Let C be a finite complete code. Then there exist polynomials X, Y, Z in $\mathbb{Z}\langle A\rangle$ such that*

$$\underline{C} - 1 = X(d(\underline{A} - 1) + (\underline{A} - 1)Z(\underline{A} - 1))Y \qquad (4.1)$$

and

(i) *d is the degree of C,*
(ii) *C is prefix (suffix) if and only if $Y = 1$ ($X = 1$).*

Example 4.1 We have

$$a^2 + a^2b + ab + ab^2 + b^2 - 1 = (1 + a)(a + b - 1)(1 + b)\,.$$

The corresponding code is neither prefix nor suffix, but *synchronizing* (that is of degree 1).

Example 4.2 Let C be the square of the code of Example 4.1. Then C is of degree 2 and

$$\underline{C} - 1 = (1 + a)(2(a + b - 1) + (a + b - 1)(1 + b)(1 + a)(a + b - 1))(1 + b)\,.$$

Example 4.3 We have

$$a^3 + a^2ba + a^2b^2 + ab + ba^2 + baba + bab^2 + b^2a + b^3 - 1$$
$$= 3(a + b - 1) + (a + b - 1)(2 + a + b + ab)(a + b - 1)\,.$$

The corresponding code is a bifix code and has degree 3.

The following corollary (which also uses Theorem 2.7) characterizes completely finite complete codes.

Corollary 4.2 (Reutenauer 1985) *Let C be a language not containing the empty word. Then the following conditions are equivalent:*

(i) *C is a complete finite code.*
(ii) *There exist polynomials P, S in $\mathbb{Z}\langle A\rangle$ such that*

$$\underline{C} - 1 = P(\underline{A} - 1)S\,. \qquad \square$$

In order to prove Theorem 4.1, we need the following lemma.

Lemma 4.3 *Let C be a finite complete code of degree d. Then there exist words $u_1, \ldots, u_d, v_1, \ldots, v_d$, with $u_1, v_1 \in C^*$, such that for any i, $1 \le i \le d$:*

$$\underline{A}^* = \sum_{1\le j\le d} u_i^{-1}(\underline{C}^*)v_j^{-1}$$

and for any j, $1 \le j \le d$:

$$\underline{A}^* = \sum_{1\le i\le d} u_i^{-1}(\underline{C}^*)v_j^{-1}\,.$$

Proof. By Theorem 3.1 there exist a finite monoid M and a surjective morphism $\phi : A^* \to M$ such that $C^* = \phi^{-1}\phi(C^*)$; moreover, there exists an idempotent e in $D \cap \phi(C^*)$, where D is the minimal ideal of M, $G = eMe$ is a finite group and $H = eMe \cap \phi(C^*)$ is a subgroup of G of index d.

Let $u_1, \ldots, u_d, v_1, \ldots, v_d$ be words in $\phi^{-1}(G)$ such that

$$G = \bigcup_{1\le i\le d} \phi(v_i)H \tag{4.2}$$

and

$$G = \bigcup_{1 \le j \le d} H\phi(u_j)$$

(disjoint unions). By elementary group theory, we may assume that $\phi(u_1) = \phi(v_1) = e$ (hence $u_1, v_1 \in \phi^{-1}(e) \subset \phi^{-1}\phi(C^*) = C^*$) and that $\phi(u_i)$ is the inverse of $\phi(v_i)$ in G.

Let $1 \le j \le d$ and w be a word. Then there exists one and only one i, $1 \le i \le d$, such that $w \in u_i^{-1}(C^*)v_j^{-1}$, that is $u_i w v_j \in C^*$. Indeed, the element $e\phi(wv_j)$ of G is in some $\phi(v_i)H$ by Equation (4.2). Consequently, $\phi(u_i w v_j) = \phi(u_i)e\phi(wv_j) \in \phi(u_i)\phi(v_i)H = eH = H$, which implies that $u_i w v_j \in \phi^{-1}(H) \subset \phi^{-1}\phi(C^*) = C^*$. Conversely, $u_i w v_j \in C^*$ implies $\phi(u_i w v_j) \in eMe \cap \phi(C^*) = H$, because $\phi(u_i w v_j) = e\phi(u_i w v_j)e$ is in eMe. It follows that $e\phi(wv_j) = \phi(v_i)\phi(u_i w v_j) \in \phi(v_i)H$, and i is completely determined by j and w.

We have shown that one has the disjoint union, for any j, $1 \le j \le d$:

$$A^* = \bigcup_{1 \le i \le d} u_i^{-1}(C^*)v_j^{-1}\,.$$

But this is equivalent to the last relation of the lemma. By symmetry, we have also the first. □

We easily derive the following lemma.

Lemma 4.4 *Let C be a finite complete code of degree d. Then there exist polynomials $P, P_1, S, S_1, Q, G_1, D_1$ with coefficients 0, 1 such that:*

(i) $d\underline{A}^* - Q = S\underline{C}^*P$;
(ii) $\underline{A}^* - G_1 = S\underline{C}^*P_1$;
(iii) $\underline{A}^* - D_1 = S_1\underline{C}^*P$;
(iv) P_1, S_1 *have constant term* 1;
(v) G_1, D_1 *have constant term* 0;
(vi) *if C is a prefix (suffix) code, then $S_1 = 1$ ($P_1 = 1$).*

Proof. We use Lemma 4.3 and the notation of Section 11.1. We have, by Proposition 1.6, $u_i^{-1}(\underline{C}^*)v_j^{-1} = S_{u_i}\underline{C}^*P_{v_j} + F_{u_i,v_j}$; moreover, by Lemma 1.5 and Proposition 1.6, S_{u_i}, P_{v_j} and F_{u_i,v_j} are polynomials with nonnegative coefficients.

Now, by Lemma 4.3, for any i

$$\underline{A}^* = \sum_{1 \le j \le d} S_{u_i}\underline{C}^*P_{v_j} + \sum_{1 \le j \le d} F_{u_i,v_j}$$

and for any j

$$\underline{A}^* = \sum_{1 \le i \le d} S_{u_i}\underline{C}^*P_{v_j} + \sum_{1 \le i \le d} F_{u_i,v_j}\,.$$

Let

$$P = \sum_{1\le j\le d} P_{v_j}\,, \quad S = \sum_{1\le i\le d} S_{u_i}\,, \quad P_1 = P_{v_1}\,, \quad S_1 = S_{u_1}\,,$$

$$G_1 = \sum_i F_{u_i,v_1}\,, \quad D_1 = \sum_j F_{u_1,v_j}\,, \quad Q = \sum_{i,j} F_{u_i,v_j}\,.$$

Then we obtain

$$d\underline{A}^* = S\underline{C}^*P + Q, \quad \underline{A}^* = S\underline{C}^*P_1 + G_1, \quad \underline{A}^* = S_1\underline{C}^*P + D_1\,, \tag{4.3}$$

which proves (i), (ii) and (iii).

Since $u_1 \in C^*$ by Lemma 4.3, $u_1^{-1}(C^*)$ contains 1, hence $u_1^{-1}(\underline{C}^*)$ has constant term 1. As $u_1^{-1}(\underline{C}^*) = S_{u_1}\underline{C}^*$ by Lemma 1.5, $S_1 = S_{u_1}$ must have constant term 1. The same holds for P_1 by symmetry, and proves (iv).

Since $S = \sum_i S_{u_i}$, the S_{u_i}'s are nonnegative and as S_{u_1} has constant term 1, S has positive constant term. Moreover, P_1 has constant term 1. Hence, because $\underline{A}^*$ has constant term 1 and by Equation (4.3), G_1 has constant term 0. Similarly, D_1 has constant term 0. This proves (v).

Suppose now that C is prefix. Then, by Exercise 1.4, $u_1^{-1}(C^*) = C^*$ (because $u_1 \in C^*$). Hence $u_1^{-1}(\underline{C}^*) = \underline{C}^*$. Since by Lemma 1.5, $u_1^{-1}(\underline{C}^*) = S_{u_1}\underline{C}^*$, we obtain $S_1 = S_{u_1} = 1$. Similarly, if C is suffix, then $P_1 = 1$. This proves (vi). □

Given a Bernoulli morphism π, define a mapping λ for each word w by

$$\lambda(w) = \pi(w)\,|w|\,.$$

For each language X, define $\lambda(X)$ by

$$\lambda(X) = \sum_{w\in X} \lambda(w) \in \mathbb{R}_+ \cup \infty\,.$$

This is called the *average length* of X. On the other hand λ extends to a linear mapping $\mathbb{Z}\langle A\rangle \to \mathbb{R}$ by

$$\lambda(P) = \sum_w (P, w)\lambda(w)\,.$$

Lemma 4.5 *Let $P_1, \ldots, P_n$ be polynomials. Then*

$$\lambda(P_1\cdots P_n) = \sum_{1\le i\le n} \pi(P_1)\cdots\pi(P_{i-1})\lambda(P_i)\pi(P_{i+1})\cdots\pi(P_n)\,.$$

Proof. For $n = 2$, it is enough, by linearity, to prove the lemma when $P_1 = u$, $P_2 = v$ are words. But in this case

$$\begin{aligned}\lambda(uv) &= \pi(uv)\,|uv| = \pi(u)\pi(v)(|u| + |v|)\\ &= \pi(u)|u|\pi(v) + \pi(u)\pi(v)|v| = \lambda(u)\pi(v) + \pi(u)\lambda(v)\,.\end{aligned}$$

The general case is easily proved by induction. □

Proof of Theorem 4.1. 1. We use the notation of Lemma 4.4. We have $\underline{A}^* - G_1 = (1-\underline{A})^{-1} - G_1 = (1-\underline{A})^{-1}(1-(1-\underline{A})G_1)$. Since $\underline{A}^* - G_1 = S\underline{C}^* P_1$ and P_1 has constant term 1 (Lemma 4.4), P_1 is invertible in $\mathbb{Z}\langle A\rangle$ and we obtain from

$$S\underline{C}^* P_1 = (1-\underline{A})^{-1}(1-(1-\underline{A})G_1)\,,$$

by multiplying by $1-\underline{A}$ on the left and by P_1^{-1} on the right,

$$(1-\underline{A})S\underline{C}^* = (1-(1-\underline{A})G_1)P_1^{-1}\,. \tag{4.4}$$

Multiply the relation (i) of Lemma 4.4 by $1-\underline{A}$ on the left. This yields

$$d-(1-\underline{A})Q = (1-\underline{A})S\underline{C}^* P\,.$$

Hence, by Equation (4.4),

$$d-(1-\underline{A})Q = (1-(1-\underline{A})G_1)P_1^{-1}P\,.$$

Note that, because G_1 has no constant term, $1-(1-\underline{A})G_1$ is invertible in $\mathbb{Z}\langle\langle A\rangle\rangle$, so that we obtain, by multiplying the previous relation by $P_1(1-(1-\underline{A})G_1)^{-1}$ on the left

$$P = P_1(1-(1-\underline{A})G_1)^{-1}(d-(1-\underline{A})Q)\,.$$

2. We apply Corollary 10.4.3 to the last equality: there exist E, F, G, H in $\mathbb{Z}\langle A\rangle$ such that

$$\begin{aligned} P_1 &= EF, \quad 1-(1-\underline{A})G_1 = GF \\ d-(1-\underline{A})Q &= GH, \quad P = EH\,. \end{aligned} \tag{4.5}$$

By Proposition 10.4.4 applied to the second equality (with $1-\underline{A}$ instead of Y), we obtain

$$G \equiv \pm 1 \mod (1-\underline{A})\mathbb{Z}\langle A\rangle\,.$$

Replacing if necessary E, F, G, H by their opposites, we may suppose that $G \equiv +1$, and hence we obtain, again by Proposition 10.4.4, and by the third equality in Equation (4.5), that $H = d + (\underline{A}-1)R$, with $R \in \mathbb{Z}\langle A\rangle$. This implies, by the fourth equality in Equation (4.5),

$$P = E(d+(\underline{A}-1)R)\,. \tag{4.6}$$

3. We have $\underline{A}^* - D_1 = (1-\underline{A})^{-1}(1-(1-\underline{A})D_1)$ so that by Lemma 4.4 (iii),

$$S_1\underline{C}^* P = (1-\underline{A})^{-1}(1-(1-\underline{A})D_1)\,.$$

Since D_1 has constant term 0, $1-(1-\underline{A})D_1$ is invertible in $\mathbb{Z}\langle\langle A\rangle\rangle$; moreover S_1 is also invertible because it has constant term 1. So we obtain, by multiplying by $(1-\underline{C})S_1^{-1}$ on the left and by $(1-(1-\underline{A})D_1)^{-1}(1-\underline{A})$ on the right,

$$(1-\underline{C})S_1^{-1} = P(1-(1-\underline{A})D_1)^{-1}(1-\underline{A})\,.$$

Now we use Equation (4.6) and multiply by $-S_1$ on the right, thus obtaining

$$\underline{C}-1 = E(d+(\underline{A}-1)R)(1-(1-\underline{A})D_1)^{-1}(\underline{A}-1)S_1\,.$$

4. By Corollary 10.4.3, there exist $E', F', G', H' \in \mathbb{Z}\langle A\rangle$ such that

$$\begin{aligned} E(d+(\underline{A}-1)R) &= E'F', \quad 1-(1-\underline{A})D_1 = G'F' \\ (\underline{A}-1)S_1 &= G'H', \quad \underline{C}-1 = E'H'. \end{aligned} \tag{4.7}$$

Let π be any Bernoulli morphism. Replacing if necessary E', F', G', H' by their opposites, we may assume that

$$\pi(F') \geq 0.$$

Thus, by Equation (4.7) and Proposition 10.4.4, we obtain (since $\pi(\underline{A}-1)=0$)

$$G' = 1+(\underline{A}-1)G'', \quad F' = 1+(\underline{A}-1)F'' \tag{4.8}$$

for some $G'', F'' \in \mathbb{Z}\langle A\rangle$. This and Equation (4.7) imply that

$$(\underline{A}-1)S_1 = (1+(\underline{A}-1)G'')H' = H' + (\underline{A}-1)G''H'.$$

Thus, we have

$$H' = (\underline{A}-1)H'', \quad H'' \in \mathbb{Z}\langle A\rangle. \tag{4.9}$$

Now, Eqs. (4.7) and (4.8) imply also

$$E(d+(\underline{A}-1)R) = E'(1+(\underline{A}-1)F'').$$

5. We now apply Theorem 10.2.2 to this equality and denote by p_i the continuant polynomial $p(a_1, \ldots, a_i)$ and $\tilde{p}_i = p(a_i, \ldots, a_1)$. Thus there exist polynomials $U, V \in \mathbb{Q}\langle A\rangle$ such that

$$\begin{aligned} E &= Up_n, \quad d+(\underline{A}-1)R = \tilde{p}_{n-1}V, \\ E' &= Up_{n-1}, \quad 1+(\underline{A}-1)F'' = \tilde{p}_n V. \end{aligned} \tag{4.10}$$

Applying Corollary 10.1.3 to the second and the last equalities (with $X \to \tilde{p}_{n-1}$ or $\tilde{p}_n$, $Y \to \underline{A}-1$, $Q_1 \to 0$, $P \to V$), we obtain that $\underline{A}-1$ is a weak left divisor of $\tilde{p}_{n-1}$ and $\tilde{p}_n$, that is $\tilde{p}_{n-1}$ and $\tilde{p}_n$ are both congruent to a scalar mod$(\underline{A}-1)\mathbb{Q}\langle A\rangle$. This implies, by Proposition 10.2.3, that

$$p_{n-1} \quad \text{and} \quad \tilde{p}_{n-1}\ (p_n \text{ and } \tilde{p}_n) \tag{4.11}$$

are congruent to the same scalar mod $(\underline{A}-1)\mathbb{Q}\langle A\rangle$. Moreover, by Corollary 10.4.2, they have the same content

$$c(p_{n-1}) = c(\tilde{p}_{n-1}), \quad c(p_n) = c(\tilde{p}_n). \tag{4.12}$$

6. Since D_1 has coefficients 0, 1, the polynomial $1-(\underline{A}-1)D_1$ is primitive. Hence, by Equation (4.7) and by Gauss's Lemma, G' and F' are primitive. Since by Eqs. (4.10) and (4.8)

$$\tilde{p}_n V = 1+(\underline{A}-1)F'' = F',$$

we obtain by Gauss's Lemma

$$c(\tilde{p}_n)c(V) = 1$$

and

$$\bar{\bar{p}}_n \overline{V} = F' .$$

This equality, Proposition 10.4.4 and Equation (4.8) imply that

$$\overline{V} = \varepsilon + (\underline{A} - 1)V', \quad \varepsilon = \pm 1, \ V' \in \mathbb{Z}\langle A\rangle . \tag{4.13}$$

Furthermore, $\underline{C} - 1$ is primitive, and so is E' by Gauss' Lemma and Equation (4.7). As $E'F' = E(d + (\underline{A} - 1)R)$ by Equation (4.7) and E', F' are primitive, we obtain by Gauss's Lemma that $d + (\underline{A} - 1)R$ is primitive. Thus by Equation (4.10) and Gauss's Lemma again

$$d + (\underline{A} - 1)R = \bar{\bar{p}}_{n-1}\overline{V} .$$

This implies, by Proposition 10.4.4 and Equation (4.13),

$$\bar{\bar{p}}_{n-1} = \varepsilon d + (\underline{A} - 1)L, \quad L \in \mathbb{Z}\langle A\rangle .$$

By Eqs. (4.11) and (4.12), we obtain that $\bar{p}_{n-1}$ and $\bar{\bar{p}}_{n-1}$ are congruent to the same scalar $\mathrm{mod}(\underline{A} - 1)\mathbb{Q}\langle A\rangle$. Therefore

$$\bar{p}_{n-1} = \varepsilon d + (\underline{A} - 1)M$$

with $M \in \mathbb{Q}\langle A\rangle$. Now $\bar{p}_{n-1} - \varepsilon d = (\underline{A} - 1)M$ and $\underline{A} - 1$ is primitive, so that $c(M) = c(\bar{p}_{n-1} - \varepsilon d) \in \mathbb{N}$ and $M \in \mathbb{Z}\langle A\rangle$, by Equation (4.2) in Chapter 10.

We have seen that E' is primitive, so that by Gauss's Lemma and Equation (4.10), we have

$$E' = \overline{U}\bar{p}_{n-1}$$

which implies

$$E' = \overline{U}(\varepsilon d + (\underline{A} - 1)M) .$$

Hence, by Eqs. (4.7) and (4.9),

$$\underline{C} - 1 = \overline{U}(\varepsilon d + (\underline{A} - 1)M)(\underline{A} - 1)H'' ,$$

where all polynomials are in $\mathbb{Z}\langle A\rangle$ and where $\varepsilon = \pm 1$. This shows that we have a relation of the form

$$\underline{C} - 1 = X(\varepsilon' d + (\underline{A} - 1)D)(\underline{A} - 1)Y ,$$

where

$$X = \pm\overline{U}, \ Y = \pm H'', \ \varepsilon' d + (\underline{A} - 1)D = \pm(\varepsilon d + (\underline{A} - 1)M)$$

are chosen in such a way that, for some Bernoulli morphism π, one has

$$\pi(X) \geq 0, \ \pi(Y) \geq 0 .$$

7. Apply Lemma 4.5 to this relation, using the fact that $\pi(\underline{A} - 1) = 0$; we obtain

$$\lambda(\underline{C} - 1) = \pi(X)\varepsilon' d\lambda(\underline{A} - 1)\pi(Y) .$$

Now $\lambda(1) = 0$, $\lambda(\underline{C}) > 0$, $\lambda(\underline{A}) > 0$, and we obtain

$$\varepsilon' d\pi(X)\pi(Y) > 0\,.$$

This shows that $\varepsilon' = 1$ and proves Equation (4.1) and (i).

8. First, note that the "if" part of (ii) is a consequence of Theorem 2.7. Now, if C is a prefix code, we have by Lemma 4.4 (vi) that $S_1 = 1$. Hence, by Equation (4.7), $\underline{A} - 1 = G'H'$, which implies by Equation (4.9) $\underline{A} - 1 = G'(\underline{A} - 1)H''$. Thus $H'' = \pm 1$, and we obtain $Y = \pm 1$. Now $\pi(Y) \geq 0$, and consequently $Y = 1$.

On the other hand, if C is suffix, then $P_1 = 1$ by Lemma 4.4 (vi). Then, by Equation (4.5), $E = \pm 1$ which implies by Equation (4.10) and Gauss's Lemma that $\overline{U} = \pm 1$. Thus $X = \pm 1$. Since $\pi(X) \geq 0$, we obtain $X = 1$. This proves the theorem. □

Exercises for Chapter 11

1.1 Show that a submonoid of A^* is of the form C^*, C a code, if and only if it is free (that is isomorphic to some free monoid). Show that a submonoid M of A^* is free if and only if for any words u, v, w

$$u, uv, vw, w \in M \implies v \in M\,.$$

1.2 Show that, given rational languages K, L, it is decidable whether their union (their product, the star of K if $1 \notin K$) is unambiguous.

1.3 Show that S_u (P_u, $F_{u,v}$) as defined in Section 11.1 is a sum of proper suffixes (prefixes, factors) of words of C.

1.4 Show that for a prefix code C and $v \in C^*$, one has $v^{-1}C^* = C^*$.

1.5 Show that for any series S and words u, v, one has $(u^{-1}S)v^{-1} = u^{-1}(Sv^{-1})$.

2.1 Show that for a finite code C the three following conditions are equivalent:

(i) C is a complete and prefix code.

(ii) For any word w, $wA^* \cap CA^*$ is not empty.

(iii) For any word w, $wA^* \cap C^*$ is not empty.

2.2 Let C be a finite complete language. Show that for any word w, there exists some power of a conjugate of w which is in C^* (two words w, w' are *conjugate* if $w = uv$, $w' = vu$ for some words u, v).

2.3 Deduce from Theorem 2.4 an algorithm to show that a finite set C is a complete code. (*Hint*: It is decidable whether C is complete, since the set of factors of a rational language is rational.)

3.1 Let C be a finite complete code. Show that C is synchronizing (that is of degree 1) if and only if for some word w, one has $wA^*w \subset C^*$.

4.1 Let C be a finite complete code which is bifix. Let n be such that $a^n \in C$ for some letter a.

(a) Show that for any i, $1 \leq i \leq n$, $C_i = a^{-i}C$ is a prefix set such that $C_i A^* \cap wA^*$ is not empty for any word w.

(b) Show that the set of proper suffixes of C is the disjoint union of the C_i's.

(c) Deduce that $\underline{C_i} - 1 = P_i(\underline{A} - 1)$ and that

$$\underline{C} - 1 = n(\underline{A} - 1) + (\underline{A} - 1)\left(\sum_{i=1}^{n} P_i\right)(\underline{A} - 1).$$

Show that n is the degree of C. Show that it is also equal to the average length of C (cf. Perrin 1977).

Notes to Chapter 11

Theorem 4.1 is a noncommutative generalization of a theorem due to Schützenberger (1965). Corollary 4.2 is a partial answer to the main conjecture in the theory of finite codes, the *factorization conjecture* which states that P and S may be chosen to have nonnegative coefficients (or equivalently coefficients 0 and 1).

Finite complete codes are maximal codes, and conversely, every maximal code is complete. Most of the general results on codes are stated here in the finite case. However, they hold for rational and even for *thin* codes (a language X is *thin* if there exists a word which is not a factor of any word in X). For a general exposition of the theory of codes, see the book by Berstel *et al.* (2009).

12

Semisimple syntactic algebras

It is shown that the syntactic algebra of the characteristic series of a rational language L is semisimple in the following two cases: L is a free submonoid generated by a bifix code, or L is a cyclic language.

This chapter has two appendices, one on semisimple algebras (without proofs) and another on simple semigroups, with concise proofs. We use the symbols A1 and A2 to refer to them.

12.1 Bifix codes

Let E be a set of endomorphisms of a finite dimensional vector space V. Recall that E is called *irreducible* if there is no subspace of V other than 0 and V itself which is invariant under all endomorphisms in E. Similarly, we say that E is *completely reducible* if V is a direct sum $V = V_1 \oplus \cdots \oplus V_k$ of subspaces such that for each i, the set of induced endomorphisms $e|V_i$ of V_i, for $e \in E$, is irreducible.

A set of matrices in $K^{n\times n}$ (K being a field) is *irreducible* (resp. *completely reducible*) if it is so, viewed as a set of endomorphisms acting at the right on $K^{1\times n}$, or equivalently at the left on $K^{n\times 1}$ (for this equivalence, see Exercises 1.1 and 1.2).

A linear representation (λ, μ, γ) of a series $S \in K\langle\langle A\rangle\rangle$ is *irreducible* (resp. *completely reducible*) if the set of matrices $\{\mu a \mid a \in A\}$ (or equivalently the sets μA^* or $\mu(K\langle A\rangle)$) is so. By a change of basis, we see that (λ, μ, γ) is completely reducible if and only if it is similar to a linear representation which has a block diagonal form

$$\lambda = (\lambda_1, \ldots, \lambda_k), \quad \mu = \begin{pmatrix} \mu_1 & 0 & \cdots & \cdots & 0 \\ 0 & \mu_2 & \ddots & & \vdots \\ \vdots & \ddots & \ddots & \ddots & \vdots \\ \vdots & & \ddots & \mu_{k-1} & 0 \\ 0 & \cdots & \cdots & 0 & \mu_k \end{pmatrix}, \quad \gamma = \begin{pmatrix} \gamma_1 \\ \vdots \\ \gamma_k \end{pmatrix}$$

where each representation $(\lambda_i, \mu_i, \gamma_i)$ is irreducible.

Recall that codes, bifix codes and complete codes have been defined in Sections 11.1 and 11.2. We assume that K is a field of characteristic 0.

Theorem 1.1 *Let C be a rational code and let S be the characteristic series of C^*. Let $\rho = (\lambda, \mu, \gamma)$ be a minimal representation of S.*

(i) *If C is bifix, then ρ is completely reducible.*
(ii) *If C is complete and ρ is completely reducible, then C is bifix.*

An equivalent formulation of this result is the following. For semisimple algebras, see Appendix A1.

Corollary 1.2 *Let C and S be as in the theorem and let $\mathfrak{A}$ be the syntactic algebra of S.*

(i) *If C is bifix, then $\mathfrak{A}$ is semisimple.*
(ii) *If C is complete and $\mathfrak{A}$ is semisimple, then C is bifix.*

We thus obtain that a complete rational code C is bifix if and only if the syntactic algebra of $\underline{C}^*$ is semisimple.

Proof. Let $\rho = (\lambda, \mu, \gamma)$ be as in the theorem. Then $\mathfrak{A} = \mu(K\langle A\rangle)$ is isomorphic to the syntactic algebra of S by Corollary 2.2.2. Evidently, $\mathfrak{A}$ acts on $K^{1\times n}$, and it acts faithfully. Thus statement (i) follows from Theorem 1.1(i) and from A1.5. For (ii), we use Theorem 1.1(ii) and A1.6. □

For the proof of Theorem 1.1 we need a lemma.

Lemma 1.3 *Let C, S, ρ be as in the theorem. Then in the finite monoid $M = \mu(A^*)$, there is a finite group G, with neutral element e, such that $e \in \mu(C^*)$ and that*

- *if M has no zero, then $eMe = G$;*
- *if M has a zero, then $e \neq 0$ and $eMe = G \cup 0$.*

Proof. The language C^* is rational. Therefore, by Propositions 3.3.1 and 3.3.2, M is the syntactic monoid of C^* and is finite. If M has no zero, let D be its minimal ideal. If M has a zero, let D be a 0-minimal ideal. For these notions, see A2.1 and A2.2. In both cases, Card $D \geq 2$. Consequently $\mu(C^*)$ intersects D since otherwise we obtain a coarser congruence than the syntactic congruence by taking $\mu^{-1}(D)$ as a single equivalence class, contradicting the fact that M is the syntactic monoid of C^*.

If M has a zero, $\mu(C^*)$ does not contain it. Indeed, if $0 = \mu(w)$ for some $w \in C^*$, then for any letter a, one has $0 = \mu(aw) = \mu(wa)$, hence $w, wa, aw \in C^*$ and by Exercise 1.4, $a \in C^*$. Thus $C = A$ and $M = \{1\}$ which would yield $1 = 0$ in M, a contradiction with the definition of a zero in A2.1.

We conclude that in both cases (zero or not) some element and its powers are in $\mu(C^*) \cap D$ and are nonzero. It follows that there is some nonzero idempotent e in $\mu(C^*) \cap D$ and the lemma is a consequence of A2.5(ii). □

Proof of Theorem 1.1. (i) Let the algebra $\mathfrak{A} = \mu(K\langle A\rangle)$ act on the right on $V = K^{1\times n}$. In view of Exercise 1.3, it is enough to show that each subspace W of V which is invariant under $\mathfrak{A}$ has a supplementary space W' which is also invariant.

With the notations of Lemma 1.3, in particular $M = \mu(A^*)$, define the subspace $E = \{ve \mid v \in V\}$ of V. Set $F = W \cap E$. If $g \in G$, then $Wg \subset W$ (W being invariant under $\mathfrak{A}$) and $g = ge$, hence $Eg = Ege \subset E$. This implies that F is invariant under G. By Maschke's theorem A1.7, there exists a G-invariant subspace F' of E such that E is the direct sum over K of F and F'. Let

$$W' = \{v \in V \mid vMe \subset F'\}.$$

We show that W' is a subspace of V, supplementary of W and invariant under $\mathfrak{A}$. First, it is invariant, since for m in M, the inclusion $vMe \subset F'$ implies $vmMe \subset F'$.

We claim that $\lambda \in E$. This will imply that $\lambda = t + t'$ for some $t \in F$, $t' \in F'$. Since $F \subset W$ and $F' \subset W'$ (indeed, $t' \in F'$ implies $t' \in E$, and therefore $t' = t'e$ from which follows $t'Me = t'eMe \subset F'G \subset F'$, hence $t' \in W'$), we obtain $\lambda \in W + W'$. Since these two subspaces are invariant and since $\lambda\mathfrak{A} = V$ (Proposition 2.2.1), we obtain that $V = W + W'$.

In order to prove the claim, it suffices to show that $\lambda = \lambda e$. We know that $e = \mu(w)$ for some $w \in C^*$. Since C is a prefix code, we have $u \in C^* \iff wu \in C^*$ for any word $u \in A^*$ (see Exercise 1.5). Thus $(S, u) = (S, wu)$ and therefore $(S, (1 - w)u) = 0$. This implies that for any P in $K\langle A\rangle$, one has $0 = (S, (1 - w)P) = (S \circ (1 - w), P)$. We obtain that $1 - w$ is in the right syntactic ideal of S (Proposition 2.1.4) and therefore $\lambda\mu(1 - w) = 0$ (Proposition 2.2.1), and finally $\lambda = \lambda e$.

It remains to show that $W \cap W' = 0$. For this, consider a vector in $W \cap W'$. By Proposition 2.2.1, it is of the form $\lambda\mu P$ for some P in $K\langle A\rangle$. If $m \in M$, then $\lambda\mu Pme \in E \cap W = F$ since W is stable and by definition of E. Moreover, $\lambda\mu Pme \in F'$ since $\lambda\mu P \in W'$ and by definition of W'. It follows that $\lambda\mu Pme \in F \cap F' = 0$.

Now, since C is a suffix code, we have symmetrically $(S, u) = (S, uw)$ for any word u, and w as above. Consequently, for $Q \in K\langle A\rangle$, we have $(S, Q) = (S, Qw)$ or equivalently $\lambda\mu Q\gamma = \lambda\mu Q\mu w\gamma$. We deduce that for any word u,

$$\lambda\mu P\mu u\gamma = \lambda\mu P\mu u\mu w\gamma = \lambda\mu Pme\gamma = 0$$

by the preceding argument and with $m = \mu u$. Since the $\mu u\gamma$ span $K^{n\times 1}$ by Proposition 2.2.1, we conclude that $\lambda\mu P = 0$.

(ii) It is enough, by left-right symmetry, to show that C is prefix. By Lemma 1.3, we know that $M = \mu(A^*)$ is a finite monoid. Since C is complete, C^* intersects each ideal in A^*, hence $\mu(C^*)$ intersects the minimal ideal D of M.

Let $V = K^{1\times n}$, with its right action of $\mathfrak{A} = \mu(K\langle A\rangle)$. Let W be the subspace of V composed of the elements v in V such that $v\underline{H}\gamma = v\underline{K}\gamma$ for any maximal subgroups H, K in D contained in the same minimal left ideal of M, where we write $\underline{H}$ for $\sum_{m\in H} m$. The subspace W is invariant under M, hence under $\mathfrak{A}$. Indeed, if $v \in W$ and $m \in M$, then for any H, K as above, mH and mK are maximal subgroups of the same minimal left ideal and the mapping $h \mapsto mh$ is a bijection $H \to mH$ by A2.5(iv).

Consequently

$$vm\underline{H}\gamma = v\underline{mH}\gamma = v\underline{mK}\gamma = vm\underline{K}\gamma\,,$$

which implies that $vm \in W$.

Observe that for any m in D and v in V, one has $vm \in W$. This is because for any maximal subgroups H, K contained in the same minimal left ideal of M, one has $mH = mK$ (see A2.5(iv)).

Since V is completely reducible, we know by A1.3 that $V = W \oplus W'$ for some stable subspace W'. Let $\lambda = v + v'$ with $v \in W$, $v' \in W'$. Let H, K be as before. Then

$$\lambda\underline{H}\gamma - \lambda\underline{K}\gamma = v\underline{H}\gamma - v\underline{K}\gamma + v'\underline{H}\gamma - v'\underline{K}\gamma = v'\underline{H}\gamma - v'\underline{K}\gamma$$

since v is in W. By our previous observation, $v'\underline{H}$ and $v'\underline{K}$ are in W. Since they are also in W', they vanish, hence $\lambda\underline{H}\gamma = \lambda\underline{K}\gamma$. Note that for m in M, $\lambda m\gamma = 1$ if $m \in \mu(C^*)$, and $= 0$ otherwise. This shows that if $\mu(C^*)$ intersects some maximal subgroup of a minimal left ideal, then it intersects each such maximal subgroup.

In other words, $\mu(C^*)$ intersects each minimal right ideal of M (see A2.5(ii) and (iii)). Applying A2.4, we have $D = I \times G \times D$ and by Exercise 1.4, $D \cap \mu(C^*) = I_1 \times H \times J_1$, where H is a subgroup of G and $I_1 \subset I$, $J_1 \subset J$. In fact, by what we have just said, we must have $I = I_1$. Moreover, $p_{j,i} \in H$ for $j \in J_1, i \in I_1$.

By Exercise 1.5, C is a prefix code if we establish that for any words u, v,

$$u, uv \in C^* \implies v \in C^*\,.$$

Since the syntactic congruence of C^* saturates C^*, and in view of Proposition 3.3.2, it suffices to show that for any m, n in M, $m, mn \in \mu(C^*) \implies n \in \mu(C^*)$. By multiplying m on the left by some element in $D \cap \mu(C^*)$, we may assume that $m \in D$. We may write $m = (i, h, j)$ for some $i \in I$, $h \in H$, $j \in J_1$ and we have $mn \in D \cap \mu(C^*)$. Now $nm \in D$ and is a left multiple of m; hence it is in the same minimal left ideal as m and therefore, by A2.5.(iii), $nm = (i', g, j)$ with $i' \in I$, $g \in G$. It follows that

$$(i, hp_{j,i'}g, j) = (i, h, j)(i', g, j) = mnm \in D \cap \mu(C^*)\,.$$

Thus $hp_{j,i'}g \in H$, which implies $g \in H$. We conclude that m, mn and nm are all in $\mu(C^*)$ and consequently $n \in \mu(C^*)$ by Exercise 1.4. □

12.2 Cyclic languages

A language $L \subset A^*$ is called *cyclic* if it has the following two properties:

(i) for any words $u, v \in A^*$, $uv \in L \iff vu \in L$.

(ii) for any nonempty word w and any integer $n \geq 1$, $w \in L \iff w^n \in L$.

Given a finite deterministic automaton $\mathcal{A}$ over A, we call *character* of $\mathcal{A}$, denoted by $\chi_{\mathcal{A}}$, the formal series

$$\chi_{\mathcal{A}} = \sum_{w \in A^*} \alpha_w \, w \,,$$

where α_w is the number of closed paths labeled w in $\mathcal{A}$.

Recall that a 0, 1-*matrix* is a matrix with entries equal to 0 or 1, and that a *row-monomial matrix* is a matrix having at most one nonzero entry in each row. A series is the character of some finite deterministic automaton if and only if there is a representation μ of A^* by row-monomial 0, 1-matrices such that this series is equal to $\sum_{w \in A^*} \operatorname{tr}(\mu w) w$. This follows from the equivalence between automata and linear representations, see Section 1.6.

Theorem 2.1 *The characteristic series of a rational cyclic language is a $\mathbb{Z}$-linear combination of characters of finite deterministic automata.*

Corollary 2.2 *The syntactic algebra over a field K of a rational cyclic language is semisimple.*

This will follow from the theorem and the next lemma.

Lemma 2.3 *Let $\mu_1, \ldots, \mu_k$ be linear representations of A^*, let $\alpha_1, \ldots, \alpha_k \in K$ and let S be the series defined by*

$$S = \sum_{1 \le i \le k} \alpha_i \operatorname{tr}(\mu_i \, w) \,.$$

Then the syntactic algebra of S is semisimple.

Proof. We may assume that each representation is irreducible. Indeed, if μ_i is reducible, we put it, by an appropriate change of basis, into block-triangular form with each block irreducible, and then, keeping only the diagonal blocks, into block-diagonal form. These transformations do not change the trace. Since the trace of a diagonal sum is the sum of the traces of the blocks, we obtain the desired form.

Consider now the algebra

$$\mathfrak{A} = \{(\mu_1 P, \ldots, \mu_k P) \mid P \in K\langle A \rangle\} \,.$$

It acts faithfully on the right on $K^{1 \times n}$, where n is of the appropriate size; moreover $K^{1 \times n}$ is completely reducible under this action. Thus $\mathfrak{A}$ is semisimple by A1.5.

There is a surjective algebra morphism $\mu : K\langle A \rangle \to \mathfrak{A}$, namely $\mu = (\mu_1, \ldots, \mu_k)$, and a linear mapping $\varphi : \mathfrak{A} \to K$ such that $(S, w) = \varphi(\mu w)$, namely $\varphi(\mu_1 P, \ldots, \mu_k P) = \sum_{1 \le i \le k} \alpha_i \operatorname{tr}(\mu_i P)$. Consequently, by Exercise 2.1.4, the syntactic algebra of S is a quotient of $\mathfrak{A}$, hence is semisimple by A1.1. □

Corollary 2.2 follows from Lemma 2.3 because of the trace form of the character of an automaton seen above.

Let L be a language and let a_n be the number of words of length n in L. The *zeta function* of L is the series

$$\zeta_L = \exp\Big(\sum_{n\geq 1} a_n \frac{x^n}{n}\Big).$$

Corollary 2.4 *Let L be a rational cyclic language. Then its zeta function is rational.*

Proof. Let $\mathcal{A}$ be a finite deterministic automaton with associated representation $\mu : A^* \to \mathbb{Z}^{n\times n}$, see the remark before Theorem 2.1. Then the character of $\mathcal{A}$ is

$$\sum_{w\in A^*} \mathrm{tr}(\mu w)\, w .$$

Setting $a_n = \sum_{|w|=n} \mathrm{tr}(\mu w)$, we obtain $a_n = \mathrm{tr}(M^n)$, where $M = \sum_{a\in A} \mu a$. It follows that

$$\zeta_{\mathcal{A}} := \exp\Big(\sum_{n\geq 1} a_n \frac{x^n}{n}\Big) = \exp\Big(\sum_{n\geq 1} \frac{\mathrm{tr}(M^n)}{n} x^n\Big) = \exp\Big(\sum_{n\geq 1}\sum_{i=1}^{k} \frac{\lambda_i^n}{n} x^n\Big)$$

where $\lambda_1, \ldots, \lambda_k$ are the eigenvalues of M with multiplicities. Thus this series is equal to

$$\prod_{i=1}^{k} \exp\Big(\sum_{n\geq 1} \frac{\lambda_i^n x^n}{n}\Big) = \prod_{i=1}^{k} \exp\Big(\log \frac{1}{1-\lambda_i x}\Big)$$

$$= \prod_{i=1}^{k} \frac{1}{1-\lambda_i x} = \det(1 - Mx)^{-1}.$$

Since by Theorem 2.1, $\underline{L}$ is a $\mathbb{Z}$-linear combination of characters of finite deterministic automata $\mathcal{A}_j$ for $j \in J$, we have $\underline{L} = \sum_{j\in J} \alpha_j \chi_{\mathcal{A}_j}$ for some α_j in $\mathbb{Z}$. Then it is easily verified that $\zeta_L = \prod_{j\in J} \zeta_{\mathcal{A}_j}^{\alpha_j}$, which concludes the proof. □

In view of the proof of Theorem 2.1 we establish two lemmas. For this, we call *permutation character* of a group G a function $\chi : G \to \mathbb{N}$, where $\chi(g)$ is the number of fixpoints of g in some action of G on a finite set. Equivalently, $\chi(g) = \mathrm{tr}(\mu(g))$, where $\mu : G \to \mathbb{Z}^{n\times n}$ is a representation of G such that each matrix $\mu(g)$ is a permutation matrix.

Lemma 2.5 *Let G be a group and let $\theta : G \to \mathbb{Z}^{n\times n}$ be a multiplicative monoid morphism such that each matrix $\theta(g)$ is a row-monomial $0, 1$-matrix. Then $g \mapsto \mathrm{tr}(\theta(g))$ is a permutation character.*

Observe that it is not assumed that $\theta(g)$ is an invertible matrix for any $g \in G$.

Proof. The row vector e_i of the canonical basis of $\mathbb{Z}^{1\times n}$ is mapped by each g in G onto some e_j or onto 0. Thus each $g \in G$ induces a partial function from $\{1, \ldots, n\}$

into itself. These partial functions have all the same image E. The restriction of g to E is a bijection and the number of fixpoints of this bijection is $\mathrm{tr}(\theta(g))$. □

Recall that two elements in a semigroup are *conjugate* if, for some elements x, y in it, they may be written xy and yx.

Lemma 2.6 *Let D be a 0-minimal ideal of a finite monoid M and let G be a maximal subgroup in $D \setminus 0$. Any element $x \in D$ with $x^2 \neq 0$ is conjugate to some element in G.*

Proof. We use the Rees matrix semigroup form for D, see A2.4. We may by A2.5(ii) assume that the maximal subgroup is $\{(i, g, j) \mid g \in G\}$ and that $x = (i', g', j')$. Since $x^2 \neq 0$, we have $p_{j',i'} \neq 0$. Similarly $p_{j,i} \neq 0$. Let $u = (i', g', j)$ and $v = (i, p_{j,i}^{-1}, j')$. Then $uv = (i', g' p_{j,i} p_{j,i}^{-1}, j') = x$ and $vu = (i, p_{j,i}^{-1} p_{j',i'} g', j)$, which proves the lemma. □

We call a formal series $S = \sum_{w \in A^*} (S, w)$ *cyclic* if it has the following properties:

(i) There is a finite monoid M, a surjective monoid morphism $\mu : A^* \to M$ and a function $\varphi : M \to \mathbb{Z}$ such that for any word w, $(S, w) = \varphi(\mu w)$. Moreover, for any group G in M, the restriction of φ to G is a $\mathbb{Z}$-linear combination of permutation characters of G.

(ii) For any words u and v, one has $(S, uv) = (S, vu)$.

(iii) For any word w, the sequence $u_n = (S, w^{n+1})$, $n \in \mathbb{N}$, satisfies a strict linear recurrence relation (see Section 6.1).

Observe that a $\mathbb{Z}$-linear combination of cyclic series is a cyclic series (take the product monoid and use Exercise 2.2). Moreover, the character of a finite deterministic automaton is a cyclic series: this follows from Lemma 2.5 for condition (i); condition (ii) is evident, and condition (iii) follows from Theorem 6.2.1 and the equality $(S, w^{n+1}) = \sum \lambda^{n+1}$, where the sum is over all nonzero eigenvalues of μw (μ is the representation given after the definition of the character of an automaton).

Proof of Theorem 2.1. The proof is in two parts. First, we show that the characteristic series of a rational cyclic language is a cyclic series. Next, we prove that each cyclic series satisfies the conclusion of the theorem. This implies the theorem.

1. Let S be the characteristic series of a rational cyclic language L. Since L is recognizable by Theorem 3.1.1, there is some monoid morphism $\mu : A^* \to M$, where M is a finite monoid, and a subset P of M such that $L = \mu^{-1}(P)$. We may assume that μ is surjective. Define $\varphi : M \to \mathbb{Z}$ by $\varphi(m) = 1$ if $m \in P$, and $\varphi(m) = 0$ otherwise. Then $(S, w) = \varphi(\mu w)$.

If G is a group in M, then either $\varphi(G) = 1$ or $\varphi(G) = 0$. Indeed, any two elements in G have a positive power in common, namely the neutral element e of G, and we conclude according to $e \in P$ or not, since L is cyclic and μ is surjective. Hence condition (i) is satisfied for S.

Moreover, condition (ii) is satisfied since L is cyclic, and (iii) follows also, since u_n is constant, for the same reason. This proves that S is cyclic.

2. It remains to prove that each cyclic series S is a $\mathbb{Z}$-linear combination of characters of finite deterministic automata. We take the notations of conditions (i),(ii) and (iii) above and prove the claim by induction on the cardinality of M. If M has a 0, we may assume that $\varphi(0) = 0$ by replacing φ by $\varphi - \varphi(0)$ and S by $S - \varphi(0)\underline{A}^*$, since $\underline{A}^*$ is evidently the character of some finite deterministic automaton.

Now, let D be some 0-minimal ideal of M if M has a zero, and the minimal ideal of M if M has no zero. Note that Card $D \geq 2$.

Suppose first that $D^2 = 0$. Then $x^2 = 0$ for each element x of D. Hence the sequence $\varphi(x^{n+1})$ is $\varphi(x), 0, 0, \ldots$, and therefore by (iii) we have $\varphi(x) = 0$. Hence φ vanishes on D and we may replace M by the quotient M/D and conclude by induction.

Suppose now that $D^2 \neq 0$. Then by A2.4, D contains some maximal group G. By A2.6 there exists a monoid representation $\theta : M \to (G_0)^{r\times r}$ where G_0 is G with a zero adjoined, where each matrix is row-monomial, and where the restriction of θ to G is of the form

$$\theta(g) = \begin{pmatrix} g & 0 & \cdots & 0 \\ * & 0 & \cdots & 0 \\ \vdots & \vdots & & \vdots \\ * & 0 & \cdots & 0 \end{pmatrix}$$

and moreover $\theta(0) = 0$.

Let $\beta : G \to \mathbb{Z}^{d\times d}$ be a representation of G by permutation matrices. Replacing in each matrix $\theta(m)$, for $m \in M$, each nonzero entry $g \in G$ by $\beta(g)$, we obtain a representation $\psi : M \to \mathbb{Z}^{dr\times dr}$ by row-monomial 0, 1-matrices. Hence

$$\sum_{w\in A^*} \operatorname{tr}(\psi(\mu w))\, w$$

is the character of some finite deterministic automaton. If H is a group in M, then the function $H \to \mathbb{Z}$, $h \mapsto \operatorname{tr}(\psi(h))$ is a permutation character of H by Lemma 2.5.

Since $\varphi|G$ is a $\mathbb{Z}$-linear combination of permutation characters of G, the previous construction shows that for some $\mathbb{Z}$-linear combination T of characters of finite deterministic automata, the series $S' = S - T$ vanishes on G. Then S' is a cyclic series. By Lemma 2.6 it vanishes on D. Indeed, let $x \in D$. If $x^2 \neq 0$, we use this lemma and the cyclicity of S'. On the contrary, if $x^2 = 0$, we use property (iii) of cyclic series together with the fact that $\theta(0) = 0$. Thus we may replace M by the quotient M/D and conclude by induction. □

Appendix 1: Semisimple algebras

Here, all algebras are finite dimensional over the field K. Likewise the modules over these algebras that we consider will be finite dimensional over K.

A1.1 An algebra is called *simple* if it has no two-sided ideal other than 0 and itself. An algebra is called *semisimple* if it is a finite direct product of simple algebras. It follows that a quotient of a semisimple algebra is semisimple (see Exercise A1.1).

A1.2 A right module M over an algebra $\mathfrak{A}$ is *faithful* if, whenever $Ma = 0$ for some a in $\mathfrak{A}$, then $a = 0$. Similarly for left modules.

A1.3 A module is *irreducible*, or *simple*, if it has no submodules other than 0 and itself. It is *completely reducible* if it is a finite direct sum of irreducible modules. A module is completely reducible if and only if each submodule has a supplementary submodule.

A1.4 If an algebra has a faithful irreducible module, then this algebra is simple.

A1.5 If an algebra has a faithful completely reducible module, then this algebra is semisimple.

A1.6 Each module over a semisimple algebra is completely reducible and this property characterizes semisimple algebras.

A1.7 If K is a field of characteristic 0 and G is a finite group, then the group algebra KG is semisimple. In other words, a finite group of endomorphisms of a vector space is completely reducible (Maschke's theorem).

A1.8 Each simple algebra is isomorphic to a matrix algebra $D^{n\times n}$, where D is a skew field containing K in its center and finite dimensional over K. In particular, if K is algebraically closed, then each simple algebra is a matrix algebra $K^{n\times n}$ (theorem of Wedderburn).

Appendix 2: Minimal ideals in finite monoids

All monoids and semigroups considered here are finite.

A2.1 An *ideal* in a monoid M is a nonempty subset D of M such that for all $m \in M$, $t \in D$, the elements mt and tm are in D. A *zero* in M is an element 0 such that $M \neq \{0\}$ and such that $\{0\}$ is an ideal. It is necessarily unique. Note that the neutral element is $\neq 0$.

A2.2 The *minimal ideal* of a monoid M is the smallest ideal in M. It always exists, since it necessarily contains the product, in some order, of the elements in M. If M has a zero, a 0-*minimal ideal* of M is an ideal in M strictly containing 0, and minimal for this property.

Observe that if m is an element (resp. a nonzero element) of the minimal (resp. of a 0-minimal) ideal D of M, then $MmM = D$.

A2.3 A *right ideal* in a monoid M is a nonempty subset R of M such that for all $m \in M$ and all $r \in R$, $rm \in M$. *Minimal right ideals* are defined appropriately. If M has a 0, a 0-*minimal right ideal* of M is a right ideal strictly containing 0 and which is minimal among all right ideals having this property. Observe that if m is an element (resp. a nonzero element) of a minimal (resp. 0-minimal) right ideal R of M, then $mM = R$.

Similar definitions and properties hold for *left ideals*.

A2.4 A *Rees matrix semigroup with* 0 (resp. *without* 0) is a semigroup denoted $\mathcal{M}_0(G, I, J, P)$ (resp. $\mathcal{M}(G, I, J, P)$), where G is a finite group, I, J are finite sets, and P is a $J \times I$ matrix over $G_0 = G \cup \{0\}$ (resp. over G), called the *sandwich matrix*, with the property that P has at least one nonzero element in each row and each column (this property holds automatically in the case "without 0"). The elements of the Rees matrix are the triples (i, g, j), $i \in I$, $g \in G$, $j \in J$, together with 0 in the case "with 0". The product is

$$(i, g, j)(i', g', j') = \begin{cases} (i, gp_{j,i'}g', j') & \text{if } p_{j,i'} \neq 0\,, \\ 0 & \text{if } p_{j,i'} = 0\,. \end{cases}$$

Note that in the case "without 0", only the first case occurs.

The fact that this product is associative is easily deduced from the fact that (i, g, j) may be represented by the $I \times J$ matrix, denoted $(g)_{i,j}$, over G_0, whose only nonzero element is g in position i, j. Then the product above is represented by the matrix $(g)_{i,j}P(g')_{i',j'}$.

Theorem (Rees–Suschkewitsch). *Let M be a finite monoid (resp. monoid with* 0*) and D be the minimal (resp. a* 0*-minimal ideal) of M (resp. such that $D^2 \neq 0$). Then D is, as semigroup, isomorphic to a Rees matrix semigroup without* 0 *(resp. with* 0*).*

We prove the theorem only in the case where M has a zero. The other case is easily deduced (with some new arguments) from this case by adjoining a zero to M.

(a) D is the disjoint union of the 0-minimal right (resp. left) ideals of M that it contains.

Indeed, if R is a 0-minimal right ideal, then for any $m \in M$, mR is a right ideal. Assuming that $mR \neq 0$, we show that mR is 0-minimal. For this, let R' be a nonzero right ideal with $R' \subseteq mR$. Let $R_1 = \{r \in R \mid mr \in R'\}$. Then $R' = mR_1$ and $R_1 \neq 0$ since $R' \neq 0$. Clearly, R_1 is a right ideal contained in R, hence $R_1 = R$ by 0-minimality. Consequently $R' = mR$ and mR is 0-minimal.

Let D' be the union of all 0-minimal right ideals contained in D. Then $D' \subseteq D$ and D' is a right ideal; it is also a left ideal, by the previous discussion, since the inclusion $R \subseteq D'$ implies $mR \subseteq D$. Thus $D' = D$, since D is a 0-minimal ideal.

(b) From (a), it follows that for each nonzero $s \in D$, the set sM (resp. Ms) is a 0-minimal right (resp. left) ideal. Using A2.3, we see that for each nonzero $s \in D$, if $st \neq 0$ (resp. $ts \neq 0$), then $sM = stM$ (resp. $Ms = Mts$).

(c) Let R be a 0-minimal right ideal and $s, t \in R \setminus 0$. By (b), $t = sa$ and $s = tb$ for some $a, b \in M$. Denote by ρ_m the mapping representing multiplication on the right by m. Then ρ_a and ρ_b induce inverse bijections $Ms \to Mt$ and $Mt \to Ms$ such that $xM = \rho_a(x)M$ for any $x \in Ms$.

Let indeed $x \in Ms$. Then $xa \in Msa = Mt$; since $s = tb = sab$, we have, for any m in M, the equality $ms = msab$ and therefore $x = xab$ for any x in Ms. Thus $x \mapsto xa$ is a mapping $Ms \to Mt$, with left inverse $y \to yb$. Similarly, the latter maps Mt into Ms, and has left inverse $x \to xa$, which implies that they are inverse bijections.

Finally, for $x \in Ms$, $x = xab$; if $x \neq 0$, then $xa \neq 0$. If on the other hand $x = 0$, then clearly $xM = \rho_a(x)M$.

(d) Let $s, t \in D$. Then $st \neq 0$ if and only if $Ms \cap tM$ contains a nonzero idempotent.

Indeed, let $e \in Ms \cap tM$ with $e^2 = e \neq 0$. Then $e = ns = tm$ for some $m, n \in M$. Hence $0 \neq e = e^2 = nstm$, so that $st \neq 0$.

Conversely, suppose that $st \neq 0$. Then by (b), $sM = stM$ and $Mt = Mst$. By (c), $x \mapsto xt$ is a bijection $Ms \to Mst$. Since $t \in Mst$, it has an inverse image that we denote by e. Then $et = t$ and $e \neq 0$, since $t \neq 0$ (because $st \neq 0$). By (b) $tM = eM$, hence $e = tm$ for some $m \in M$. Thus $e^2 = etm = tm = e$.

(e) Let e be an idempotent in $D \setminus 0$. Then $G = eM \cap Me \setminus 0$ is a group with neutral element e.

If $s, t \in G$, then $Ms = Me$ and $tM = eM$ by (b) and A2.3. Therefore $Ms \cap tM$ contains e, a nonzero idempotent. Thus $st \neq 0$ by (d) and $st \in G$. We conclude that G is a monoid contained in D with neutral element e. Let $s = ea \in G$ with $a \in M$. Then $s = es$. Hence $sM = eM$ by (b). By (c), $x \to xs$ is a bijection from Me onto Ms. The latter is equal to Me, since $s \in Me \setminus 0$. Hence $e \in Ms$. Therefore there exists t in Me such that $ts = e$. Thus $t \neq 0$. By (b), $tM = eM$, hence $t \in G$. This shows that s has a left inverse, and similarly a right inverse.

(f) Let R (resp. L) be a 0-minimal right (resp. left) ideal contained in D. Then $R \cap L$ is nonzero.

Indeed, let $x \in R$, $y \in L$, $x, y \neq 0$. Then by A2.2, $D = MxM = MyM$. Hence $x = ayb$ and $y = a'xb'$ for some $a, b, a', b' \in M$. Then for some $n \geq 1$, the power $(aa')^n$ is idempotent (see Exercise A2.1); denote by e this idempotent. One has $x = ayb = aa'xb'b = \cdots = (aa')^n x(b'b)^n = ex(b'b)^n$. Then $xb' \neq 0$ and $ex = x$. Thus $(aa')^n x = x$, which implies by (b) that $Ma'x = Mx$. This in turn implies that $My = Ma'xb' = Mxb'$. Moreover, by A2.3 we have $xb'M = xM$. Since by (b), $xM = R$ and $My = L$, we see that $xb' \in R \cap L \setminus 0$.

(g) Each 0-minimal right (resp. left) ideal contained in D is generated by an idempotent.

Indeed, let R be some 0-minimal right ideal. Since $D^2 \neq 0$, we know by (d) that there is a nonzero idempotent e in D; let $e \in R_0$, some 0-minimal right ideal contained in D. By (f), there is some nonzero n in $Me \cap R$. Then $n = ve$, $e = v'n$, since $Mn = Me$ by (b). Let $m = ev'$. Then $mn = ev'n = ee = e$. Thus $mn \neq 0$ and we conclude by (d) that nM contains a nonzero idempotent. Since $nM = R$ by (b), R contains a nonzero idempotent, which by (b) generates it.

(h) We now prove the theorem. By (d), since $D^2 \neq 0$, there exists a nonzero idempotent e in D. Let G be the group of (e), and let $G_0 = G \cup \{0\}$. Let I (resp. J) be the set of 0-minimal right (resp. left) ideals contained in D and $R_0 = eM$, $L_0 = Me$. For each $L \in J$, $R_0 \cap L$ is nonzero by (f) and we take a nonzero u_L in $R_0 \cap L$. Similarly for $R \in I$, let $v_R \in L_0 \cap R$ with $v_R \neq 0$. Let $p_{L,R} = u_L v_R$. Then $p_{L,R} \in R_0 \cap L_0 = G_0$.

Define $\Phi : \mathcal{M}_0(G, I, J, P) \to D$ by $\Phi(R, g, L) = v_R g u_L$ and $\Phi(0) = 0$. We show that Φ induces a bijection from $\{R\} \times G \times \{L\}$ onto $R \cap L \setminus 0$. This will imply that Φ is a bijection.

Since $v_R \in L_0 = Me$, we have $v_R e = v_R$. Since $v_R, e \in L$, by the symmetric statement of (c), $x \mapsto v_R x$ is a bijection from $eM = R_0$ onto $v_R M = R$ and $Mx = Mv_R x$. It induces by restriction a bijection from $R_0 \cap L_0$ onto $R \cap L_0$, hence also from $G = R_0 \cap L_0 \setminus 0$ onto $R \cap L_0 \setminus 0$. Now, since $eu_L = u_L \in R_0 = eM$, we have by (c) that $x \mapsto xu_L$ is a bijection $L_0 \to L$ such that $xM = xu_L M$. Consequently we obtain by restriction a bijection from $R \cap L_0 \setminus 0$ onto $R \cap L \setminus 0$. Composing the two bijections, we see that $g \mapsto v_R g u_L$ is a bijection from G onto $R \cap L \setminus 0$.

We must show that Φ is a semigroup morphism. This is reduced to show that

$$\Phi((R, g, L)(R', g', L')) = \Phi(R, g, L)\Phi(R', g', L').$$

The right-hand side is $v_R g u_L v_{R'} g' u_{L'}$. The left-hand side is $\Phi(R, g p_{L,R'} g', L')$ if $p_{L,R'} \neq 0$, and $\Phi(0)$ if $p_{L,R'} = 0$; that is, $v_R g p_{L,R'} g' u_{L'}$ in the first case and 0 in the second. Since $p_{L,R'} = u_L v_{R'}$, both sides are equal.

Finally, let $L \in J$. Then by (g), one has $L = Mf$ for some idempotent f. Let $f \in R$, some 0-minimal right ideal contained in D. Then $0 \neq f \in L \cap R = Mu_L \cap v_R M$. Thus, by (d), $p_{L,R} = u_L v_R \neq 0$. This shows that each row of P contains a nonzero element, and similarly for the columns.

A2.5 Let M be a monoid and D its minimal ideal if M has no zero, and some 0-minimal ideal if M has a zero, with the assumption $D^2 \neq 0$. In the first case, D may be identified with a Rees matrix semigroup without zero $\mathcal{M}(G, I, J, \Lambda)$, and in the second case with a Rees matrix semigroup with zero $\mathcal{M}_0(G, I, J, \Lambda)$. Then this matrix representation has the following properties, which we state in the case "with zero" only (the case "without zero" is easily deduced):

(i) the idempotents in $D \setminus 0$ are the elements $(i, p_{j,i}^{-1}, j)$ where $i \in I$, $j \in J$ are such that $p_{j,i} \neq 0$;

(ii) the maximal subgroups in $D \setminus 0$ are the subsets

$$G_{i,j} = \{(i, g, j) \mid g \in G\}, \quad i \in I, j \in J, p_{j,i} \neq 0.$$

They are isomorphic to G. If e is the neutral element of $G_{i,j}$, then $e = (i, p_{j,i}^{-1}, j)$ and $eMe = G_{i,j} \cup 0$.

(iii) The 0-minimal right (resp. left) ideals of M contained in D are the subsets $R_i = \{0\} \cup \{(i, g, j) \mid g \in G, j \in J\}$ for $i \in I$ (resp. $L_j = \{0\} \cup \{(i, g, j) \mid i \in I, g \in G\}$ for $j \in J$).

Here one uses the fact that R_i, which is clearly a right ideal of D, is also a right ideal of M. This follows from the fact that for $s \in D$, one has $se = s$ for some idempotent e in D by (g), hence $sm = s(em)$ for any $m \in M$.

(iv) For $m \in M$, $i \in I$, $j \in J$, with $p_{j,i} \neq 0$, the function $x \mapsto xm$ either maps $G_{i,j}$ onto 0, or is a bijection from $G_{i,j}$ onto $G_{i,j'}$ for some $j' \in J$.

Indeed, $xe = x$ for the idempotent in $G_{i,j}$; thus replacing m by em we are reduced to the case where $m \in D$. Then, if $m \neq 0$, we have $m = (i', g', j')$. For $x = (i, g, j)$, we have $xm = (i, gp_{j,i'}g', j')$ if $p_{j,i'} \neq 0$ and we obtain the desired bijection, or it is 0 for any x in $G_{i,j}$.

A2.6 Let M be a monoid and let D be its minimal ideal if M has no zero, and a 0-minimal ideal if M has a zero.

Suppose that $D^2 \neq 0$. Then D contains an idempotent $e \neq 0$. Then D has a maximal subgroup G containing e, which is the neutral element of G. There exists a representation of M by square row-monomial matrices over $G \cup \{0\}$ such that the image of each g in G has nonzero coefficients only in the first column, and such that the image of 0 is the zero matrix.

Indeed, we may assume that M has a zero. We identify D with $\mathcal{M}_0(G, I, J, P)$ and write $e = (i_0, p_{j_0,i_0}^{-1}, j_0)$. For each $j \in J$, choose $a_j, b_j \in M$ such that $x \mapsto xa_j$ and $y \mapsto yb_j$ are inverse bijections from G_{i_0,j_0} onto $G_{i_0,j}$ and from $G_{i_0,j}$ onto G_{i_0,j_0}: we may indeed take $i_1 \in I$ such that $p_{j,i_1} \neq 0$; then we choose $a_j = (i_0, p_{j_0,i_0}^{-1}, j)$ and $b_j = (i_1, p_{j,i_1}^{-1}, j_0)$.

Let $m \in M$. For $j \in J$, $x \mapsto xm$ is either a bijection $G_{i_0,j} \to G_{i_0,j'}$ for some $j' \in J$, or $xm = 0$ for any $x \in G_{i_0,j}$ (see A2.5(iv)). In the first case, we put $j \cdot m = j'$, and in the second case $j \cdot m = \emptyset$. Then $(j, m) \mapsto j \cdot m$ is a partial right action of M on J. Define

$$j * m = \begin{cases} ea_jmb_{j'} & \text{if } j \cdot m = j', \\ 0 & \text{if } j \cdot m = \emptyset. \end{cases}$$

Note that $ea_j \in G_{i_0,j}$, and that if $j \cdot m = j'$, then $ea_jm \in G_{i_0,j'}$ and therefore $j * m = ea_jmb_{j'}$ is in G_{i_0,j_0} which may be identified with G.

Now define for any $m \in M$ the row-monomial matrix $\theta(m) \in G_0^{J \times J}$ by

$$\theta(m)_{j,j'} = \begin{cases} j * m & \text{if } j \cdot m = j', \\ 0 & \text{if } j \cdot m \neq j' \text{ or } j \cdot m = \emptyset. \end{cases}$$

In order to show that θ is a monoid morphism, it is enough to show that $j * (mn) = (j * m)((j \cdot m) * n)$. Now, the left-hand side is nonzero if and only if $G_{i_0,j}mn$ is nonzero, and the right-hand side is nonzero if and only if $G_{i_0,j}m$ is nonzero and $(G_{i_0,j}m)n$ is nonzero. So we may assume that both sides are nonzero. Set $j' = j \cdot m$ and $j'' = j' \cdot n$. Then $j * (mn) = ea_jmnb_{j''}$ and $(j * m)((j \cdot m) * n) = ea_jmb_{j'}ea_{j'}nb_{j''}$.

Since $ea_jmb_{j'} \in G_{i_0,j_0}$ (as we saw above), $ea_jmb_{j'}e = ea_jmb_{j'}$. Now, $ea_jm \in G_{i_0,j'}$ and $y \mapsto yb_{j'}$, $G_{i_0,j'} \to G_{i_0,j_0}$ and $x \mapsto xa_{j'}$, $G_{i_0,j_0} \to G_{i_0,j'}$ are inverse bijections. Hence $ea_jmb_{j'}ea_{j'} = ea_jmb_{j'}a_{j'} = ea_jm$ and $(j * m)((j \cdot m) * n) = ea_jmnb_{j''} = j * mn$.

Exercises for Chapter 12

1.1 Show that a set M of square matrices of order n is reducible (that is, not irreducible) if and only if for some invertible matrix g and some $i, j \geq 1$ with $i + j = n$, the matrices gmg^{-1}, for $m \in M$, have all the block triangular form $\begin{pmatrix} a & b \\ 0 & c \end{pmatrix}$, where a (resp. b) is square of order i (resp. j). Show that equivalently the form may be $\begin{pmatrix} a & 0 \\ b & c \end{pmatrix}$.

1.2 Show that a set M of square matrices is completely reducible if and only if for some invertible matrix g, the matrices gmg^{-1} have all the block diagonal matrix form of the same size

$$\begin{pmatrix} a_1 & 0 & . & . & 0 \\ 0 & a_2 & . & & . \\ . & . & . & . & . \\ . & & & . & 0 \\ 0 & . & . & 0 & a_k \end{pmatrix}$$

where for each $i = 1, \ldots, k$ the induced set of matrices a_i is irreducible.

1.3 Show that a set of endomorphisms of a finite dimensional vector space is completely reducible if and only if for each subspace which is invariant under these endomorphisms, there is a supplementary subspace which is also invariant. (*Hint*: Use A1.3.)

1.4 Let C be a code. Show that if $u, uv, vu \in C^*$, then $v \in C^*$ (consider uvu or use Exercise 11.1.1).

1.5 Let C be a code. Show that C is prefix if and only if for any words u and v, $u, uv \in C^*$ implies $v \in C^*$.

1.6 Let D be a Rees matrix semigroup as in A2.4. Let E be a subsemigroup of D not containing 0. Show that for some subgroup H of G, some subsets I_1 of I and J_1

of J, one has

$$E = \{(i, h, j) \mid i \in I_1, h \in H, j \in J_1\},$$

together with the condition $p_{j,i} \in H$ for all $i \in I_1$, $j \in J_1$.

1.7 Let G be a finite group and take as alphabet $A = G$. Let $\mu : A^* \to G$ be the natural monoid morphism which is the identity on G. Show that $\mu^{-1}(1) = C^*$ for some rational bifix code C. Show that the syntactic algebra of C^* is isomorphic to the group algebra KG.

2.1 Let L be a rational language such that for any w in L, one has $w^n \in L$ for all $n \geq 1$. Show that the *cyclic closure* of L (that is the smallest cyclic language containing L) is rational.

2.2 Show that the sum of two cyclic series is a cyclic series. (*Hint*: Show that if G is a group in the product monoid $M_1 \times M_2$, then G is a subgroup of $G_1 \times G_2$, for some subgroup G_1 (G_2) of M_1 (M_2).)

A1.1 Let $\mathfrak{A}$, $\mathfrak{B}$ be two algebras with $\mathfrak{A}$ simple. Show that if $\mathfrak{I}$ is a two-sided ideal of $\mathfrak{A} \times \mathfrak{B}$, then either $\mathfrak{I} = \mathfrak{A} \times \mathfrak{J}$ or $\mathfrak{I} = 0 \times \mathfrak{J}$ for some ideal $\mathfrak{J}$ of $\mathfrak{B}$. Deduce that each quotient of $\mathfrak{A} \times \mathfrak{B}$ is either a quotient of $\mathfrak{B}$ or of the form $\mathfrak{A} \times$ (a quotient of $\mathfrak{B}$). Deduce that a quotient of a semisimple algebra is semisimple.

A1.2 Let $\mathfrak{A}$ be a subalgebra of $K^{n \times n}$. Show that it acts faithfully at the right on $K^{1 \times n}$.

A1.3 Let $\mathfrak{A}_1, \ldots, \mathfrak{A}_n$ be simple algebras and let $\mathfrak{A}$ be a subalgebra of $\mathfrak{A}_1 \times \cdots \times \mathfrak{A}_n$ such that the projections $\mathfrak{A} \to \mathfrak{A}_i$ are surjective. Show that $\mathfrak{A}$ is semisimple. (*Hint*: Let $\mathfrak{B}$ be the projection of $\mathfrak{A}$ onto $\mathfrak{A}_1 \times \cdots \times \mathfrak{A}_{n-1}$. It is semisimple by induction. If $\mathfrak{A} \to \mathfrak{B}$ is not injective, then $(0, \ldots, 0, a) \in \mathfrak{A}$ for some $a \neq 0$ in $\mathfrak{A}_n$. Then $0 \times \cdots \times 0 \times \mathfrak{A}_n \subset \mathfrak{A}$ and finally $\mathfrak{A} = \mathfrak{B} \times \mathfrak{A}_n$.)

A1.4 Let $\mathfrak{A}$ act faithfully on a completely reducible module M. Using A1.4 and the previous exercise, prove A1.5.

A2.1 Show that for any element s of a finite semigroup S, there is some $n \geq 1$ such that s^n is idempotent. (*Hint*: For some $1 \leq i \leq j$, one has $s^i = s^{i+j}$, then s^j is idempotent.)

Notes to Chapter 12

Corollary 1.2 is from Reutenauer (1981). For the proof of the equivalent Theorem 1.1, we have followed Berstel and Perrin (1985), Section VIII.7. Theorem 2.1 and Corollary 2.4 are from Berstel and Reutenauer (1990). Concerning Corollary 2.4, it is shown in Reutenauer (1997) that the zeta function of a rational cyclic language is even ℕ-rational, and this is extended to rational cyclic relations. Corollary 2.4 may be used to show that the zeta function of each sofic system (in the sense of symbolic dynamics) is rational, see Berstel and Reutenauer (1990); for other proofs of this result, see Manning (1971) whose proof may be adapted, Fried (1987), Béal (1995) and Lind and Marcus (1995).

For Appendix 1, see Lang (1984) and for Appendix 2, see Lallement (1979) or Clifford and Preston (1961).

Open problems and conjectures

In this appendix we collect, for the convenience of the reader, several open problems and conjectures already mentioned at various places in the book. None of them are easy, and all deserve to be studied.

1. Field dependence problem of supports. Does there exist a language which is the support of an $\mathbb{R}$-rational series without being the support of a $\mathbb{Q}$-rational series? Page 64
2. Rational separation of disjoint supports: Let L and K be disjoint languages which are both support of some rational series. Does there exist two disjoint rational languages L' and K' such that

$$K \subset K', \ L \subset L'$$

(that is K and L are *rationally separated*)? Page 60
3. (Un)decidability of star height over the rationals: Is the the star height over $\mathbb{Q}$ of a rational series in $\mathbb{Q}\langle\langle A\rangle\rangle$ effectively computable? Page 78
4. Conjecture: Each rational Pólya series over $\mathbb{Q}$ is an unambiguous rational series. Page 128
5. Decidability of a zero in a linear recurrence series: Is it decidable, for a rational series $\sum a_n x^n$, whether there exists an n such that $a_n = 0$? Page 128
6. Characterization of strong Fatou rings. Page 142
7. Characterization of polynomials whose inverse is an $\mathbb{N}$-rational or $\mathbb{R}_+$-rational series. Page 159
8. Decidability of a common stable subspace for matrices over $\mathbb{Q}$. Page 175
9. Factorization conjecture of finite complete codes. Page 216
10. Hadamard quotient. Let S and T be two series with integer coefficients, such that $(T, w) = 0$ whenever $(S, w) = 0$. The *Hadamard quotient conjecture* is that if S and $S \odot T$ are $\mathbb{Z}$-rational series, then T is $\mathbb{Z}$-rational. This conjecture has been stated by C. Pisot for the case of one variable, and has been solved positively. In this case, see (Everest *et al.* 2003, page 69, Theorem 4.4). Partial cases have been solved by Lamèche (1973), Jacob (1980) and Reutenauer (1980a).

References

Allouche, J.-P. (1987). Automates finis en théorie des nombres. *Exposition. Math.*, **5** (3): 239–266. 104

Allouche, J.-P. and Shallit, J. O. (1992). The ring of k-regular sequences. *Theoret. Comput. Sci.*, **98**: 163–197. 104

Allouche, J.-P. and Shallit, J. O. (2003a). *Automatic Sequences: Theory, Applications, Generalizations*. Cambridge University Press. 104

Allouche, J.-P. and Shallit, J. (2003b). The ring of k-regular sequences. II. *Theoret. Comput. Sci.*, **307**(1): 3–29. 104

Amice, Y. (1975). *Les nombres p-adiques*. Presses Universitaires de France, Paris. Préface de Ch. Pisot, Collection SUP: Le Mathématicien, No. 14. 114

Bacher, R. (2008). The special subgroup of invertible non-commutative rational power series as a metric group. Technical report, Institut Fourier, Laboratoire de Mathématiques, St Martin d'Hères. 27, 45, 46

Bacher, R. (2009). On exponentials of exponential generating functions. Preprint. 27

Barcucci, E., Del Lungo, A., Frosini, A. and Rinaldi, S. (2001). A technology for reverse-engineering a combinatorial problem from a rational generating function. *Adv. in Appl. Math.*, **26**(2): 129–153. 158, 159

Bassino, F. (1997). Nonnegative companion matrices and star-height of **N**-rational series. *Theoret. Comput. Sci.*, **180**(1–2): 61–80. 158

Béal, M.-P. (1995). Puissance extérieure d'un automate déterministe, application au calcul de la fonction zêta d'un système sofique. *RAIRO Inform. Théor. Appl.*, **29**(2): 85–103. 232

Béal, M.-P. and Perrin, D. (2003). On the generating sequences of regular languages on k-symbols. *J. ACM*, **50**(6): 955–980 (electronic). 63

Bell, J. P. and Gerhold, S. (2007). On the positivity set of a linear recurrence sequence. *Israel J. Math.*, **157**: 333–345. 158

Benzaghou, B. (1970). Algèbres de Hadamard. *Bull. Soc. Math. France*, **98**: 209–252. 112, 114

Berger, R. (2008). Gerasimov's theorem and n-Koszul algebras. Technical report, Available at `arXiv:0801.3383v2`. 159

Bergman, G. M. (1968). Commuting elements in free algebras and related topics in ring theory. Thesis, Harvard University. 193

Berstel, J. (1971). Sur les pôles et le quotient de Hadamard de séries N-rationnelles. *C. R. Acad. Sci. Paris Sér. A-B*, **272**: A1079–A1081. 143, 155

Berstel, J. and Mignotte, M. (1976). Deux propriétés décidables des suites récurrentes linéaires. *Bull. Soc. Math. France*, **104**(2): 175–184. 128

Berstel, J. and Perrin, D. (1985). *Theory of Codes*, volume 117 of *Pure and Applied Mathematics*. Academic Press. 201, 231

Berstel, J. and Reutenauer, C. (1990). Zeta functions of formal languages. *Trans. American Math. Soc.*, **321**: 533–546. 231, 232

Berstel, J. and Reutenauer, C. (2008a). Another proof of Soittola's theorem. *Theoret. Comput. Sci.*, **393**(1-3): 196–203. 158

Berstel, J. and Reutenauer, C. (2008b). Extension of Brzozowski's derivation calculus of rational expressions to series over the free partially commutative monoids. *Theoret. Comput. Sci.*, **400**: 144–158. 26

Berstel, J., Perrin, D., and Reutenauer, C. (2009). *Codes and Automata*. Cambridge University Press. 201, 208, 216

Bézivin, J.-P. (1984). Factorisation de suites récurrentes linéaires et applications. *Bull. Soc. Math. France*, **112**(3): 365–376. 128

Boë, J. M., de Luca, A. and Restivo, A. (1980). Minimal complete sets of words. *Theoret. Comput. Sci.*, **12**(3): 325–332. 205

Bourbaki, N. (1964). *Éléments de mathématique. Fasc. XXX. Algèbre commutative. Chapitre 5: Entiers. Chapitre 6: Valuations*. Actualités Scientifiques et Industrielles, No. 1308. Hermann, Paris. 142

Brown, T. C. (1971). An interesting combinatorial method in the theory of locally finite semigroups. *Pacific J. Math.*, **36**: 285–289. 179

Brzozowski, J. A. (1964). Derivatives of regular expressions. *J. Assoc. Comput. Mach.*, **11**: 481–494. 79

Cahen, P.-J. and Chabert, J.-L. (1975). Éléments quasi-entiers et extensions de Fatou. *J. Algebra*, **36**(2): 185–192. 142

Carlyle, J. W. and Paz, A. (1971). Realizations by stochastic finite automata. *J. Comput. System Sci.*, **5**: 26–40. 33, 45

Chabert, J.-L. (1972). Anneaux de Fatou. *Enseignement Math.*, **18**: 141–144. 131

Christol, G. (1979). Ensembles presque périodiques k-reconnaissables. *Theoret. Comput. Sci.*, **9**: 141–145. 93, 104

Christol, G., Kamae, T., Mendès France, M. and Rauzy, G. (1980). Suites algébriques, automates et substitutions. *Bull. Soc. Math. France*, **108**: 401–419. 93, 104

Clifford, A. H. and Preston, G. B. (1961). *The Algebraic Theory of Semigroups*, volume 1. American Mathematical Society. 232

Cobham, A. (1969). On the base-dependence of sets of numbers recognizable by finite automata. *Math. Systems Th.*, **3**: 186–192. 104

Cobham, A. (1972). Uniform tag sequences. *Math. Systems Th.*, **6**: 164–192. 104

Cobham, A. (1978). Representation of a word function as the sum of two functions. *Math. Systems Th.*, **12**: 373–377. 46

Cohen, R. S. (1970). Star height of certain families of regular events. *J. Comput. System Sci.*, **4**: 281–297. 79

Cohn, P. M. (1961). On a generalization of the Euclidean algorithm. *Proc. Cambridge Philos. Soc.*, **57**: 18–30. 184, 192, 193, 197, 198

Cohn, P. M. (1969). Free associative algebras. *Bull. London Math. Soc.*, **1**: 1–39. 40, 187

Cohn, P. M. (1982). The universal field of fractions of a semifir. I. Numerators and denominators. *Proc. London Math. Soc. (3)*, **44**(1): 1–32. 193

Cohn, P. M. (1985). *Free Rings and Their Relations*, volume 19 of London Mathematical Society Monographs. Academic Press. 27, 198

Cohn, P. M. (2006). *Free Ideal Rings and Localization in General Rings*, volume 3 of New Mathematical Monographs. Cambridge University Press. 198

Cohn, P. M. and Reutenauer, C. (1999). On the construction of the free field. *Internat. J. Algebra Comput.*, **9**(3-4): 307–323. Dedicated to the memory of Marcel-Paul Schützenberger. 46

Connes, A. (1994). *Noncommutative Geometry*. Academic Press. 46

Conway, J. H. (1971). *Regular Algebra and Finite Machines*. Chapman and Hall. 26

Cox, D., Little, J. and O'Shea, D. (1997). *Ideals, Varieties, and Algorithms: An Introduction to Computational Algebraic Geometry and Commutative Algebra*. Springer-Verlag. 78

Derksen, H. (2007). A Skolem–Mahler–Lech theorem in positive characteristic and finite automata. *Invent. Math.*, **168**(1): 175–224. 128

Drensky, V. (2000). *Free Algebras and PI-Algebras*. Springer-Verlag. Graduate course in algebra. 136, 137, 141

Droste, M. and Gastin, P. (1999). The Kleene-Schützenberger theorem for formal power series in partially commuting variables. *Inform. and Comput.*, **153**(1): 47–80. 26

Droste, M. and Gastin, P. (2007). Weighted automata and weighted logics. *Theor. Comput. Sci.*, **380**(1-2): 69–86. 27

Droste, M. and Gastin, P. (2008). On aperiodic and star-free formal power series in partially commuting variables. *Theory Comput. Syst.*, **42**(4): 608–631. 27

Droste, M. and Kuich, W. (2009). Semirings and formal power series. In Droste, M., Kuich, W. and Vogler, H., editors, *Handbook of Weighted Automata*, EATCS Monographs in Theoretical Computer Science. Springer-Verlag. 27

Droste, M., Kuich, W., and Rahonis, G. (2008). Multi-valued MSO logics over words and trees. *Fundam. Inform.*, **84**(3-4): 305–327. 27

Droste, M., Kuich, W. and Vogler, H., editors (2009). *Handbook of Weighted Automata*, Monographs in Theoretical Computer Science. An EATCS Series. Springer-Verlag. 27

Duboué, M. (1983). Une suite récurrente remarquable. *European J. Combin.*, **4**(3): 205–214. 128

Duchamp, G. and Reutenauer, C. (1997). Un critère de rationalité provenant de la géométrie non commutative. *Invent. Math.*, **128**(3): 613–622. 27, 46

Eggan, L. C. (1963). Transition graphs and the star-height of regular events. *Michigan Math. J.*, **10**: 385–397. 79

Ehrenfeucht, A., Parikh, R. and Rozenberg, G. (1981). Pumping lemmas for regular sets. *SIAM J. Comput.*, **10**(3): 536–541. 58

Eilenberg, S. (1974). *Automata, Languages, and Machines. Vol. A*. Academic Press. 26, 104, 155

Eilenberg, S. and Schützenberger, M.-P. (1969). Rational sets in commutative monoids. *J. Algebra*, **13**: 173–191. 133

Ésik, Z. and Kuich, W. (2003). Formal tree series. *J. Automat. Lang. Comb.*, **8**(2): 219–285. 27

Everest, G., van der Poorten, A., Shparlinski, I. and Ward, T. (2003). *Recurrence Sequences*. American Mathematical Society. 128, 233

Fagnot, I. (1996). Langage de Lukasiewicz et diagonales de séries formelles. *J. Théor. Nombres Bordeaux*, **8**(1): 31–46. 104

Fatou, P. (1904). Sur les séries entières à coefficients entiers. *Comptes Rendus Acad. Sci. Paris*, **138**: 342–344. 130

Fliess, M. (1970). Sur le plongement de l'algèbre des séries rationelles non commutatives dans un corps gauche. *C. R. Acad. Sci. Paris Sér. A-B*, **271**: A926–A927. 27

Fliess, M. (1971). Formal languages and formal power series. In IRIA., editor, *Séminaire Logique et Automates*, pages 77–85. 55, 62

Fliess, M. (1974a). Matrices de Hankel. *J. Math. Pures Appl. (9)*, **53**: 197–222. 26, 33, 35, 45, 46, 129, 132, 142

Fliess, M. (1974b). Sur divers produits de séries formelles. *Bull. Soc. Math. France*, **102**: 181–191. 27, 104

Fliess, M. (1975). Séries rationnelles positives et processus stochastiques. *Ann. Inst. H. Poincaré Sect. B (N.S.)*, **11**: 1–21. 132, 158

Fliess, M. (1981). Fonctionnelles causales non linéaires et indéterminées non commutatives. *Bull. Soc. Math. France*, **109**(1): 3–40. 27

Fried, D. (1987). Finitely presented dynamical systems. *Ergod. Theor. Dyn. Syst.*, **7**(4): 489–507. 232

Furstenberg, H. (1967). Algebraic functions over finite fields. *J. Algebra*, **7**: 271–277. 97, 104

Gelfand, I., Gelfand, S., Retakh, V. and Wilson, R. L. (2005). Quasideterminants. *Adv. in Math.*, **193**: 56–141. 159

Gessel, I. M. (1981). Two theorems of rational power series. *Utilitas Math.*, **19**: 247–254. 104

Gessel, I. M. (2003). Rational functions with nonnegative integer coefficients. In *The 50th séminaire Lotharingien de Combinatoire*. Unpublished, available at Gessel's homepage. 158

Haiman, M. (1993). Noncommutative rational power series and algebraic generating functions. *European J. Combin.*, **14**(4): 335–339. 104

Halava, V., Harju, T., and Hirvensalo, M. (2006). Positivity of second order linear recurrent sequences. *Discrete Applied Math.*, **154**: 447–451. 158

Hansel, G. (1986). Une démonstration simple du théorème de Skolem–Mahler–Lech. *Theoret. Comput. Sci.*, **43**(1): 91–98. 128

Harrison, M. A. (1978). *Introduction to Formal Language Theory*. Addison-Wesley. 59

Hashiguchi, K. (1982). Limitedness theorem on finite automata with distance functions. *J. Comput. System Sci.*, **24**(2): 233–244. 181

Hashiguchi, K. (1991). Algorithms for determining relative inclusion star height and inclusion star height. *Theoret. Comput. Sci.*, **91**(1): 85–100. 79

Herstein, I. N. (1968). *Noncommutative Rings*. The Carus Mathematical Monographs, No. 15. Published by The Mathematical Association of America. 24

Inagaki, Y., Hiroyuki, M. and Fukumura, T. (1972). Some aspects of linear space automata. *Inform. Control*, : 439–479. 45

Jacob, G. (1975). Représentations et substitutions matricielles dans la théorie algébrique des transductions. Thesis, University of Paris. 26

Jacob, G. (1978). La finitude des représentations linéaires des semi-groupes est décidable. *J. Algebra*, **52**(2): 437–459. 63, 163, 166

Jacob, G. (1980). Un théorème de factorisation des produits d'endomorphismes de k^n. *J. Algebra*, **63**: 389–412. 57, 233

Karhumäki, J. (1977). Remarks on commutative N-rational series. *Theoret. Comput. Sci.*, **5**(2): 211–217. 142

Katayama, T., Okamoto, M. and Enomoto, H. (1978). Characterization of the structure-generating functions of regular sets and the DOL growth functions. *Inform. Control*, **36**(1): 85–101. 157, 158

Kirsten, D. (2005). Distance desert automata and the star height problem. *Theor. Inform. Appl.*, **39**(3): 455–509. 79

Kleene, S. C. (1956). Representation of events in nerve nets and finite automata. In Shannon, C. E. and McCarthy, J., editors, *Automata Studies*, Annals of Mathematics Studies, no. 34, pages 3–41. Princeton University Press. 26, 47

Koblitz, N. (1984). *p-adic Numbers, p-adic Analysis, and zeta-Functions*, volume 58 of Graduate Texts in Mathematics. Springer-Verlag. 114

Konvalinka, M. and Pak, I. (2007). Non-commutative extensions of the MacMahon master theorem. *Adv. in Math.*, **216**: 29–61. 159

Koutschan, C. (2005). Regular languages and their generating functions: The inverse problem. Diplomarbeit Informatik, Friedrich-Alexander-Universität Erlangen-Nürnberg. 158

Koutschan, C. (2008). Regular languages and their generating functions: the inverse problem. *Theoret. Comput. Sci.*, **391**: 65–74. 158

Krob, D. (1991). Expressions rationnelles sur un anneau. In *Topics in Invariant Theory (Paris, 1989/1990)*, volume 1478 of Lecture Notes in Mathematics, pages 215–243. Springer-Verlag. 79

Krob, D. (1994). The equality problem for rational series with multiplicities in the tropical semiring is undecidable. *Internat. J. Algebra Comput.*, **4**(3): 405–425. 181

Kuich, W. and Salomaa, A. (1986). *Semirings, Automata, Languages*, volume 5 of EATCS Monographs on Theoretical Computer Science. Springer-Verlag. 27

Lallement, G. (1979). *Semigroups and Combinatorial Applications*. John Wiley & Sons. 201, 232

Lamèche, K. (1973). Quelques propriétés des "séries rationnelles en variables non commutatives". *J. Combinatorial Theory Ser. A*, **14**: 128–135. 233

Lang, S. (1984). *Algebra*. Addison-Wesley, second edition. First edition in 1965. 14, 100, 170, 232

Laohakosol, V. and Tangsupphathawat, P. (2009). Positivity of third order linear recurrence sequences. *Discrete Applied Math.*, **157**: 3239–3248. 158

Lascoux, A. (1986). Suites récurrentes linéaires. *Adv. in Appl. Math.*, **7**(2): 228–235. 128

Lavallée, S. (2008). ℕ-rationality of a certain class of formal series. *Inform. Process. Lett.* **108**(3): 124–126. 158

Lavallée, S. (2009). Séries rationnelles et matrices génériques non commutatives. PhD thesis, Université du Québec à Montréal. 159

Lech, C. (1953). A note on recurring series. *Ark. Mat.*, **2**: 417–421. 118, 128

Leung, H. (1988). On the topological structure of a finitely generated semigroup of matrices. *Semigroup Forum*, **37**(3): 273–287. 181

Lewin, J. (1969). Free modules over free algebras and free group algebras: The Schreier technique. *Trans. Amer. Math. Soc.*, **145**: 455–465. 40

Lewin, J. (1974). Fields of fractions for group algebras of free groups. *Trans. Amer. Math. Soc.*, **192**: 339–346. 27

Lidl, R. and Niederreiter, H. (1983). *Finite Fields*. Encyclopedia of Mathematics and its Applications. Addison-Wesley. 126, 128

Lind, D. A. and Marcus, B. H. (1995). *An Introduction to Symbolic Dynamics and Coding*. Cambridge University Press. 232

Lothaire, M. (1983). *Combinatorics on Words*, volume 17 of Encyclopedia of Mathematics and its Applications. Addison-Wesley. 137, 141

Lothaire, M. (2002). *Algebraic Combinatorics on Words*, volume 90 of Encyclopedia of Mathematics and its Applications. Cambridge University Press. 198

Lyndon, R. C. and Schupp, P. E. (1977). *Combinatorial Group Theory*. Springer-Verlag. Ergebnisse der Mathematik und ihrer Grenzgebiete, Band 89. 45

Mahler, K. (1935). Eine arithmetische Eigenschaft der Taylor-Koeffizienten rationaler Funktionen. *Akad. Wetensch. Amsterdam Proc.*, **38**: 50–60. 118, 128

Mandel, A. and Simon, J. (1977). On finite semigroups of matrices. *Theoret. Comput. Sci.*, **5**: 101–111. 163

Manin, Y. I. (1977). *A Course in Mathematical Logic*. Springer-Verlag. 63

Manning, A. (1971). Axiom A diffeomorphisms have rational zeta functions. *Bull. London Math. Soc.*, **3**: 215–220. 232

McNaughton, R. and Zalcstein, Y. (1975). The Burnside problem for semigroups. *J. Algebra*, **34**: 292–299. 166

Okniński, J. (1998). *Semigroups of Matrices*, volume 6 of Series in Algebra. World Scientific Publishing Co. Inc. 64, 181

Perrin, D. (1977). Codes asynchrones. *Bull. Soc. Math. France*, **105**(4): 385–404. 216

Perrin, D. (1992). On positive matrices. *Theoret. Comput. Sci.*, **94**(2): 357–366. Discrete mathematics and applications to computer science (Marseille, 1989). 157, 158

Pólya, G. (1921). Arithmetische Eigenschaften der Reihenentwicklungen rationaler Funktionen. *J. reine angew. Math.*, **151**: 1–31. 114

Procesi, C. (1973). *Rings with Polynomial Identities*. Marcel Dekker. Pure and Applied Mathematics, 17. 136, 141

Restivo, A. and Reutenauer, C. (1984). On cancellation properties of languages which are supports of rational power series. *J. Comput. System Sci.*, **29**(2): 153–159. 58

Reutenauer, C. (1978). Variétés d'algèbres et de séries rationnelles. In *1er Congrès Math. Appl. AFCET-SMF*, volume 2, pages 93–102. AFCET, Paris. 30, 45

Reutenauer, C. (1980a). Séries formelles et algèbres syntactiques. *J. Algebra*, **66**(2): 448–483. 30, 43, 45, 142, 233

Reutenauer, C. (1980b). An Ogden-like iteration lemma for rational power series. *Acta Inform.*, **13**(2): 189–197. 64

Reutenauer, C. (1981). Semisimplicity of the algebra associated to a biprefix code. *Semigroup Forum*, **23**(4): 327–342. 231

Reutenauer, C. (1982). Sur les éléments inversibles de l'algèbre de Hadamard des séries rationnelles. *Bull. Soc. Math. France*, **110**(3): 225–232. 128

Reutenauer, C. (1985). Noncommutative factorization of variable-length codes. *J. Pure Appl. Algebra*, **36**(2): 167–186. 207, 208, 209

Reutenauer, C. (1996). Inversion height in free fields. *Selecta Math. (N.S.)*, **2**(1): 93–109. 79

Reutenauer, C. (1997). $\mathbb{N}$-rationality of zeta functions. *Adv. in Appl. Math.*, **18**(1): 1–17. 159, 231

Reutenauer, C. (1999). Malcev-Neumann series and the free field. *Exposition. Math.*, **17**(5): 469–478. 27

Rowen, L. H. (1980). *Polynomial Identities in Ring Theory*, volume 84 of Pure and Applied Mathematics. Academic Press. 136, 137, 141

Ryser, H. J. (1963). *Combinatorial Mathematics*. The Carus Mathematical Monographs, No. 14. Published by The Mathematical Association of America. 59

Sakarovitch, J. (2009a). *Elements of Automata Theory*. Cambridge University Press. 27, 79

Sakarovitch, J. (2009b). Rational and recognisable power series. In Droste, M., Kuich, W. and Vogler, H., editors, *Handbook of Weighted Automata*, Monographs in Theoretical Computer Science. An EATCS Series, pages 105–174. Springer-Verlag. 27

Salomaa, A. and Soittola, M. (1978). *Automata-Theoretic Aspects of Formal Power Series*. Springer-Verlag. 26, 27, 64, 135, 142, 158

Schützenberger, M.-P. (1961a). On the definition of a family of automata. *Inform. Control*, **4**: 245–270. 18, 26, 35, 36, 45, 51, 55, 63

Schützenberger, M.-P. (1961b). On a special class of recurrent events. *Ann. Math. Statist.*, **32**: 1201–1213. 45, 57

Schützenberger, M.-P. (1962a). On a theorem of R. Jungen. *Proc. Amer. Math. Soc.*, **13**: 885–890. 15, 26

Schützenberger, M.-P. (1962b). Finite counting automata. *Inform. Control*, **5**: 91–107. 26, 164, 165, 169, 171, 172, 173

Schützenberger, M.-P. (1965). Sur certains sous-monoïdes libres. *Bull. Soc. Math. France*, **93**: 209–223. 206, 216

Schützenberger, M.-P. (1970). Parties rationnelles d'un monoïde libre. In *Proc. Intern. Math. Conf.*, volume 3, pages 281–282. 135

Schützenberger, M.-P. and Marcus, R. S. (1959). Full decodable code-word sets. *IRE Trans.*, **IT-5**: 12–15. 205

Simon, I. (1978). Limited subsets of a free monoid. In *19th Annual Symposium on Foundations of Computer Science (Ann Arbor, Mich., 1978)*, pages 143–150. IEEE. 177, 181

Simon, I. (1988). Recognizable sets with multiplicities in the tropical semiring. In *Mathematical Foundations of Computer Science, 1988 (Carlsbad, 1988)*, volume 324 of Lecture Notes in Computer Science, pages 107–120. Springer-Verlag. 181

Simon, I. (1994). On semigroups of matrices over the tropical semiring. *RAIRO Inform. Théor. Appl.*, **28**(3-4): 277–294. 181

Skolem, T. (1934). Ein Verfahren zur Behandlung gewisser exponentialer Gleichungen und diophantischer Gleichungen. *C. R. 8e Congr. Scand. Stockholm*: 163–188. 118, 128

Soittola, M. (1976). Positive rational sequences. *Theoret. Comput. Sci.*, **2**(3): 317–322. 147, 158

Sontag, E. D. (1975). On some questions of rationality and decidability. *J. Comput. System Sci.*, **11**(3): 375–381. 51, 63

Sontag, E. D. and Rouchaleau, Y. (1977). Sur les anneaux de Fatou forts. *C. R. Acad. Sci. Paris Sér. A-B*, **284**(5): A331–A333. 142

Turakainen, P. (1972). On the minimization of linear space automata. *Ann. Acad. Sci. Fennicae AI*, **506**: 1–14. 45

Turakainen, P. (1985). A note on test sets for ℕ-rational languages. *Bull Europ. Assoc. Theor. Comput. Sci.*, **25**: 40–42. 62

Wedderburn, H. M. (1932). Noncommutative domains of integrity. *J. reine angew. Math.*, **167**: 129–141. 187

Index of notation

Index